SINGULAR PERTURBATION THEORY

MATHEMATICAL AND ANALYTICAL TECHNIQUES
WITH APPLICATIONS TO ENGINEERING

MATHEMATICAL AND ANALYTICAL TECHNIQUES WITH APPLICATIONS TO ENGINEERING

Alan Jeffrey, *Consulting Editor*

Published:

Inverse Problems
A. G. Ramm

Singular Perturbation Theory
R. S. Johnson

Forthcoming:

Methods for Constructing Exact Solutions of Partial Differential Equations with Applications
S. V. Meleshko

The Fast Solution of Boundary Integral Equations
S. Rjasanow and O. Steinbach

Stochastic Differential Equations with Applications
R. Situ

SINGULAR PERTURBATION THEORY

MATHEMATICAL AND ANALYTICAL TECHNIQUES WITH APPLICATIONS TO ENGINEERING

R. S. JOHNSON

Springer
the language of science

Library of Congress Cataloging-in-Publication Data

Johnson, R. S. (Robin Stanley), 1944–
Singular perturbation theory: mathematical and analytical techniques with applications to engineering/R. S. Johnson
p. cm.
Includes bibliographical references and index.
ISBN 0-387-23200-1 E-ISBN 0-387-23217-6
1. Perturbation (Mathematics) 2. Engineering mathematics I. Title

QA871.J54 2004
515′.392—dc22 2004058790

Printed in the United States of America.

9 8 7 6 5 4 3 2 1 SPIN 11325628

springeronline.com

To Ros, who still, after nearly 40 years,
sometimes listens when I extol the wonders of
singular perturbation theory, fluid mechanics or water waves
—usually on a long trek in the mountains.

CONTENTS

FOREWORD

The importance of mathematics in the study of problems arising from the real world, and the increasing success with which it has been used to model situations ranging from the purely deterministic to the stochastic, is well established. The purpose of the set of volumes to which the present one belongs is to make available authoritative, up to date, and self-contained accounts of some of the most important and useful of these analytical approaches and techniques. Each volume provides a detailed introduction to a specific subject area of current importance that is summarized below, and then goes beyond this by reviewing recent contributions, and so serving as a valuable reference source.

The progress in applicable mathematics has been brought about by the extension and development of many important analytical approaches and techniques, in areas both old and new, frequently aided by the use of computers without which the solution of realistic problems would otherwise have been impossible.

A case in point is the analytical technique of singular perturbation theory which has a long history. In recent years it has been used in many different ways, and its importance has been enhanced by it having been used in various fields to derive sequences of asymptotic approximations, each with a higher order of accuracy than its predecessor. These approximations have, in turn, provided a better understanding of the subject and stimulated the development of new methods for the numerical solution of the higher order approximations. A typical example of this type is to be found in the general study of nonlinear wave propagation phenomena as typified by the study of water waves.

Elsewhere, as with the identification and emergence of the study of inverse problems, new analytical approaches have stimulated the development of numerical techniques for the solution of this major class of practical problems. Such work divides naturally into two parts, the first being the identification and formulation of inverse problems, the theory of ill-posed problems and the class of one-dimensional inverse problems, and the second being the study and theory of multidimensional inverse problems.

On occasions the development of analytical results and their implementation by computer have proceeded in parallel, as with the development of the fast boundary element methods necessary for the numerical solution of partial differential equations in several dimensions. This work has been stimulated by the study of boundary integral equations, which in turn has involved the study of boundary elements, collocation methods, Galerkin methods, iterative methods and others, and then on to their implementation in the case of the Helmholtz equation, the Lamé equations, the Stokes equations, and various other equations of physical significance.

A major development in the theory of partial differential equations has been the use of group theoretic methods when seeking solutions, and in the introduction of the comparatively new method of differential constraints. In addition to the useful contributions made by such studies to the understanding of the properties of solutions, and to the identification and construction of new analytical solutions for well established equations, the approach has also been of value when seeking numerical solutions. This is mainly because of the way in many special cases, as with similarity solutions, a group theoretic approach can enable the number of dimensions occurring in a physical problem to be reduced, thereby resulting in a significant simplification when seeking a numerical solution in several dimensions. Special analytical solutions found in this way are also of value when testing the accuracy and efficiency of new numerical schemes.

A different area in which significant analytical advances have been achieved is in the field of stochastic differential equations. These equations are finding an increasing number of applications in physical problems involving random phenomena, and others that are only now beginning to emerge, as is happening with the current use of stochastic models in the financial world. The methods used in the study of stochastic differential equations differ somewhat from those employed in the applications mentioned so far, since they depend for their success on the Ito calculus, martingale theory and the Doob-Meyer decomposition theorem, the details of which are developed as necessary in the volume on stochastic differential equations.

There are, of course, other topics in addition to those mentioned above that are of considerable practical importance, and which have experienced significant developments in recent years, but accounts of these must wait until later.

Alan Jeffrey
University of Newcastle
Newcastle upon Tyne
United Kingdom

PREFACE

The theory of singular perturbations has been with us, in one form or another, for a little over a century (although the term 'singular perturbation' dates from the 1940s). The subject, and the techniques associated with it, have evolved over this period as a response to the need to find approximate solutions (in an analytical form) to complex problems. Typically, such problems are expressed in terms of differential equations which contain at least one small parameter, and they can arise in many fields: fluid mechanics, particle physics and combustion processes, to name but three. The essential hallmark of a singular perturbation problem is that a simple and straightforward approximation (based on the smallness of the parameter) does not give an accurate solution throughout the domain of that solution. Perforce, this leads to *different* approximations being valid in *different* parts of the domain (usually requiring a 'scaling' of the variables with respect to the parameter). This in turn has led to the important concepts of breakdown, matching, and so on.

Mathematical problems that make extensive use of a small parameter were probably first described by J. H. Poincaré (1854–1912) as part of his investigations in celestial mechanics. (The small parameter, in this context, is usually the ratio of two masses.) Although the majority of these problems were not obviously 'singular'—and Poincaré did not dwell upon this—some are; for example, one is the earth-moon-spaceship problem mentioned in Chapter 2. Nevertheless, Poincaré did lay the foundations for the technique that underpins our approach: the use of asymptotic expansions. The notion of a singular perturbation problem was first evident in the seminal work of L. Prandtl (1874–1953) on the viscous boundary layer (1904). Here, the small parameter is the inverse Reynolds number and the equations are based on the classical Navier-Stokes equation of fluid mechanics. This analysis, coupled with small-Reynolds-number approximations that were developed at about the same time (1910), prepared the ground for a century of singular perturbation work in fluid mechanics. But other fields over the century also made important contributions, for example: integration of differential equations, particularly in the context of quantum mechanics; the theory of nonlinear oscillations; control theory; the theory of semiconductors. All these, and many others, have helped to develop the mathematical study of singular perturbation theory, which has, from the mid-1960s, been supported and made popular by a range of excellent text books and research papers. The subject is now quite familiar to postgraduate students in applied mathematics (and related areas) and, to some extent, to undergraduate students who specialise in applied mathematics. Indeed, it is an essential tool of the modern applied mathematician, physicist and engineer.

This book is based on material that has been taught, mainly by the author, to MSc and research students in applied mathematics and engineering mathematics, at the University of Newcastle upon Tyne over the last thirty years. However, the presentation of the introductory and background ideas is more detailed and comprehensive than has been offered in any particular taught course. In addition, there are many more worked examples and set exercises than would be found in most taught programmes. The style adopted throughout is to explain, with examples, the essential techniques, but without dwelling on the more formal aspects of proof, *et cetera*; this is for two reasons. Firstly, the aim of this text is to make all the material readily accessible to the reader who wishes to learn and use the ideas to help with research problems and who (in all likelihood) does not have a strong mathematical background (or who is not that concerned about these niceties). And secondly, many of the results and solutions that we present cannot be recast to provide anything that resembles a routine proof of existence or asymptotic correctness. Indeed, in many cases, no such proof is available, but there is often ample evidence that the results are relevant, useful and probably correct.

This text has been written in a form that should enable the relatively inexperienced (or new) worker in the field of singular perturbation theory to learn and apply all the essential ideas. To this end, the text has been designed as a learning tool (rather than a reference text, for example), and so could provide the basis for a taught course. The numerous examples and set exercises are intended to aid this process. Although it is assumed that the reader is quite unfamiliar with singular perturbation theory, there are many occasions in the text when, for example, a differential equation needs to be solved. In most cases the solution (and perhaps the method of solution) are quoted, but some readers may wish to explore this aspect of mathematical analysis; there are many good texts that describe methods for solving (standard) ordinary and partial differential equations. However, if the reader can accept the given solution, it will enable the main theme of singular perturbation theory to progress more smoothly.

Chapter 1 introduces all the mathematical preliminaries that are required for the study of singular perturbation theory. First, a few simple examples are presented that highlight some of the difficulties that can arise, going some way towards explaining the need for this theory. Then notation, definitions and the procedure of finding asymptotic expansions (based on a parameter) are described. The notions of uniformity and breakdown are introduced, together with the important concepts of scaling and matching. Chapter 2 is devoted to routine and straightforward applications of the methods developed in the previous chapter. In particular, we discuss how these ideas can be used to find the roots of equations and how to integrate functions represented by a number of matched asymptotic expansions. We then turn to the most significant application of these methods: the solution of differential equations. Some simple *regular* (i.e. not singular) problems are discussed first—these are rather rare and of no great importance—followed by a number of examples of *singular* problems, including some that exhibit boundary or transition layers. The rôle of scaling a differential equation is given some prominence.

In Chapter 3, the techniques of singular perturbation theory are applied to more sophisticated problems, many of which arise directly from (or are based upon) important examples in applied mathematics or mathematical physics. Thus we look at nonlinear wave propagation, supersonic flow past a thin aerofoil, solutions of Laplace's equation, heat transfer to a fluid flowing through a pipe and an example taken from gas dynamics. All these are classical problems, at some level, and are intended to show the efficacy of these techniques. The chapter concludes with some applications to ordinary differential equations (such as Mathieu's equation) and then, as an extension of some of the ideas already developed, the method of strained coordinates is presented.

One of the most general and most powerful techniques in the armoury of singular perturbation theory is the *method of multiple scales*. This is introduced, explained and developed in Chapter 4, and then applied to a wide variety of problems. These include linear and nonlinear oscillations, classical ordinary differential equations (such as Mathieu's equation—again—and equations with turning points) and the propagation of dispersive waves. Finally, it is shown that the method of multiple scales can be used to great effect in boundary-layer problems (first mentioned in Chapter 2).

The final chapter is devoted to a collection of worked examples taken from a wide range of subject areas. It is hoped that each reader will find something of interest here, and that these will show—perhaps more clearly than anything that has gone before—the relevance and power of singular perturbation theory. Even if there is nothing of immediate interest, the reader who wishes to become more skilled will find these a useful set of additional examples. These are listed under seven headings: mechanical & electrical systems; celestial mechanics; physics of particles & light; semi- and superconductors; fluid mechanics; extreme thermal processes; chemical & biochemical reactions.

Throughout the text, worked examples are used to explain and describe the ideas, which are reinforced by the numerous exercises that are provided at the end of each of the first four chapters. (There are no set exercises in Chapter 5, but the extensive references can be investigated if more information is required.) Also at the end of each of Chapters 1–4 is a section of further reading which, in conjunction with the references cited in the body of the chapter, indicate where relevant reference material can be found. The references (all listed at the end of the book) contain both texts and research papers. Sections in each chapter are numbered following the decimal pattern, and equations are numbered according to the chapter in which they appear; thus equation (2.3) is the third (numbered) equation in Chapter 2. The worked examples follow a similar pattern (so E3.3 is the third worked example in Chapter 3) and each is given a title in order to help the reader—perhaps—to select an appropriate one for study; the end of a worked example is denoted by a half-line across the page. The set exercises are similarly numbered (so Q3.2 is the second exercise at the end of Chapter 3) and, again, each is given a title; the answers (and, in some cases, hints and intermediate steps) are given at the end of the book (where A3.2 is the answer to Q3.2). A detailed and comprehensive subject index is provided at the very end of the text.

I wish to put on record my thanks to Professor Alan Jeffrey for encouraging me to write this text, and to Kluwer Academic Publishers for their support throughout. I must also record my heartfelt thanks to all the authors who came before me (and most are listed in the References) because, without their guidance, the selection of material for this text would have been immeasurably more difficult. Of course, where I have based an example on something that already exists, a suitable acknowledgement is given, but I am solely responsible for my version of it. Similarly, the clarity and accuracy of the figures rests solely with me; they were produced either in Word (as was the main text), or as output from Maple, or using SmartDraw.

1. MATHEMATICAL PRELIMINARIES

Before we embark on the study of *singular perturbation theory*, particularly as it is relevant to the solution of differential equations, a number of introductory and background ideas need to be developed. We shall take the opportunity, first, to describe (without being too careful about the formalities) a few simple problems that, it is hoped, explain the need for the approach that we present in this text. We discuss some elementary differential equations (which have simple exact solutions) and use these—both equations and solutions–to motivate and help to introduce some of the techniques that we shall present. Although we will work, at this stage, with equations which possess known solutions, it is easy to make small changes to them which immediately present us with equations which we cannot solve exactly. Nevertheless, the approximate methods that we will develop are generally still applicable; thus we will be able to tackle far more difficult problems which are often important, interesting and physically relevant.

Many equations, and typically (but not exclusively) we mean differential equations, that are encountered in, for example, science or engineering or biology or economics, are too difficult to solve by standard methods. Indeed, for many of them, it appears that there is no realistic chance that, even with exceptional effort, skill and luck, they could ever be solved. However, it is quite common for such equations to contain parameters which are small; the techniques and ideas that we shall present here aim to take advantage of this special property.

The second, and more important plan in this first chapter, is to introduce the ideas, definitions and notation that provide the appropriate language for our approach. Thus we will describe : order, asymptotic sequences, asymptotic expansions, expansions with parameters, non-uniformities and breakdown, matching.

1.1 SOME INTRODUCTORY EXAMPLES

We will present four simple ordinary differential equations–three second-order and one first-order. In each case we are able to write down the exact solution, and we will use these to help us to interpret the difficulties that we encounter. Each equation will contain a small parameter, ε, which we will always take to be positive; the intention is to obtain, directly from the equation, an approximate solution which is valid for small ε.

E1.1 An oscillation problem

We consider the constant coefficient equation

$$\frac{d^2x}{dt^2} + (1+\varepsilon)x = 0, \quad t \geq 0, \tag{1.1}$$

with $x(0) = 0$, $\dot{x}(0) = 1$ (where the dot denotes the derivative with respect to t); this is an initial-value problem. Let us assume that there is a solution which can be written as a power series in ε:

$$x = x_0(t) + \varepsilon x_1(t) + \varepsilon^2 x_2(t) + \cdots\cdot\cdot \tag{1.2}$$

where each of the x_ns is *not* a function of ε. The equation (1.1) then gives

$$\ddot{x}_0 + \varepsilon\ddot{x}_1 + \varepsilon^2\ddot{x}_2 + \cdots + (1+\varepsilon)(x_0 + \varepsilon x_1 + \varepsilon^2 x_2 + \cdots) = 0, \tag{1.3}$$

where we again use, for convenience, the dot to denote derivatives. We write (1.3) in the form

$$(\ddot{x}_0 + x_0) + \varepsilon(\ddot{x}_1 + x_1 + x_0) + \cdots\cdots = 0$$

and, since the right-hand side is precisely zero, all the ε-dependence must vanish; thus we require

$$\ddot{x}_0 + x_0 = 0; \quad \ddot{x}_1 + x_1 + x_0 = 0 \quad \text{and so on.} \tag{1.4}$$

(Remember that each $x_n(t)$ does not depend on ε.)

The two initial conditions give

$$x_0(0) + \varepsilon x_1(0) + \cdots = 0; \quad \dot{x}_0(0) + \varepsilon\dot{x}_1(0) + \cdots = 1$$

and, using the same argument as before, we must choose

$$x_0(0) = x_1(0) = \cdots\cdot = 0; \quad \dot{x}_0(0) = 1; \quad \dot{x}_1(0) = \cdots\cdot = 0,$$

where the '1' in the second condition is accommodated by $\dot{x}_0(0) = 1$. (If the initial conditions were, say, $x(0) = \varepsilon$, $\dot{x}(0) = 1 - \varepsilon$, then we would have to select $x_1(0) = 1$, $x_0(0) = x_2(0) = \cdots\cdot = 0$; $\dot{x}_0(0) = 1$, $\dot{x}_1(0) = -1$, $\dot{x}_2(0) = \cdots\cdot = 0$.)

Thus the first approximation is represented by the problem

$$\ddot{x}_0 + x_0 = 0 \quad \text{with} \quad x_0(0) = 0, \quad \dot{x}_0(0) = 1;$$

the general solution is

$$x_0 = A \sin t + B \cos t,$$

where A and B are arbitrary constants which, to satisfy the initial conditions, must take the values $A = 1$, $B = 0$. The solution is therefore

$$x_0 = \sin t. \tag{1.5}$$

The problem for the second term in the series becomes

$$\ddot{x}_1 + x_1 = -x_0 = -\sin t \quad \text{with} \quad x_1(0) = \dot{x}_1(0) = 0.$$

The solution of this equation requires the inclusion of a particular integral, which here is $\frac{1}{2}t \cos t$; the complete general solution is therefore

$$x_1 = C \sin t + D \cos t + \tfrac{1}{2}t \cos t,$$

where C and D are arbitrary constants. (The particular integral can be found by any one of the standard methods e.g. variation of parameters, or simply by trial-and-error.) The given conditions then require that $C = -\frac{1}{2}$ and $D = 0$ i.e.

$$x_1 = \tfrac{1}{2}(t \cos t - \sin t) \tag{1.6}$$

and so our series solution, at this stage, reads

$$x(t) = \sin t + \tfrac{1}{2}\varepsilon(t \cos t - \sin t) + \cdots\cdots. \tag{1.7}$$

Let us now review our results.

The original differential equation, (1.1), should be recognised as the harmonic oscillator equation for all $\varepsilon > -1$ and, as such, it possesses bounded, periodic solutions. The first term in our series, (1.5), certainly satisfies both these properties, whereas the second *fails* on *both* counts. Thus the series, (1.7), also fails: our approximation procedure has generated a solution which is not periodic and for which the amplitude grows without bound as $t \to \infty$. Yet the exact solution is simply

$$x_e = \frac{\sin\left(t\sqrt{1+\varepsilon}\right)}{\sqrt{1+\varepsilon}} \tag{1.8}$$

which is easily obtained by scaling out the '$1+\varepsilon$' factor, by working with $t\sqrt{1+\varepsilon}$ rather than t. (The 'e' subscript here is used to denote the exact solution.) It is now an elementary exercise to check that (1.8) and (1.7) agree, in the sense that the expansion of (1.8), for small ε and fixed t, reproduces (1.7). (A few examples of expansions are set as exercises in Q1.1, 1.2.) This process immediately highlights one of our difficulties, namely, taking $\varepsilon \to 0$ first and then allowing $t \to \infty$; this is a classic case of a *non-uniform limiting process* i.e. the answer depends on the order in which the limits are taken. (Examples of simple limiting processes can be found in Q1.4.) Clearly, any approximate methods that we develop must be able to cope with this type of behaviour. So, for example, if it is known (or expected) that bounded, periodic solutions exist, the approach that we adopt must produce a suitable approximation to this solution.

We have taken some care in our description of this first example because, at this stage, the approach and ideas are new; we will present the other examples with slightly less detail. However, before we leave this problem, there is one further observation to make. The original equation, (1.1), can be solved easily and directly; an associated problem might be

$$\ddot{x} + (1+\varepsilon x^2)x = 0, \quad t \geq 0, \tag{1.9}$$

with appropriate initial data. This describes an oscillator for which the frequency depends on the value of $x(t)$ at that instant—it is a nonlinear problem. Such equations are much more difficult to solve; our techniques have got to be able to make some useful headway with equations like (1.9).

E1.2 A first-order equation

We consider the equation

$$\frac{dy}{dx} = (1-2\varepsilon x)y, \quad x \geq 0, \tag{1.10}$$

with $y(0) = 1$. Again, let us seek a solution in the form

$$y = y_0(x) + \varepsilon y_1(x) + \varepsilon^2 y_2(x) + \cdots\cdot.$$

and then obtain

$$y_0' + \varepsilon y_1' + \cdots\cdot - (1-2\varepsilon x)(y_0 + \varepsilon y_1 + \cdots\cdot) = 0$$

or

$$(y_0' - y_0) + \varepsilon(y_1' - y_1 + 2xy_0) + \cdots\cdot\cdot = 0;$$

we use the prime to denote the derivative. Thus we require

$$y_0' - y_0 = 0; \quad y_1' - y_1 = -2xy_0, \quad \text{and so on,}$$

with the boundary conditions

$$y_0(0) = 1, \quad y_1(0) = y_2(0) = \cdots\cdots = 0.$$

The solution for y_0 is immediately

$$y_0 = e^x \quad (= 1 \text{ on } x = 0); \tag{1.11}$$

but this result is clearly unsatisfactory: the solution for y_0 grows exponentially, whereas the solution of equation (1.10) must decay for $x > 1/(2\varepsilon)$ (because then $y'/y < 0$). Perhaps the next term in the series will correct this behaviour for large enough x; we have

$$y_1' - y_1 = -2xy_0 = -2xe^x \quad \text{with} \quad y_1(0) = 0.$$

Thus $$\left(y_1 e^{-x}\right)' = -2x \quad \text{or} \quad y_1 = Ae^x - x^2 e^x$$

and we require $A = 0$; the series solution so far is therefore

$$y = e^x - \varepsilon x^2 e^x + \cdots\cdots. \tag{1.12}$$

However, this is no improvement; now, for sufficiently large x, the second term dominates and the solution grows towards $-\infty$! Let us attempt to clarify the situation by examining the exact solution.

We write equation (1.10) as

$$\frac{y'}{y} = 1 - 2\varepsilon x \quad \text{or} \quad \ln|y| = x - \varepsilon x^2 + \text{constant};$$

the general solution is therefore

$$y = C\exp(x - \varepsilon x^2)$$

and, with $C = 1$ to satisfy the given condition at $x = 0$, this yields

$$y_e = \exp(x - \varepsilon x^2). \tag{1.13}$$

Clearly the series, (1.12), is recovered directly by expanding the exact solution, (1.13), in ε for fixed x, so that we obtain

$$y_e = e^x(1 - \varepsilon x^2 + \cdots\cdot).$$

Equally clearly, this procedure will give a very poor approximation for large x; indeed, for x about the size of ε^{-1}, the approximation altogether fails. A neat way to see this is to redefine x as $x = X/\varepsilon$; this is called *scaling* and will play a crucial rôle in what we describe in this text. If we now consider ε small, for X fixed, the size of x is now proportional to ε^{-1}, and the results are very different:

$$y_e = \exp(x - \varepsilon x^2) = \exp\{\varepsilon^{-1}(X - X^2)\}, \tag{1.14}$$

indeed, in this example, we cannot even write down a suitable approximation of (1.14) for small ε! The expression in (1.14) attains a maximum at $X = 1/2$, and for larger X the function tends to zero.

We observe that any techniques that we develop must be able to handle this situation; indeed, this example introduces the important idea that the function of interest may take different (approximate) forms for different sizes of x. This, ultimately, is not surprising, but the significant ingredient here is that 'different sizes' are measured in terms of the small parameter, ε. We shall be more precise about this concept later.

E1.3 Another simple second-order equation

This time we consider

$$\frac{d^2y}{dx^2} = \varepsilon^2 y, \quad x \geq 0, \tag{1.15}$$

with

$$y(0) = 1 \quad \text{and} \quad y \to 0 \quad \text{as} \quad x \to \infty.$$

(The use of ε^2 here, rather than ε, is simply an algebraic convenience, as will become clear; obviously any small positive number could be represented by ε or ε^2—or anything equivalent, such as $\sqrt{\varepsilon}$ or ε^3, *et cetera*.) Presumably—or so we will assume—a first approximation to equation (1.15), for small ε, is just

$$y'' = 0 \quad \text{with} \quad y(0) = 1, \quad y \to 0 \quad \text{as } x \to \infty,$$

but this problem has no solution. The general solution is $y = Ax + B$, where A and B are the two arbitrary constants, and no choice of them can satisfy *both* conditions. In a sense, this is a more worrying situation than that presented by either of the two previous examples: we cannot even get started this time.

The exact solution is

$$y_e = \exp(-\varepsilon x)$$

and the difficulties are immediately apparent: $\varepsilon \to 0$, with x fixed, gives $y_e \to 1$, but then how do we accommodate the condition at infinity? Correspondingly, with $x \to \infty$ and ε fixed, we obtain $y_e \to 0$, and now how can we obtain the dependence on ε? As we can readily see, to treat ε and x separately is not appropriate here—we need to work with a scaled version of x (i.e. $X = \varepsilon x$). The choice of such a variable avoids the non-uniform limiting process: $\varepsilon \to 0$ and $x \to \infty$.

E1.4 A two-point boundary-value problem

Our final introductory example is provided by

$$\varepsilon \frac{d^2 y}{dx^2} + (1+\varepsilon)\frac{dy}{dx} + y = 0, \quad 0 \le x \le 1, \tag{1.16}$$

with $y(0)$ and $y(1)$ given. This equation contains the parameter ($\varepsilon > 0$) in two places: multiplying the higher derivative, which is critical here (as we will see), and adjusting the coefficient of the other derivative by a small amount. This latter appearance of the parameter is altogether unimportant—the coefficient is certainly close to unity—and serves only to make more transparent the calculations that we present.

Once again, we will start by seeking a solution which can be represented by the series

$$y = y_0(x) + \varepsilon y_1(x) + \cdots,$$

so that we obtain

$$y_0' + y_0 + \varepsilon(y_1' + y_1 + y_0' + y_0'') + \cdots = 0;$$

the shorthand notation for derivatives is again being employed. Thus we have the set of differential equations

$$y_0' + y_0 = 0; \quad y_1' + y_1 + y_0' + y_0'' = 0, \quad \text{and so on,} \tag{1.17}$$

with boundary conditions written as

$$y(0) = \alpha; \quad y(1) = \beta, \tag{1.18}$$

where α and β are given (but we will assume that they are not functions of ε). The general solution for y_0 is

$$y_0 = A\mathrm{e}^{-x} \quad (A \text{ is an arbitrary constant}),$$

but it is not at all clear how we can determine A. The difficulty that we have in this example is that we must apply *two* boundary conditions, which is patently impossible (unless some special requirement is satisfied). So, if we use $y_0(0) = \alpha$ we obtain

$$y_0 = \alpha\mathrm{e}^{-x} \quad \text{and then} \quad y_0(1) = \alpha\mathrm{e}^{-1}; \tag{1.19}$$

if, by extreme good fortune, we have $\beta = \alpha\mathrm{e}^{-1}$, then we also satisfy the second boundary condition (on $x = 1$). Of course, in general, this will not be the case; let us proceed with the problem for which $\alpha \ne \beta\mathrm{e}$. Thus the solution using $y_0(0) = \alpha$ does not satisfy $y_0(1) = \beta$, and the solution

$$y_0 = \beta e^{1-x} \tag{1.20}$$

does not satisfy $y_0(0) = \alpha$. Indeed, we have no way of knowing which, if either, is correct; thus there is little to be gained by solving the y_1 problem:

$$y_1' + y_1 + y_0'' + y_0' = 0 \quad \text{with} \quad y_1(0) = y_1(1) = 0. \tag{1.21}$$

(We note that, since $y_0' + y_0 = 0$, we must have $y_0'' + y_0' = 0$, and then there is, exceptionally, a solution of the complete problem: $y_1 = 0$, for $0 \le x \le 1$. But we still do not know y_0.)

As in our previous examples, let us construct and examine the exact solution. Equation (1.16) is a second order, constant coefficient, ordinary differential equation and so we may seek a solution in the form

$$y = e^{\lambda x} \quad \text{where} \quad \varepsilon\lambda^2 + (1+\varepsilon)\lambda + 1 = 0$$

i.e. $$(\varepsilon\lambda + 1)(\lambda + 1) = 0 \quad \text{so} \quad \lambda = -1 \text{ or } \lambda = -1/\varepsilon.$$

The general solution is therefore

$$y = Be^{-x} + Ce^{-x/\varepsilon}$$

and, imposing the two boundary conditions, this becomes

$$y_e = \frac{(\beta - \alpha e^{-1/\varepsilon})e^{-x} + (\alpha e^{-1} - \beta)e^{-x/\varepsilon}}{e^{-1} - e^{-1/\varepsilon}}. \tag{1.22}$$

(We can note here that the contribution from the term $\exp(-x/\varepsilon)$ is absent in the special case $\alpha = \beta e$; we proceed with the problem for which $\alpha \ne \beta e$.)

This solution, (1.22), is defined for $0 \le x \le 1$ and with $\varepsilon > 0$; let us select any $x \in (0, 1]$ and, for this x fixed, allow $\varepsilon \to 0^+$ (where $\to 0^+$ denotes tending to zero through the positive numbers). We observe that the terms $\exp(-1/\varepsilon)$ and $\exp(-x/\varepsilon)$ vanish rapidly in this limit, leaving

$$y_e \to \frac{\beta e^{-x}}{e^{-1}} = \beta e^{1-x}; \tag{1.23}$$

this is our approximate solution given in (1.20). (Some examples that explore the relative sizes of x^n, $\exp(x)$ and $\ln(x)$ can be found in Q1.5.) Thus one of the possible options for $y_0(x)$, introduced above, is indeed correct. However, this solution is, as already noted, incorrect on $x = 0$ (although, of course, $y_e(0) = \alpha$). The difficulty is plainly with the term $\exp(-x/\varepsilon)$: for any $x > 0$ fixed, as $\varepsilon \to 0^+$, this vanishes exponentially, but on $x = 0$ this takes the value 1 (one). In order to examine the rôle of this term, as $\varepsilon \to 0$, we need to retain it (but not to restrict ourselves to $x = 0$); as we have seen in earlier examples, a suitable rescaling of x is useful. In this case we set $x = \varepsilon X$, and so obtain

$$y_e = \frac{(\beta - \alpha e^{-1})e^{-\varepsilon X} + (\alpha e^{-1} - \beta)e^{-X}}{e^{-1} - e^{-1/\varepsilon}}, \quad 0 \leq X \leq 1/\varepsilon, \tag{1.24}$$

and now, for any X fixed, as $\varepsilon \to 0$, we have

$$y_e \to \frac{\beta + (\alpha e^{-1} - \beta)e^{-X}}{e^{-1}} = \beta e + (\alpha e^{-1} - \beta)e^{1-X}. \tag{1.25}$$

This is a second, and different, approximation to y_e, valid for xs which are proportional to ε; note that on $X = 0$, (1.25) gives the value α, which is the correct boundary value.

In summary, therefore, we have (from (1.23))

$$y_e \approx \beta e^{1-x} \quad \text{for } x \text{ not too small} \tag{1.26}$$

and (from (1.25))

$$y_e \approx \beta e + (\alpha e^{-1} - \beta)e^{1-x/\varepsilon} \quad \text{for } x \text{ about the size of } \varepsilon. \tag{1.27}$$

These two together constitute an approximation to the exact solution, each valid for an appropriate size of x. Further, these two expressions possess the comforting property that they describe a smooth—not discontinuous—transition from one to the other, in the following sense. The approximation (1.26) is not valid for small x, but as x decreases we have

$$\beta e^{1-x} \to \beta e \quad (\text{as } x \to 0) \tag{1.28}$$

(which we already know is incorrect because $\alpha \neq \beta e$); correspondingly, (1.27) is not valid for large $X = x/\varepsilon$, but we see that

$$\beta e + (\alpha e^{-1} - \beta)e^{1-x/\varepsilon} \to \beta e \quad \text{as } x/\varepsilon \to \infty: \tag{1.29}$$

results (1.28) and (1.29) *agree precisely*. This is clearly demonstrated in figure 1, where we have plotted the exact solution for $\alpha = 0,\ \beta = 1$ (as an example) i.e.

$$y_e = \frac{e^{-x} - e^{-x/\varepsilon}}{e^{-1} - e^{-1/\varepsilon}}, \tag{1.30}$$

for various ε. As ε decreases, the dramatically different behaviours for x not too small, and x small, are very evident. (Note that the solution for x not too small is

$$y_e \approx \exp(1 - x) \to e \quad \text{as } x \to 0.)$$

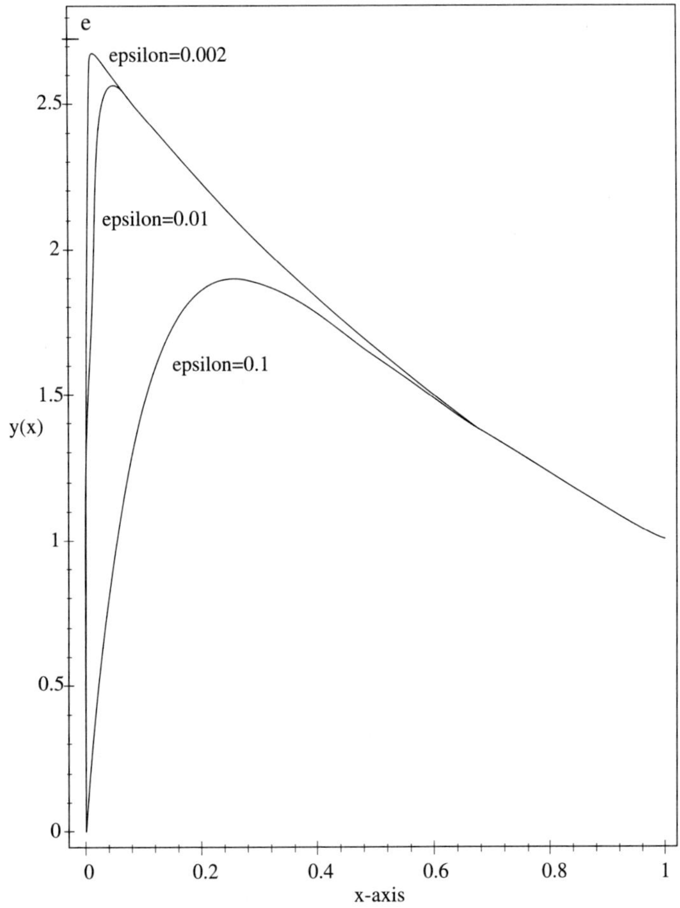

Figure 1. Plot of $y_\varepsilon = (e^{-x} - e^{-x/\varepsilon})/(e^{-1} - e^{-1/\varepsilon})$ for $0 \le x \le 1$, with $\varepsilon = 0.1, 0.01, 0.002$; the maximum value attained (e) is marked on the y-axis.

In these four simple examples, we have described some difficulties that are encountered when we attempt to construct approximate solutions, valid as $\varepsilon \to 0^+$, directly from given differential equations; a number of other examples of equations with exact solutions can be found in Q1.3. We must now turn to the discussion of the ideas that will allow a systematic study of such problems. In particular, we first look at the notation that will help us to be precise about the expansions that we write down.

1.2 NOTATION

We need a notation which will accurately describe the behaviour of a function in a limit. To accomplish this, consider a function $f(x)$ and a limit $x \to a$; here a may be any finite value (and approached either from the left or the right) or infinite. Further, it is convenient to compare $f(x)$ against another, simpler, function, $g(x)$; we call $g(x)$ a *gauge* function. The three definitions, and associated notation, that we introduce are based on the result of finding the limit

$$L = \lim_{x \to a} \left[\frac{f(x)}{g(x)} \right]. \tag{1.31}$$

We consider three cases in turn.

(a) ***Little-oh***

We write

$$f(x) = o[g(x)] \quad \text{as } x \to a$$

if the limit, (1.31), is $L = 0$; we say that 'f is little-oh of g as $x \to a$'. Clearly, this property of a function does not provide very useful information; essentially all it says is that $f(x)$ is smaller than $g(x)$ (as $x \to a$). So, for example, we have

$$\frac{1}{1+x^2} = o(1/x) \quad \text{as } |x| \to \infty$$

but also

$$\frac{1}{1+x^2} = o(1) \quad \text{as } |x| \to \infty;$$

and

$$\sin(x^2) = o(x) \quad \text{as } x \to 0.$$

It is an elementary exercise to show that each satisfy the definition $L = 0$ from (1.31), by using familiar ideas that are typically invoked in standard 'limit' problems. For example, the last example above involves

$$\frac{\sin(x^2)}{x} = \frac{x^2 - x^6/6 + \cdots}{x} = x - \frac{x^5}{6} + \cdots \to 0 \quad \text{as } x \to 0,$$

confirming that the limit is zero. (Note that, in the above examples, the gauge function which is a non-zero constant is conventionally taken to be $g(x) = 1$; note also that the limit under consideration should always be quoted, or at least understood.)

(b) ***Big-oh***

We write

$$f(x) = O[g(x)] \quad \text{as } x \to a$$

if the limit, (1.31), is finite and non-zero; this time we say that 'f is big-oh of g as $x \to a$' or simply 'f is order g as $x \to a$'. As examples, we offer

$$\frac{1}{3+2x^2} = O(1) \quad \text{as } x \to 0$$

but

$$\frac{1}{3+2x^2} = O(x^{-2}) \quad \text{as } |x| \to \infty;$$

also

$$\sin(3x) = O(x) \quad \text{as } x \to 0,$$

$$\tan(2x^2) = O(x^2) \quad \text{as } x \to 0;$$

finally $$\frac{1}{4-3x-2e^{-x}} = O(x^{-1}) \quad \text{as } x \to +\infty$$

but $$\frac{1}{4-3x-2e^{-x}} = O(e^{x}) \quad \text{as } x \to -\infty.$$

(Little-oh and big-oh–o and O—are usually called the *Landau symbols*.)

(c) *Asymptotically equal to or behaves like*

Finally, we write

$$f(x) \sim g(x) \quad \text{as } x \to a$$

if the limit L, in (1.31), is precisely $L = 1$; then we say that 'f is asymptotically equal to g as $x \to a$' or 'f behaves like g as $x \to a$'. Some examples are

$$\frac{1}{3+2x^2} \sim \frac{1}{2x^2} \quad \text{as } |x| \to \infty;$$

$$\frac{5}{1-x-e^{-x}} \sim -\frac{10}{x^2} \quad \text{as } x \to 0;$$

$$\sin(3x) \sim 3x \quad \text{as } x \to 0$$

and then we may also write

$$\sin(3x) - 3x \sim -\frac{(3x)^3}{3!} \quad \text{as } x \to 0. \tag{1.32}$$

Finally, it is not unusual to use '=' in place of '$\sim$', but in conjunction with a measure of the error. So, with '$\sim$', 'O' and 'o' as defined above, we write

$$\sin(3x) = 3x + O(x^3) \quad \text{as } x \to 0$$

or $$\sin(3x) = 3x + o(x) \quad \text{as } x \to 0,$$

but such statements should be regarded as no more than equivalents to some of the statements given earlier. Some exercises that use o, O and $\sim$ are given in Q1.6, 1.7 and 1.8.

We should comment that other definitions exist for O, for example, although what we have presented is, we believe, the most straightforward and most directly useful. An alternative, in particular, is to define $f(x) = O[g(x)]$ as $x \to a$ if $\exists$ positive constants C and R s.t.

$$|f(x)| \le C\,|g(x)| \quad \text{for } 0 < |x-a| < R;$$

our limit definition follows directly from this.

1.3 ASYMPTOTIC SEQUENCES AND ASYMPTOTIC EXPANSIONS

First we recall example (1.32), which epitomises the idea that we will now generalise. We already have

$$\sin(3x) - 3x \sim -\frac{(3x)^3}{3!} \quad \text{as } x \to 0$$

and this procedure can be continued, so

$$\sin(3x) - 3x + \frac{(3x)^3}{3!} \sim \frac{(3x)^5}{5!} \quad \text{as } x \to 0, \tag{1.33}$$

(and the correctness of this follows directly from the Maclaurin expansion of $\sin(3x)$). The result in (1.33), and its continuation, produces progressively better approximations to $\sin(3x)$, in that we may write

$$\sin(3x) = 3x - \frac{(3x)^3}{3!} + \frac{(3x)^5}{5!} + \mathrm{O}(x^7) \quad \text{as } x \to 0,$$

and then

$$\sin(3x) = 3x - \frac{(3x)^3}{3!} + \frac{(3x)^5}{5!} - \frac{(3x)^7}{7!} + \mathrm{O}(x^9) \quad \text{as } x \to 0. \tag{1.34}$$

At each stage, we perform a 'varies as' calculation (as in (1.33), *via* the definition of '$\sim$'); in this example we have used the *set* of gauge functions $\{x^{1+2n}\}$ for $n = 0, 1, 2, \ldots$; such a set is called an *asymptotic sequence*. In order to proceed, we need to define a general set of functions which constitute an asymptotic sequence.

Definition (*asymptotic sequence*)

The set of functions $\{\phi_n(x)\}$, $n = 0, 1, 2, \ldots$, is an asymptotic sequence as $x \to a$, if

$$\phi_{n+1}(x) = \mathrm{o}\left[\phi_n(x)\right] \quad \text{as } x \to a \tag{1.35}$$

for every n.

As examples, we have

$$\{x^n\} \quad \text{as } x \to 0; \quad \left\{x^{-n}\right\} \quad \text{as } x \to \infty;$$

$$\left\{\mathrm{e}^{-nx}\right\} \quad \text{as } x \to +\infty; \quad \{\ln[1 + (1 - x)^n]\} \quad \text{as } x \to 1.$$

(In each case, it is simply a matter of confirming that $\lim_{x\to a}[\phi_{n+1}(x)/\phi_n(x)] = 0$.) Some further examples are given in Q1.9.

Now, with respect to an asymptotic sequence (that is, using the chosen sequence), we may write down a set of terms, such as (1.34); this is called an *asymptotic expansion*. We now give a formal definition of an asymptotic expansion (which is usually credited to Henri Poincaré (1854–1912)).

Definition (*asymptotic expansion*)

The series of terms written as

$$\sum_{n=0}^{N} c_n \phi_n(x) + \mathrm{O}(\phi_{n+1}),$$

where the c_n are constants, is an *asymptotic expansion* of $f(x)$, with respect to the asymptotic sequence $\{\phi_n(x)\}$ if, for every $N \geq 0$,

$$f(x) - \sum_{n=0}^{N} c_n \phi_n(x) = \mathrm{o}[\phi_N(x)] \quad \text{as } x \to a.$$

If this expansion exists, it is unique in that the coefficients, c_n, are completely determined.

There are some comments that we should add in order to make clear what this definition says and implies—and what it does not.

First, given only a function and a limit of interest (i.e. $f(x)$ and $x \to a$), the asymptotic expansion is not unique; it is unique (if it exists—we shall comment on this shortly) only if the asymptotic sequence is also prescribed. To see that this is the case, let us consider our function $\sin(3x)$ again; we will demonstrate that this can be represented, as $x \to 0$, in any number of different ways, by choosing different asymptotic sequences (although, presumably, we would wish to use the sequence which is the simplest). So, for example,

$$\{x^n\} : \sin(3x) \sim 3x - \frac{9}{2}x^3 + \frac{81}{40}x^5;$$

$$\{\ln(1+x^n)\} : \sin(3x) \sim 3\ln(1+x) + \frac{3}{2}\ln(1+x^2) - \frac{11}{2}\ln(1+x^3);$$

$$\left\{\frac{x^n}{(1+x^2)^{3/2}}\right\} : \sin(3x) \sim 3\frac{x}{(1+x^2)^{3/2}} - \frac{18}{5}\frac{x^5}{(1+x^2)^{3/2}};$$

$$\{(\sin x)^n\} : \sin(3x) \sim 3\sin x - 4(\sin x)^3;$$

indeed, this last example, is a familiar *identity* for $\sin(3x)$. (Another simple example of this non-uniqueness is discussed in Q1.10.) So, given a function and the limit, we need to select an *appropriate* asymptotic sequence—appropriate because, for some choices, the asymptotic expansion does not exist.

To see this, let us consider the function $\sin(3x)$ again, the limit $x \to 0^+$ and the asymptotic sequence $\{e^{-n/x}\}$. The first term in such an expansion, if it exists, will be a constant (corresponding to $n = 0$); but $\sin(3x) \to 0^+$ in this limit, so the constant is zero. Perhaps the first term is proportional to $e^{-n/x}$, for some $n > 0$; thus we examine

$$L = \lim_{x \to 0^+} \left[\frac{\sin(3x)}{c\,e^{-n/x}} \right].$$

If we are to have $\sin(3x) \sim c\,e^{-n/x}$ (for some n and some constant c), then this limit is to be $L = 1$. However, this limit does not exist—it is infinite—for *every* $n > 0$. Hence we are unable to represent $\sin(3x)$, as $x \to 0^+$, with the asymptotic sequence proposed (which many readers will find self-evident, essentially because $\sin(3x) \sim 3x$ as $x \to 0$). If every c_n, $n \geq 0$, in the asymptotic expansion is either zero or is undefined, then the expansion does not exist.

Let us take this one step further; if we have a function, a limit *and* an appropriate asymptotic sequence, then the coefficients, c_n, are unique. This is readily demonstrated. From the definition of an asymptotic expansion, we have

$$\lim_{x \to a} \left[\frac{f(x) - \sum_{n=0}^{N} c_n \phi_n(x)}{\phi_N(x)} \right] = 0;$$

consider

$$\frac{f(x) - \sum_{n=0}^{N-1} c_n \phi_n(x)}{\phi_N(x)} = \frac{f(x) - \sum_{n=0}^{N} c_n \phi_n(x) + c_N \phi_N(x)}{\phi_N(x)} = \frac{f(x) - \sum_{n=0}^{N} c_n \phi_n(x)}{\phi_N(x)} + c_N$$

and take the limit to give

$$c_N = \lim_{x \to a} \left[\frac{f(x) - \sum_{n=0}^{N-1} c_n \phi_n(x)}{\phi_N(x)} \right],$$

which determines each c_N, $N \geq 0$.

Finally, the terms $\sum_{n=0}^{N} c_n \phi_n(x)$ should not be regarded or treated as a series in any conventional way. This notation $\left(\sum_{n=0}^{N}\right)$ is simply a shorthand for a sequence of 'varies as' calculations (as in (1.33), for example); at no stage in our discussion have we written that these are the familiar objects called series—and certainly not *convergent* series. Indeed, many asymptotic expansions, if treated conventionally i.e. select a value $|x - a| \neq 0$ and compute the terms in the series, turn out to be *divergent* (although, exceptionally, some *are* convergent). Of course, numerical estimates are sometimes relevant, either to gain an insight into the nature of the solution or, more often, to provide a starting point for an iterative solution of the problem. Because these issues may be of some interest, we will (in §1.4) deviate from our main development and offer a few comments and observations. We must emphasise, however, that the thrust of this text is towards the introduction of methods which aid the description of the *structure* of a solution (in the limit under consideration).

Finally, before we move on, we briefly comment on functions of a complex variable. (We will present no problems that sit in the complex plane, but it is quite natural to ask if our definitions of an asymptotic expansion remain unaffected in this situation.) Given $f(z)$, $z = x + iy$, and the limit $z \to z_0$, we are able to construct asymptotic expansions exactly as described above, but with one important new ingredient. Because $z = z_0$ is a point in the complex plane, it is possible to approach z_0, i.e. take the limit, from any direction whatsoever. (For real functions, the limit can only be along the real line, either $x \to a^+$ or $x \to a^-$.) However, in general, the asymptotic correctness will hold only for certain directions and not for every direction e.g. $z \to z_0$ for $\theta_0 < \arg(z - z_0) < \theta_1$ (for some θ_0, θ_1), and for other args the asymptotic expansion (with the same asymptotic sequence, $\{\phi_n\}$) fails because $\phi_{n+1} \neq o(\phi_n)$ for some n.

1.4 CONVERGENT SERIES *VERSUS* DIVERGENT SERIES

Suppose that we have a function $f(x)$ and a series

$$f_N(x) = \sum_{n=0}^{N} a_n(x),$$

then $f_N(x)$ is a *convergent* series if $f_N(x) \to f(x)$ as $N \to \infty$ for all x satisfying $|x - a| < R$ (for some $R > 0$, the radius of convergence). This is a statement of the familiar property of the type of series that is usually encountered; so we have, for example, as $N \to \infty$, that

$$\sum_{n=0}^{N} \frac{x^n}{n!} \to e^x \quad \text{for all finite } x,$$

and

$$\sum_{n=0}^{N} x^n \to \frac{1}{1-x} \quad \text{but only for } -1 < x < 1.$$

One important consequence is that we may approximate a function, which has a convergent-series representation, to any desired accuracy, by retaining a sufficient number of terms in the series. For example

$$\sum_{n=0}^{10} \frac{1}{2^n} \approx 1.9990 \quad \text{and} \quad \sum_{n=0}^{25} \frac{1}{2^n} \approx 1.99999997$$

where the limit as $N \to \infty$ is 2. With these ideas in mind, we turn to the challenge of working with divergent series.

In this case, $f_N(x)$ has no limit as $N \to \infty$ for any x (except, perhaps, at the one value $x = a$, which alone is not useful). Usually f_N diverges—the situation that is typical of asymptotic expansions—but it may remain finite and oscillate. In either case, this suggests that any attempt to use a divergent series as the basis for numerical estimates is doomed to failure; this is not true. A divergent series can be used to estimate $f(x)$

for a given x, but the error in this case cannot be made as small as we wish. However, we are able to *minimise* the error, for a given x, by retaining a precise number of terms in the series–one term more or one less will increase the error. The number of terms retained will depend on the value of x at which $f(x)$ is to be estimated. This important property can be seen in the case of a (divergent) series which has alternating signs—a quite common occurrence—*via* a general argument.

Consider the identity

$$f(x) = \sum_{n=0}^{N} a_n(x) + R_N(x),$$

where N is finite; $R_N(x)$ is the remainder. Suppose that $a_N > 0$ and $a_{N-1} < 0$, with $R_N < 0$, $R_{N-1} > 0$ (and, correspondingly, a reversal of all the signs if $a_N < 0$); this describes the alternating-sign property of the series. Let us write

$$f(x) = \sum_{n=0}^{N-1} a_n(x) + a_N(x) + R_N(x)$$

then $$R_{N-1} = a_N + R_N \quad \text{or} \quad a_N = R_{N-1} - R_N.$$

But the remainders are of *opposite* sign, so they always *add* (not cancel, approximately), which we may express as

$$|a_N| = |R_{N-1}| + |R_N| \quad \text{or} \quad |a_N| > |R_N|; \tag{1.36a}$$

similarly $$|a_{N+1}| = |R_N| + |R_{N+1}| \quad \text{or} \quad |a_{N+1}| > |R_N|. \tag{1.36b}$$

Hence the magnitude of the remainder—the error in using the series—is less than the magnitude of the last term retained and also less than that of the first term omitted. It is important to observe that, provided N remains finite, it is immaterial to this argument whether the series is convergent or divergent. Thus, for a given x, we stop the series at the term with the smallest value of a_N (which, if the series is convergent, arises at infinity and is zero); the sum of the terms selected will then provide the best estimate for the function value. Let us investigate how this idea can be implemented in a classical example.

E1.5 The exponential integral

A problem which exhibits the behaviour that we have just described, and for which the calculations are particularly straightforward, is the *exponential integral*:

$$\mathrm{Ei}(x) = \int_x^{\infty} \frac{e^{-t}}{t}\,\mathrm{d}t, \quad x > 0. \tag{1.37}$$

We are interested, here, in evaluating $\mathrm{Ei}(x)$ for large x (and we observe that $\mathrm{Ei} \to \infty$ as $x \to 0^+$; see Q1.13); of course, we cannot perform the integration, but we can generate a suitable approximation *via* the familiar technique of integration by parts. In particular we obtain

$$\mathrm{Ei}(x) = \left[-\frac{\mathrm{e}^{-t}}{t}\right]_x^\infty - \int_x^\infty e^{-t}.\frac{1}{t^2}\mathrm{d}t = \frac{\mathrm{e}^{-x}}{x} - \left\{\left[-\frac{\mathrm{e}^{-t}}{t^2}\right]_x^\infty - \int_x^\infty \mathrm{e}^{-t}.\frac{2}{t^3}\mathrm{d}t\right\}$$

$$= \mathrm{e}^{-x}\left(\frac{1}{x} - \frac{1}{x^2}\right) + 2\int_x^\infty \frac{\mathrm{e}^{-t}}{t^3}\,\mathrm{d}t$$

and so on, to give

$$\mathrm{Ei}(x) = \mathrm{e}^{-x}\left\{\frac{1}{x} - \frac{1}{x^2} + \frac{2!}{x^3} - \frac{3!}{x^4} + \cdots\cdots + \frac{(-1)^{n-1}(n-1)!}{x^n}\right\} + (-1)^n n!\int_x^\infty \frac{\mathrm{e}^{-t}}{t^{n+1}}\mathrm{d}t. \tag{1.38}$$

Note that we have used a standard mathematical procedure, which has automatically generated a sequence of terms—indeed, it has generated an asymptotic sequence, $\{x^{-n}\}$, defined as $x \to \infty$. This is another important observation: our definitions have implied a *selection* of the asymptotic sequence, but in practice a particular choice either appears naturally (as here) or is thrust upon us by virtue of the structure of the problem; we will write more of this latter point in due course. Here, for the expansion of (1.37) in the form (1.38), we might regard $\{x^{-n}\}$, $x \to \infty$, as the *natural asymptotic sequence*. It is clear that we may write, for example,

$$\mathrm{Ei}(x) \sim \mathrm{e}^{-x}\left(\frac{1}{x} - \frac{1}{x^2}\right) \quad \text{as } x \to \infty,$$

but what of the convergence, or otherwise, of this series? In order to answer this, we will use the standard *ratio test*.

We construct

$$\left|\frac{(-1)^{n-1}(n-1)!\mathrm{e}^{-x}}{x^n}\bigg/\frac{(-1)^{n-2}(n-2)!\mathrm{e}^{-x}}{x^{n-1}}\right| = \left|\frac{(-1)(n-1)}{x}\right| = \frac{n-1}{x} \tag{1.39}$$

(because $x > 0$ and $n \geq 1$) and if this expression is less than unity as $n \to \infty$ for some x, then the series converges (absolutely). But the expression in (1.39) tends to infinity as $n \to \infty$, for *all* finite x; hence the series in (1.38) diverges. To examine this series in more detail, let us write (1.38) in the form

$$\mathrm{Ei}(x) = \mathrm{e}^{-x}S_n(x) + R_n(x) \tag{1.40}$$

where the series $S_n(x)$ can be interpreted as an asymptotic expansion for $x \to \infty$; $R_n(x)$ is the remainder, given by

$$R_n(x) = (-1)^n n!\int_x^\infty \frac{\mathrm{e}^{-t}}{t^{n+1}}\mathrm{d}t. \tag{1.41}$$

It is convenient, because it simplifies the details, if we elect to work with

$$I(x) = \mathrm{e}^x \mathrm{Ei}(x) = S_n(x) + \mathrm{e}^x R_n(x),$$

and then we have

$$I(x) = \frac{1}{x} + \mathrm{e}^x R_1(x); \quad I(x) = \frac{1}{x} - \frac{1}{x^2} + \mathrm{e}^x R_2(x); \quad I(x) = \frac{1}{x} - \frac{1}{x^2} + \frac{2}{x^3} + \mathrm{e}^x R_3(x),$$

and so on. Thus, using (1.36a,b), we obtain

$$|\mathrm{e}^x R_1| < \frac{1}{x} \quad \text{and} \quad |\mathrm{e}^x R_1| < \frac{1}{x^2}; \quad |\mathrm{e}^x R_2| < \frac{1}{x^2} \quad \text{and} \quad |e^x R_2| < \frac{2}{x^3}, \text{ etc.,}$$

so that, in general,

$$|\mathrm{e}^x R_n| < \frac{(n-1)!}{x^n} \quad \text{and} \quad |\mathrm{e}^x R_n| < \frac{n!}{x^{n+1}}.$$

The best estimate, for a given x, is obtained by choosing that n which minimises the smaller of these two bounds; in this example, this is clearly $n = [x]$ (where [ ] denotes 'the integral part of'). In fact, when x is itself an integer, these two bounds for $|\mathrm{e}^x R_n|$ are identical.

As a numerical example, we seek an estimate for Ei(5)—and since our asymptotic expansion is valid as $x \to \infty$, $x = 5$ appears to be a rather bold choice. The remainder then satisfies

$$\left|e^5 R_5\right| < \frac{5!}{5^6} = \frac{4!}{5^5} \approx 0.00768$$

and

$$I(5) = \frac{1}{5} - \frac{1}{5^2} + \frac{2}{5^3} - \frac{6}{5^4} + \frac{24}{5^5} + \mathrm{e}^5 R_5 \approx 0.174 + e^5 R_5$$

i.e. $0.166 < I(5) < 0.174$, where we have re-introduced the sign of the remainder, so that $-4!/5^5 < \mathrm{e}^5 R_5 < 0$, and then we obtain $0.00112 < \mathrm{Ei}(5) < 0.00117$. The surprise, perhaps, is that a divergent asymptotic expansion, valid as $x \to \infty$, can produce tolerable estimates for xs as small as 5. Of course, for larger values of x, the estimates are more accurate e.g. $0.09155 < I(10) < 0.09158$, from which we can obtain a good estimate for Ei(10). Two further examples for you to investigate, similar to this one, can be found in Q1.11, 1.12; other asymptotic expansions of integrals are discussed in Q1.13–1.17 and finding an expansion from a differential equation is the exercise in Q1.18.

In this example, E1.5, we have used the alternating-sign property, but we could have worked directly with the remainder, $R_n(x)$. If it is possible to obtain a reasonable estimate for the remainder, there is no necessity to invoke a special property of the series (which in any event, perhaps, is not available). Here, we have (from (1.41))

$$|R_n(x)| = n! \int_x^\infty \frac{e^{-t}}{t^{n+1}}\,dt < \frac{n!}{x^{n+1}} \int_x^\infty e^{-t}\,dt$$

for $\infty > x > 0$, because $1/t^{n+1} \leq 1/x^{n+1}$ (where $\infty > t \geq x$), and so

$$|R_n(x)| < \frac{e^{-x} n!}{x^{n+1}}.$$

For any given x, this estimate for the remainder is minimised by the choice $n = [x]$, exactly as we found earlier. The only disadvantage in using this approach, for any general series, is that we may not know the sign of the remainder, and so we must content ourselves with the error $\pm|R_n|$.

Although a study of series, both convergent and divergent, is a very worthwhile undertaking and, as we have seen, it can produce results relevant to some aspects of our work, we must move on. We now turn to that most important class of asymptotic expansions: those that use a parameter as the basis for the expansion.

1.5 ASYMPTOTIC EXPANSIONS WITH A PARAMETER

We now introduce functions, $f(x;\varepsilon)$, which depend on a parameter ε, and are to be expanded as $\varepsilon \to 0$. Here, x may be either a scalar or a vector (although our early examples will involve only scalars). In the case of vectors, we might write (in longhand) $f(x, y, z, t; \varepsilon)$; note that commas separate the variables, but that a semicolon is used to separate the parameter. As we shall see, it does not much matter in this work if the function we (eventually) seek is a solution of an ordinary differential equation (x is a scalar) or a solution of a partial differential equation (x is a vector): the techniques are essentially the same. The appropriate definition of the asymptotic expansion now follows.

Definition (*asymptotic expansion with a parameter 1*)

With respect to the asymptotic sequence $\{\delta_n(\varepsilon)\}$, defined as $\varepsilon \to 0$, we write the asymptotic expansion of $f(x;\varepsilon)$ as

$$f(x;\varepsilon) \sim \sum_{n=0}^{N} a_n(x)\delta_n(\varepsilon) \qquad (1.42)$$

for $x = O(1)$ and every $N \geq 0$. The requirement that $x = O(1)$ is equivalently that x is fixed as the limit process $\varepsilon \to 0$ is imposed.

Now suppose that f is defined in some domain, D say, which will usually be prescribed by the nature of the given problem e.g. the region inside a box which contains a gas. It is at this stage that we pose a fundamental question: does the asymptotic expansion in

(1.42) hold for $\forall x \in D$? If the answer is 'yes', then the expansion is said to be *regular* or *uniform* or *uniformly valid*; if not, then the expansion is *singular* or *non-uniform* or *not uniformly valid*. Further, it is not unusual to use the terms *breakdown* or *blow up* to describe the failure of an asymptotic expansion. To explore these ideas, we introduce a first, simple example.

E1.6 An example of $f(x;\varepsilon)$

Let us consider the function

$$f(x;\varepsilon) = \sqrt{x+\varepsilon+\frac{\varepsilon^2}{1+x}}, \quad x \geq 0, \tag{1.43}$$

for $\varepsilon \to 0^+$, and use the binomial expansion to obtain the 'natural' asymptotic expansion, valid for $x = \mathrm{O}(1)$:

$$f(x;\varepsilon) = \sqrt{x}\left(1+\frac{\varepsilon}{x}+\frac{\varepsilon^2}{x(1+x)}\right)^{1/2} \sim \sqrt{x}\left\{1+\frac{\varepsilon}{2x}+\frac{\varepsilon^2}{8}\frac{(3x-1)}{x^2(1+x)}\right\}. \tag{1.44}$$

Here, the asymptotic sequence is $\{\varepsilon^n\}$ and we have taken the expansion as far as terms at $\mathrm{O}(\varepsilon^2)$. But the domain of f is given as $x \geq 0$, and clearly the expansion (1.44) is not even defined on $x = 0$ (which is more dramatic than simply not being valid near $x = 0$). Thus (1.44) is not uniformly valid–indeed, it 'blows up' at $x = 0$.

The original function can, of course, be evaluated at $x = 0$:

$$f(0;\varepsilon) = \sqrt{\varepsilon+\varepsilon^2} \sim \sqrt{\varepsilon}\left(1+\frac{\varepsilon}{2}\right) \quad \text{as } \varepsilon \to 0^+, \tag{1.45}$$

and now another complication is evident. The asymptotic sequence used in (1.44) does not include terms $\varepsilon^{1/2}, \varepsilon^{3/2}, \ldots,$ and so it could never give the correct value on $x = 0$, even if the terms were defined there. Clearly, the expansion in (1.44) has been obtained by treating x large relative to ε, but this cannot be true if x is sufficiently small. The critical size is where x is about the size of ε, which is precisely the idea that led us to the introduction of a *scaled* version of x. Let us write $x = \varepsilon X$, then

$$f(\varepsilon X;\varepsilon) \equiv F(X;\varepsilon) = \sqrt{\varepsilon(1+X)+\frac{\varepsilon^2}{1+\varepsilon X}}, \tag{1.46}$$

where we have labelled the same function, expressed in terms of X and ε, as $F(X;\varepsilon)$. The binomial expansion of (1.46), for $\varepsilon \to 0^+$ with $X = \mathrm{O}(1)$, yields

$$F(X;\varepsilon) \sim \sqrt{\varepsilon}\sqrt{1+X}\left(1+\frac{1}{2}\frac{\varepsilon}{(1+X)}\right), \tag{1.47}$$

which, on $X = 0$, recovers (1.45).

Thus we have two representations of $f(x;\varepsilon)$, one valid for $x = \mathrm{O}(1)$, (1.44), and one for $x = \mathrm{O}(\varepsilon)$, (1.47). Further, the latter expansion is defined on $X = 0$ (i.e. $x = 0$) and gives the correct value (as an expansion of $f(0;\varepsilon)$). With these observations in place, we are now in a position to discuss uniformity and breakdown more completely and more carefully.

1.6 UNIFORMITY OR BREAKDOWN

Suppose that we wish to represent $f(x;\varepsilon)$, for $x \in D$, by an asymptotic expansion

$$f(x;\varepsilon) \sim \sum_{n=0}^{N} a_n(x)\delta_n(\varepsilon) \quad \text{as } \varepsilon \to 0^+,$$

which has been constructed for $x = \mathrm{O}(1)$. This expansion is *uniformly valid* if

$$a_N(x)\delta_N(\varepsilon) = \mathrm{o}\left[a_{N-1}(x)\delta_{N-1}(\varepsilon)\right] \quad \text{as } \varepsilon \to 0^+,$$

for every $N \geq 1$ and $\forall x \in D$. Conversely, it *breaks down* (and is therefore non-uniform) if there is some $x \in D$, and some $N \geq 1$, such that

$$a_N(x)\delta_N(\varepsilon) = \mathrm{O}\left[a_{N-1}(x)\delta_{N-1}(\varepsilon)\right].$$

In other words, the expansion is said to break down if there is a size of x, in the domain of the function, for which two consecutive terms in the asymptotic expansion are the same size. On the other hand, the expansion is uniformly valid if the asymptotic ordering of the terms, as represented by the asymptotic sequence $\{\delta_n(\varepsilon)\}$, is maintained for all x in the domain.

It is an elementary exercise to apply this principle to our previous example; from (1.44) we have

$$f(x;\varepsilon) \sim \sqrt{x}\left(1 + \frac{\varepsilon}{2x}\right), \tag{1.48}$$

and the domain of the original function is $x \geq 0$. As $x \to 0^+$, the second term in the expansion, (1.48), becomes the same size as the first where $x = \mathrm{O}(\varepsilon)$: the expansion has broken down. That is, for x of this size, the expansion (1.48) is no longer valid; in order to determine the form of the expansion for $x = \mathrm{O}(\varepsilon)$, we must return to the function and use this choice i.e. write $x = \varepsilon X$—which is exactly how we generated (1.47). Thus the breakdown of an expansion can lead us to the choice of a new, scaled variable, and we note that this is based on the properties of the *expansion*, not any additional or special knowledge about the underlying function. (This point is important for what will come later: when we solve differential equations, we will not have the exact solution available—only an asymptotic expansion of the solution. But, as we shall see, the equation itself does hold information about possible scalings.) We apply this principle of breakdown and rescaling to another example.

E1.7 Another example of $f(x;\varepsilon)$

Here, we are given

$$f(x;\varepsilon)=\frac{x\sqrt{1+\varepsilon x}}{x+\varepsilon},\qquad x\geq 0,$$

with $\varepsilon\to 0^+$; for $x=\mathrm{O}(1)$ we write

$$f(x;\varepsilon)=\left(1+\frac{\varepsilon}{x}\right)^{-1}(1+\varepsilon x)^{1/2}$$

and then two applications of the binomial expansion yields

$$f(x;\varepsilon)\sim 1+\varepsilon\left(\frac{x}{2}-\frac{1}{x}\right)\quad\text{as }\varepsilon\to 0^+. \tag{1.49}$$

The domain of f is $x\geq 0$, and so we must consider $x\to 0$ and $x\to\infty$; in either case the asymptotic expansion (1.49) breaks down. For $x\to 0$, the breakdown occurs where $x=\mathrm{O}(\varepsilon)$ (from $\varepsilon/x=\mathrm{O}(1)$); for $x\to\infty$, the breakdown is where $x=\mathrm{O}(\varepsilon^{-1})$ (from $\varepsilon x=\mathrm{O}(1)$). In the former case, we introduce $x=\varepsilon X$, to give

$$f(\varepsilon X;\varepsilon)\equiv F(X;\varepsilon)=\frac{X\sqrt{1+\varepsilon^2 X}}{1+X}\sim\left(\frac{X}{1+X}\right)\left(1+\frac{1}{2}\varepsilon^2 X\right)\quad\text{for } X=\mathrm{O}(1) \tag{1.50}$$

(which, we note, recovers the correct value on $X=0$). For the other breakdown, we introduce $x=\chi/\varepsilon$, and so

$$f(\chi/\varepsilon;\varepsilon)\equiv\mathfrak{F}(\chi;\varepsilon)=\frac{\chi\sqrt{1+\chi}}{\chi+\varepsilon^2}\sim\sqrt{1+\chi}\left(1-\frac{\varepsilon^2}{\chi}\right)\quad\text{for }\chi=\mathrm{O}(1). \tag{1.51}$$

Thus the function requires three different asymptotic expansions, valid for different sizes of x, and two of these have been determined by examining the breakdown. (We note that these choices are evident from the original function, although this is not how we deduced the scalings in this example.) Furthermore, expansion (1.50) is valid as $X\to 0$, and expansion (1.51) is valid for $\chi\to\infty$: there are no further breakdowns (based on the information available in these asymptotic expansions).

Before we continue the discussion of these ideas, and their consequences, we must adjust the definition of an asymptotic expansion with a parameter; see (1.42). We have already encountered functions such as $\exp(-x/\varepsilon)$; these cannot be represented in *separable* form i.e. $a(x)\delta(\varepsilon)$, for any choice of the functions a and δ. Thus we must extend the definition of an asymptotic expansion to accommodate this:

Definition (*asymptotic expansion with a parameter 2*)

We write (cf. (1.42)

$$f(x;\varepsilon) \sim \sum_{n=0}^{N} a_n(x;\varepsilon)$$

for $x = \mathrm{O}(1)$ and every $N \geq 0$, where

$$a_N(x;\varepsilon) = \mathrm{o}[a_{N-1}(x;\varepsilon)] \quad \text{as } \varepsilon \to 0^+.$$

It is clear that the separable case is simply a special version of this more general definition; let us investigate an example which incorporates such a term.

E1.8 One more example of $f(x;\varepsilon)$

Consider the function

$$f(x;\varepsilon) = (1 + x + \varepsilon + \mathrm{e}^{-x/\varepsilon})^{-1}, \quad x \geq 0, \tag{1.52}$$

for $\varepsilon \to 0^+$; with $x = \mathrm{O}(1)$ we obtain

$$f(x;\varepsilon) \sim \frac{1}{1+x}\left(1 - \frac{\varepsilon}{1+x}\right), \tag{1.53}$$

because the term $\exp(-x/\varepsilon)$ is exponentially small. This asymptotic expansion, (1.53), as written down, is uniformly valid: there is no breakdown as $x \to 0$ and the asymptotic ordering of the terms is even reinforced as $x \to \infty$. However, from (1.52), we see that

$$f(0;\varepsilon) = (2+\varepsilon)^{-1} \sim \tfrac{1}{2}(1 - \tfrac{1}{2}\varepsilon) \tag{1.54}$$

which is not the result we obtain from (1.53): the (complete) expansion, started in (1.53), cannot be uniformly valid! Of course, it is clear that the difficulty is associated with the exponential term; it is this which contributes to the boundary value (on $x = 0$), but it is ignored in the 2-term asymptotic expansion (1.53).

The rôle of the exponential term becomes evident when we retain it, following our familiar scaling ($x = \varepsilon X$), which gives

$$f(\varepsilon X;\varepsilon) \equiv F(X;\varepsilon) = \left(1 + \mathrm{e}^{-X} + \varepsilon X + \varepsilon\right)^{-1} \sim \frac{1}{1+\mathrm{e}^{-X}}\left(1 - \frac{\varepsilon(1+X)}{1+\mathrm{e}^{-X}}\right) \tag{1.55}$$

for $X = O(1)$ as $\varepsilon \to 0^+$. Then the value of (1.55), on $X = 0$, recovers the correct boundary value, (1.54). Furthermore, the asymptotic expansion (1.55) is not uniformly valid in $X \geq 0$: as $X \to \infty$, it breaks down where $\varepsilon X = O(1)$ i.e. $x = O(1)$, which is the variable previously used to generate (1.53).

This example prompts a number of additional and important observations. For the purposes of determining a relevant scaling, from the breakdown of an expansion, it is quite sufficient for this to occur only in one direction i.e. expansion A breaks down, producing a scaling used to obtain expansion B, but B does not necessarily break down to recover the scaling used in A. This is evident here when we compare (1.53) and (1.55); expansion (1.55) breaks down, but (1.53) does not. Indeed, as we have seen, there is no clue in (1.53) that we have a problem—this is only evident when we return to the original function, (1.52), or we already have available the expansion (1.55). It *is* possible to extend the asymptotic expansion given in (1.53) and thereby make plain the nature of the breakdown; this will prove to be a useful adjunct in some of our later work.

The breakdown that must exist in the expansion of (1.52), for $x = O(1)$ as $\varepsilon \to 0^+$, arises from the exponential term. However, to include this term in the asymptotic expansion would mean, apparently, the inclusion of all terms based on the sequence $\{\varepsilon^n\}$, because $\exp(-x/\varepsilon)$ is smaller than ε^n for any n (see Q1.5). Of course, there is no need to write them down explicitly; we could indicate their presence by the use of an ellipsis (i.e. ...) or, which is the usual practice, simply to state which terms we will retain. So we might expand (1.52), for $x = O(1)$ and $\varepsilon \to 0^+$, retaining $O(1)$, $O(\varepsilon)$ and $O(e^{-x/\varepsilon})$ terms only, to give

$$f(x;\varepsilon) \sim \frac{1}{1+x}\left(1 - \frac{\varepsilon}{1+x} - \frac{e^{-x/\varepsilon}}{1+x}\right). \qquad (1.55)$$

In a sense, the omitted terms are understood, but not explicitly included and, more significantly, any further manipulation of (1.55) that we employ will use *only* the terms *written down*. It is clear that, with or without the use of ellipsis, the expansion (1.55) breaks down as $x \to 0^+$ for, eventually, the exponential term becomes $O(1)$—the same size as the first term. (The fact that there is an infinity of breakdowns, where x satisfies $e^{-x/\varepsilon} = O(\varepsilon^n)$ for each n, is immaterial; we have a well-defined breakdown the other way—from (1.55)—which is sufficient. Further, an intimate relation between different expansions of the same function, which we discuss later, shows that this infinity of breakdowns plays no rôle.)

The inclusion of the exponentially small term in (1.55) may seem superfluous, and it is in a strictly numerical sense, but it contains important information about the nature of the underlying function (and it helps us better to understand the breakdown). Because we are interested in the behaviour of functions (as $\varepsilon \to 0$), and not simply numerical estimates, we shall retain such terms when they provide useful and relevant information.

1.7 INTERMEDIATE VARIABLES AND THE OVERLAP REGION

In our examples thus far, we have expanded the given functions for $x = \mathrm{O}(1)$, $x = \mathrm{O}(\varepsilon)$ and, in one case, for $x = \mathrm{O}(\varepsilon^{-1})$. We now investigate other scalings which correspond to sizes that sit *between* those generated by the breakdown of an asymptotic expansion. This will lead us to an important and significant principle in the theory of singular perturbations.

Let us suppose that we have an asymptotic expansion of a function which is valid for $x = \mathrm{O}(1)$, and another of the same function which is valid for $x = \mathrm{O}(\varepsilon)$; further, the breakdown of at least one of these expansions produces the scaling used in the other. The line we now pursue is to examine what happens to these expansions when we allow $x = \mathrm{O}(\delta)$, where

$$\delta(\varepsilon) \to 0 \quad \text{as } \varepsilon \to 0, \qquad \text{but } \varepsilon/\delta(\varepsilon) \to 0 \quad \text{as } \varepsilon \to 0 \tag{1.56}$$

i.e. the size (scale) of x is smaller than $\mathrm{O}(1)$ but not as small as $\mathrm{O}(\varepsilon)$. Given that the expansion valid for $x = \mathrm{O}(1)$ breaks down at $x = \mathrm{O}(\varepsilon)$, the asymptotic ordering of the terms is unaltered if we use $x = \mathrm{O}(\delta)$ i.e. it is still valid for this size of x. Conversely, we are given that the expansion valid for $x = \mathrm{O}(\varepsilon)$ breaks down where $x = \mathrm{O}(1)$, but it remains valid for x smaller than $\mathrm{O}(1)$—so this is also valid for $x = \mathrm{O}(\delta)$. Hence *both* expansions are valid for $x = \mathrm{O}(\delta)$—an *intermediate variable*; furthermore, this validity holds for all $\delta(\varepsilon)$ which satisfy (1.56). In order to make plain what is happening here, let us apply this procedure to an example.

E1.9 Example with an intermediate variable

We are given the two asymptotic expansions

$$f(x;\varepsilon) \sim \sqrt{1+x}\left(1 + \frac{\varepsilon}{2x(1+x)}\right), \qquad x = \mathrm{O}(1); \tag{1.57}$$

$$f(\varepsilon X;\varepsilon) \equiv F(X;\varepsilon) \sim \sqrt{\frac{2+X}{1+X}}\left(1 + \varepsilon\frac{X(1+X)}{2(2+X)}\right), \qquad X = \mathrm{O}(1), \tag{1.58}$$

both as $\varepsilon \to 0^+$. (It is left as an exercise to show that these expansions are obtained from the function $f(x;\varepsilon) = \sqrt{1 + x + \varepsilon/(x+\varepsilon)}$, $x \geq 0$, but we do not need to know the form of the function in what follows.)

In the expansion (1.57), we write $x = \delta\chi$, where $\delta(\varepsilon)$ is defined in (1.56), and expand:

$$\sqrt{1+\delta\chi}\left(1 + \frac{\varepsilon}{2\delta\chi(1+\delta\chi)}\right) \sim 1 + \frac{1}{2}\delta\chi + \frac{\varepsilon/\delta}{2\chi}, \qquad \chi = \mathrm{O}(1), \tag{1.59}$$

where we have retained terms $\mathrm{O}(1)$, $\mathrm{O}(\delta)$ and $\mathrm{O}(\varepsilon/\delta)$. It is not clear how many terms we should retain, without being more precise about the size of $\delta(\varepsilon)$. For example,

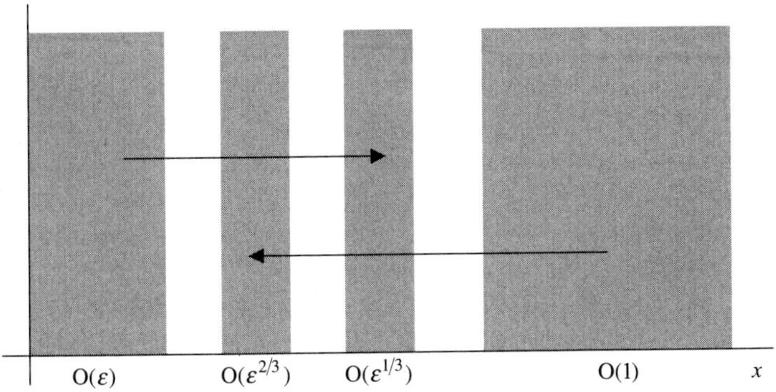

Figure 2. Diagrammatic representation of the overlap region, between $O(\varepsilon^{2/3})$ and $O(\varepsilon^{1/3})$.

if we choose $\delta = \varepsilon^{1/3}$, then $\varepsilon/\delta = \varepsilon^{2/3}$ and so, to be consistent, we should certainly include $\delta^2 = \varepsilon^{2/3}$; the issue of precisely which terms we should retain will be addressed later. Correspondingly, in the expansion (1.58), we write $X = x/\varepsilon = (\delta/\varepsilon)\chi$ and expand:

$$\sqrt{\frac{2+(\delta/\varepsilon)\chi}{1+(\delta/\varepsilon)\chi}}\left(1+\frac{\delta\chi[1+(\delta/\varepsilon)\chi]}{2[2+(\delta/\varepsilon)\chi]}\right)$$

$$= \sqrt{\frac{\chi+2(\varepsilon/\delta)}{\chi+(\varepsilon/\delta)}}\left(1+\frac{\delta\chi}{2}\left[\frac{\chi+(\varepsilon/\delta)}{\chi+2(\varepsilon/\delta)}\right]\right) \sim 1+\frac{1}{2}\delta\chi+\frac{\varepsilon/\delta}{2\chi}. \tag{1.60}$$

We have chosen to write down only the first three terms of this expansion, in order to be consistent with (1.59). We see that (1.59) and (1.60) are identical and, furthermore, this holds for all $\delta(\varepsilon)$ satisfying (1.56): we have verified, in this example, the rôle of the intermediate variable. (Again, it is left as an exercise to show that the same results are obtained directly by expanding the original function $f(\delta\chi;\varepsilon)$, for $\delta \to 0$, $\varepsilon/\delta \to 0$.)

To proceed with this discussion, we now make choices for $\delta(\varepsilon)$—one choice for the O(1) expansion (e.g. (1.59)), and another for the $x = O(\varepsilon)$ expansion (e.g. (1.60)). For example, in the former we might select $\delta = \varepsilon^{2/3}$, and in the latter we use $\delta = \varepsilon^{1/3}$; both 'expansions of expansions' are valid for these choices, because both satisfy (1.56). Thus the asymptotic expansion, constructed for $x = O(1)$, is valid also for $x = O(\varepsilon^{2/3})$; correspondingly, the expansion constructed for $x = O(\varepsilon)$, is valid for $x = O(\varepsilon^{1/3})$. Hence the expansion of expansions (e.g. (1.59) or (1.60)) is valid for xs *between* $O(\varepsilon^{2/3})$ and $O(\varepsilon^{1/3})$: the resulting expansion is now valid in an *overlap region* (which is represented in figure 2).

We have made one choice ($\varepsilon^{2/3}$, $\varepsilon^{1/3}$), but the same argument can be developed for other choices; in particular, we could use ε^{1-p}, ε^p $(0 < p < 1)$. As p decreases, so the overlap region extends; indeed, we may allow this to proceed provided $p > 0$ (because (1.56) still holds). Of course, if we permit the *limit* $p \to 0$, then conditions (1.56) will be violated, although we may allow p to be as close as we desire to zero. This obviously prompts the question: what does happen to our procedure—the expansion of expansions—if we do select $p = 0$? After all—being naïve—it would seem but a small step from p nearly zero (which is permitted) to $p = 0$ (although we are all aware that there can be big differences between $x \to a$ and $x = a$ in some contexts!). In fact, this situation here is not unfamiliar; it is analogous to the discussion that must be undertaken when the convergence of a series is investigated. Given that a series is convergent for $|x - a| < R$, and divergent for $|x - a| > R$, its status for $|x - a| = R$ (i.e. the two cases $x = a \pm R$) must be investigated *via* individual and special calculations. Here, we will employ the same philosophy, namely, to apply our procedure in the case $p = 0$, and note the results; they may, or may not, prove useful. In the event, it will transpire that the results are fundamentally important, and lead to a very significant property of asymptotic expansions.

1.8 THE MATCHING PRINCIPLE

Again, we suppose that we have two asymptotic expansions, one valid for $x = \mathrm{O}(1)$ and one for $x = \mathrm{O}(\varepsilon)$, exactly as described in the previous section. This time, however, we expand the first expansion for $x = \mathrm{O}(\varepsilon)$, and the second for $x = \mathrm{O}(1)$, i.e. the overlap region is the maximum that we can envisage (and one step beyond anything permitted so far). We know that this procedure is acceptable for the pair (ε^{1-p}, ε^p), with $0 < p < 1$, but now we set $p = 0$. Let us investigate this by returning to our previous example.

E1.10 Example with the maximum overlap

As in E1.9, we are given

$$f(x;\varepsilon) \sim \sqrt{1+x}\left(1 + \frac{\varepsilon}{2x(1+x)}\right), \quad x = \mathrm{O}(1); \tag{1.61}$$

$$f(\varepsilon X;\varepsilon) \equiv F(X;\varepsilon) \sim \sqrt{\frac{2+X}{1+X}}\left(1 + \frac{\varepsilon X(1+X)}{2(2+X)}\right), \quad X = \mathrm{O}(1), \tag{1.62}$$

and we expand (1.61) further, using $x = \varepsilon X$. We retain terms $\mathrm{O}(1)$ and $\mathrm{O}(\varepsilon)$, because we have no information about terms $\mathrm{O}(\varepsilon^2)$ and smaller. Thus we obtain

$$\sqrt{1+\varepsilon X}\left(1 + \frac{1}{2X(1+\varepsilon X)}\right) \sim 1 + \frac{1}{2X} + \frac{\varepsilon}{2}\left(X - \tfrac{1}{2}\right), \tag{1.63}$$

obtained by expanding with $X = \mathrm{O}(1)$; correspondingly, (1.62) gives

$$\sqrt{\frac{2+(x/\varepsilon)}{1+(x/\varepsilon)}}\left(1+\frac{x(1+x/\varepsilon)}{2(2+x/\varepsilon)}\right) = \sqrt{\frac{x+2\varepsilon}{x+\varepsilon}}\left(1+\frac{x}{2}\left\{\frac{x+\varepsilon}{x+2\varepsilon}\right\}\right)$$
$$\sim 1+\frac{x}{2}+\frac{\varepsilon}{2}\left(\frac{1}{x}-\frac{1}{2}\right), \qquad (1.64)$$

and we see that (1.63) and (1.64) are *identical*, when expressed in terms of the same variable (x or X). Further, these new expansions are *not* recoverable from (1.59) or (1.60) simply by writing $\delta = \varepsilon$ there: being precise about the terms to be retained has resulted in the appearance of a new term $(-\varepsilon/4)$. The two expansions, (1.61) and (1.62), are said to *match* to this order (because we can match only the terms available in the original expansions).

The *matching principle* is a fundamental tool in the techniques of singular perturbation theory; it is invoked, sometimes as a check, but more often as a means for determining arbitrary constants (or functions) that are generated in the solution of differential equations. Although we have not presented the matching principle as a *proven* property of functions—it is one reason why we call it a 'principle'—we have every confidence in its validity. For some classes of functions, it is possible to develop a proof which goes something like this. Define the operator $E_1^{(n)}$ which generates the first n terms of the asymptotic expansion of $f(x_1;\varepsilon)$, as $\varepsilon \to 0^+$ for $x_1 = \mathrm{O}(1)$ (written $E_1^{(n)} f$); correspondingly, the operator $E_2^{(m)}$ generates the first m terms of the asymptotic expansion of $F(x_2;\varepsilon)$, as $\varepsilon \to 0^+$ for $x_2 = \mathrm{O}(1)$. Here, the two functions are identical in that $F(x_2;\varepsilon) \equiv f(\delta(\varepsilon)x_2;\varepsilon)$ for some scaling $\delta(\varepsilon)$, obtained from the breakdown of the asymptotic expansion(s). Under suitable conditions—but we are able to apply the principle more widely—it can be proved that

$$E_2^{(m)}\left\{\left(E_1^{(n)} f\right)(x_2;\varepsilon)\right\} \equiv E_1^{(n)}\left\{\left(E_2^{(m)} F\right)(x_1;\varepsilon)\right\}$$

when written in the same variable i.e. x_1 or x_2. (Much more on these ideas can be found in some of the texts and references that are listed in the section on Further Reading at the end of this chapter.) Put simply, this states that the m-term expansion of the n-term expansion is identical to the n-term expansion of the m-term expansion; when presented in this form, this procedure is usually associated with the name of Milton van Dyke (1964, 1975). We present a slight variant of the principle, which we hope is transparent and readily applicable.

Definition (*matching principle*)

We are given two asymptotic expansions of a function, defined as $\varepsilon \to 0^+$, and valid for $x = O(1)$ and $X = O(1)$, where $x = \delta(\varepsilon)X$ (and either $\delta(\varepsilon) \to 0$ or $1/\delta(\varepsilon) \to 0$ as $\varepsilon \to 0^+$), in the form

$$\sum_{n=0}^{N} a_n(x;\varepsilon) \quad \text{and} \quad \sum_{m=0}^{M} b_m(X;\varepsilon),$$

respectively. These two expansions are valid in *adjacent* regions, so that the breakdown of one leads to the variable used in the other i.e. there are no other regions, and associated asymptotic expansions, between them. Here, N and M are not used simply to count the first so-many terms in the expansions; they may be used to designate the type of terms e.g. first three using the asymptotic sequence $\{\varepsilon^n\}$ and the first exponentially small term. However, N and M must retain these interpretations throughout the matching process. Now we form

$$\sum_{n=0}^{N} a_n\{\delta(\varepsilon)X;\varepsilon\} \sim \sum_{m=0}^{M} A_m(X;\varepsilon) \quad \text{for } X = O(1)$$

and

$$\sum_{m=0}^{M} b_m\left\{x/\delta(\varepsilon);\varepsilon\right\} \sim \sum_{n=0}^{N} B_n(x;\varepsilon) \quad \text{for } x = O(1);$$

the ***matching principle*** then states that

$$\sum_{m=0}^{M} A_m(X;\varepsilon) \equiv \sum_{n=0}^{N} B_n(\delta(\varepsilon)X;\varepsilon),$$

or expressed in terms of x, if preferred. We say that the expansions 'match to this order', because we can match only the terms that we have in the expansions.

Let us apply this matching principle, as we have described it, in the following example.

E1.11 An example of matching

We will show that these two expansions match:

$$f_N = \frac{1}{1+x}\left(1 - \frac{\varepsilon}{1+x} - \frac{e^{-x/\varepsilon}}{1+x} + \frac{2\varepsilon e^{-x/\varepsilon}}{(1+x)^2}\right), \quad x = O(1); \tag{1.65}$$

$$F_M = \frac{1}{1+e^{-X}}\left(1 - \frac{\varepsilon(1+X)}{1+e^{-X}}\right), \quad X = O(1) \quad \text{where } x = \varepsilon X, \tag{1.66}$$

both defined as $\varepsilon \to 0^+$. (This is based on example E1.8.) Note that, although (1.66) uses the first two terms in the sequence $\{\varepsilon^n\}$, $n = 0, 1, 2, \ldots$, the expansion (1.65) retains the first two (in size), plus the first exponentially small term and then the first which involves both the sequences $\{\varepsilon^n\}$ and $\left\{e^{-x/\varepsilon}\right\}$; these two specifications define N and M here.

To proceed, we write f_N in terms of X and expand:

$$\begin{aligned}
&\frac{1}{1+\varepsilon X}\left(1-\frac{\varepsilon}{1+\varepsilon X}-\frac{e^{-X}}{1+\varepsilon X}+\frac{2\varepsilon e^{-X}}{(1+\varepsilon X)^2}\right)\\
&\sim(1-\varepsilon X)(1-e^{-X})-\varepsilon+\varepsilon Xe^{-X}+2\varepsilon e^{-X}\\
&=1-e^{-X}+\varepsilon(2Xe^{-X}-X-1+2e^{-X}),\quad X=O(1),
\end{aligned} \tag{1.67}$$

where we have retained terms $O(1)$ and $O(\varepsilon)$ (which constitute M here). Correspondingly, we write F_M, (1.66), in terms of x and expand:

$$\begin{aligned}
\frac{1}{1+e^{-x/\varepsilon}}\left(1-\frac{x+\varepsilon}{1+e^{-x/\varepsilon}}\right)&=\left(1+e^{-x/\varepsilon}\right)^{-1}\left\{1-(x+\varepsilon)\left(1+e^{-x/\varepsilon}\right)^{-1}\right\}\\
&\sim\left(1-e^{-x/\varepsilon}\right)(1-x-\varepsilon)+(x+\varepsilon)e^{-x/\varepsilon}\\
&=1-x-\varepsilon-(1-2x)e^{-x/\varepsilon}+2\varepsilon e^{-x/\varepsilon},\quad x=O(1),
\end{aligned} \tag{1.68}$$

where we have retained terms $O(1)$, $O(\varepsilon)$, $O(e^{-x/\varepsilon})$ and $O(\varepsilon e^{-x/\varepsilon})$, which is N here. Finally, we write (1.68), say, in terms of X:

$$1-\varepsilon X-\varepsilon-(1-2\varepsilon X)e^{-X}+2\varepsilon e^{-X}=1-e^{-X}+\varepsilon(2Xe^{-X}-X-1+2e^{-X}),$$

which is identical to (1.67): the two expansions match (to this order).

This example makes clear that we need not restrict the matching to the *first* N and *first* M terms (in size)—but we must accurately identify N and M and then retain precisely these terms when the expansions are further expanded. Of course, as we have seen, there is also no requirement to work to the same number of terms in each original asymptotic expansion. However, we should offer one word of warning. In the above example, we included the term $\varepsilon e^{-x/\varepsilon}$, which arises from the two otherwise disjoint asymptotic sequences $\{\varepsilon^n\}$ and $\{e^{-nx/\varepsilon}\}$. To retain such terms, we must ensure that all terms that might contribute are also included; here, these are ε and $e^{-x/\varepsilon}$. We may not elect to keep this term alone, and ignore those in ε and $e^{-x/\varepsilon}$. To take this one step further, if we decided, in this example, to include the term $\varepsilon^2 e^{-2x/\varepsilon}$, we must also include the terms of order $1,\ \varepsilon,\ \varepsilon^2,\ e^{-x/\varepsilon},\ \varepsilon e^{-x/\varepsilon},\ e^{-2x/\varepsilon}$. (And for this same reason, it is obvious that we must retain all terms of orders $1,\ \varepsilon,\ \varepsilon^2$ when we wish to go as far as ε^3.) A number of examples of expanding and matching can be found in Q1.19, 1.20.

In our presentation of the matching principle we have described how it can be applied to functions which involve x^n and exponential terms (or functions which can be expanded in terms of these). However, when we apply this same procedure to logarithmic functions, we encounter a difficulty which requires a careful adjustment of the matching principle.

1.9 MATCHING WITH LOGARITHMIC TERMS

To see that we have a problem, let us consider an appropriate example, expand and then attempt to match in the way described above.

E1.12 A logarithmic example

We are given

$$f(x;\varepsilon) = \left(1 + x + \frac{\varepsilon}{\ln(x/\varepsilon)}\right)^{-1}, \quad x > \varepsilon,$$

and we construct the asymptotic expansions for $x = \mathrm{O}(1)$ and $x = \mathrm{O}(\varepsilon)$, as $\varepsilon \to 0^+$. So we obtain

$$f(x;\varepsilon) = \left(1 + x + \frac{\varepsilon}{\ln x - \ln\varepsilon}\right)^{-1} = \frac{1}{1+x}\left[1 - \frac{\varepsilon}{\ln\varepsilon}\frac{(1 - \ln x/\ln\varepsilon)^{-1}}{1+x}\right]^{-1}$$

$$\sim \frac{1}{1+x}\left(1 - \frac{\varepsilon/\ln\varepsilon}{1+x}\right), \quad x = \mathrm{O}(1), \tag{1.69}$$

where we have written down the first two terms. (Note that $\varepsilon/\ln\varepsilon = \mathrm{o}(\varepsilon)$ as $\varepsilon \to 0$.) Correspondingly, we have

$$f(\varepsilon X;\varepsilon) \equiv F(X;\varepsilon) = \left(1 + \varepsilon X + \frac{\varepsilon}{\ln X}\right)^{-1} \sim 1 - \varepsilon\left(X + \frac{1}{\ln X}\right), \quad X = \mathrm{O}(1), \tag{1.70}$$

where we have retained, again, the first two terms. We now match (1.69) and (1.70), and so we write (1.69) in terms of X and expand:

$$\frac{1}{1+\varepsilon X}\left(1 - \frac{\varepsilon/\ln\varepsilon}{1+\varepsilon X}\right) \sim 1 - \varepsilon X, \quad X = \mathrm{O}(1), \tag{1.71}$$

retaining terms $\mathrm{O}(1)$, $\mathrm{O}(\varepsilon)$ as required for (1.70). Similarly, from (1.70), we obtain

$$1 - x - \frac{\varepsilon}{\ln(x/\varepsilon)} = 1 - x + \frac{\varepsilon}{\ln\varepsilon - \ln x} \sim 1 - x + \frac{\varepsilon}{\ln\varepsilon}, \quad x = \mathrm{O}(1), \tag{1.72}$$

when we retain the terms O(1), O($\varepsilon/\ln\varepsilon$) used in (1.69). It is clear that (1.71) and (1.72) do *not* satisfy the matching principle: the term $\varepsilon/\ln\varepsilon$ appears in one expansion but not in the other.

However, it is easily confirmed that the two 1-term expansions [$f \sim 1/(1+x)$, $F \sim 1$] do match, as do the 1-term/2-term expansions [$f \sim 1/(1+x)$, $F \sim 1 - \varepsilon(X + 1/\ln X)$]. Further, it is easily checked that any additional terms retained in the expansion of $f(x;\varepsilon)$, by expanding the $(1 - \ln x/\ln\varepsilon)^{-1}$ contribution, will also lead to a failure of the matching principle. Perhaps if we retained *all* these terms, we might succeed; let us therefore rewrite (1.69) as

$$f(x;\varepsilon) \sim \frac{1}{1+x}\left[1 - \left(\frac{\varepsilon}{\ln x - \ln\varepsilon}\right)\frac{1}{1+x}\right], \quad x = \mathrm{O}(1),$$

$$= \frac{1}{1+\varepsilon X}\left(1 - \frac{\varepsilon}{\ln X}\frac{1}{1+\varepsilon X}\right) \sim 1 - \varepsilon\left(X + \frac{1}{\ln X}\right), \quad X = \mathrm{O}(1),$$

which immediately matches with (1.70) when we treat '$\ln X = \ln x - \ln\varepsilon$', in (1.70), in the same way.

In order to investigate the difficulty, let us consider the three simple functions x^n, $\exp(-x/\varepsilon)$ and $\ln x$ (and we may extend this to any function which is constructed from these elements, or can be represented as a series of such). Now introduce a new scaled variable, $x = \varepsilon X$ say, and so obtain, respectively,

$$\varepsilon^n X^n, \; e^{-X}, \; \ln(\varepsilon X);$$

the first two are the corresponding functions but of a different size (taking $X = \mathrm{O}(1)$ and $\varepsilon \to 0^+$) i.e. O(ε^n) and O(1), respectively. We note, however, that the logarithmic function does not follow this pattern: $\ln(\varepsilon X) = \ln\varepsilon + \ln X$, producing two terms of *different* size i.e. O($\ln\varepsilon$) and O(1), respectively. Indeed, we could write

$$\ln(\varepsilon X) = \ln\varepsilon + \ln X \sim \ln\varepsilon \quad \text{as } \varepsilon \to 0^+,$$

but in order to match any dependence in x (or X) we would need to retain the term $\ln X$ *with* $\ln\varepsilon$. (The retention of $\ln\varepsilon$, but ignoring $\ln x$, is at the heart of the problem we have with (1.71) and (1.72); but retaining $\ln\varepsilon$ *and* $\ln x$ allowed the matching principle to be applied successfully.)

The device that we therefore adopt, when log terms are present, is to treat '$\ln\varepsilon = \mathrm{O}(1)$' for the purposes of retaining the relevant terms; the matching principle, as we stated it, is then valid. Indeed, the matching principle will produce an identity (as before), with the correct identification of all the individual terms involving logarithms i.e. the interpretation '$\ln\varepsilon = \mathrm{O}(1)$' is used *only* for the retention of the appropriate terms.

Thus, for example, we must regard all the terms

$$\varepsilon,\ \varepsilon\ln\varepsilon,\ \varepsilon/\ln\varepsilon,\ \varepsilon(\ln\varepsilon)^2,\ \varepsilon\ln(\ln\varepsilon),$$

as $\mathrm{O}(\varepsilon)$ when selecting terms to use for matching. We conclude this discussion with one more example (and a few others are given in Q1.21).

E1.13 Another logarithmic example

Given

$$f(x;\varepsilon) = \sqrt{1+x+\varepsilon\ln(1+x/\varepsilon)}, \quad x \geq 0,$$

then as $\varepsilon \to 0^+$ we write

$$f(x;\varepsilon) = \sqrt{1+x}\left(1+\frac{\varepsilon[\ln x - \ln\varepsilon + \ln(1+\varepsilon/x)]}{1+x}\right)^{1/2}$$
$$\sim \sqrt{1+x}\left(1+\frac{\varepsilon}{2}\frac{(\ln x-\ln\varepsilon)}{1+x}\right), \quad x = \mathrm{O}(1), \tag{1.73}$$

keeping the first two terms in the asymptotic expansion, but treating '$\ln\varepsilon = \mathrm{O}(1)$'. For $x = \mathrm{O}(\varepsilon)$, we write $x = \varepsilon X$, to give

$$f(\varepsilon X;\varepsilon) \equiv F(X;\varepsilon) = \sqrt{1+\varepsilon X+\varepsilon\ln(1+X)}$$
$$\sim 1+\tfrac{1}{2}\varepsilon[X+\ln(1+X)], \quad X = \mathrm{O}(1), \tag{1.74}$$

if we include terms as far as $\mathrm{O}(\varepsilon)$. To match, we write (1.73) as

$$\sqrt{1+\varepsilon X}\left(1+\frac{\varepsilon}{2}\frac{\ln X}{1+\varepsilon X}\right) \sim 1+\tfrac{1}{2}\varepsilon(X+\ln X), \quad X = \mathrm{O}(1), \tag{1.75}$$

retaining terms $\mathrm{O}(1)$, $\mathrm{O}(\varepsilon)$ as required for (1.74). Similarly, we write (1.74) as

$$1+\tfrac{1}{2}x+\tfrac{1}{2}\varepsilon\ln(1+x/\varepsilon) \sim 1+\tfrac{1}{2}x+\tfrac{1}{2}\varepsilon(\ln x-\ln\varepsilon), \quad x = \mathrm{O}(1), \tag{1.76}$$

again retaining terms $\mathrm{O}(1)$, $\mathrm{O}(\varepsilon)$, as required for (1.73), but with '$\ln\varepsilon = \mathrm{O}(1)$'; it is immediately apparent that the two expansions match (to this order): (1.75) and (1.76) are identical (with $x = \varepsilon X$).

The processes of expanding, examining breakdowns, scaling and matching are the essential elements of singular perturbation theory; these will provide the basis and the framework for the rest of this text. However, there is one final aspect to which we should briefly return: the production of approximations that may prove useful in a mainly numerical context.

1.10 COMPOSITE EXPANSIONS

When we have obtained two, or more, matched expansions that represent a function, we are led to an intriguing question: is it possible to use these to produce, for example, an approximate graphical representation of the function, for all x in the domain? The main difficulty, of course, is that presumably we need to switch from one asymptotic expansion to the next at a particular value of x–which runs counter to the matching principle. One possibility might be to plot the function so that, for some values of x, both matched expansions are used and allowed to overlap, but this still involves using first one and then the other. A single expression which represents the original function, asymptotically, for $\forall x \in D$, would be far preferable. Such an expression can usually be found; it is called a *composite expansion*.

Suppose that we have a function which is described by two different asymptotic expansions, $f_N(x;\varepsilon)$ and $F_M(X;\varepsilon)$, say, valid in adjacent regions, with $x = \delta(\varepsilon)X$. (We are using the notation that was introduced in example E1.11; the extension to three or more expansions follows directly.) We now introduce a function $\mathfrak{F}\{f_N(x;\varepsilon), F_M(x/\delta;\varepsilon), x;\varepsilon)\}$ which possesses the properties that

$$\mathfrak{F} \sim f_N(x;\varepsilon) \quad \text{for } x = \mathrm{O}(1) \quad \text{and} \quad \mathfrak{F} \sim F_M(x/\delta;\varepsilon) \quad \text{for } x = \mathrm{O}(\delta),$$

both as $\varepsilon \to 0$. If such a function exists, it is called a composite expansion for the original function (for obvious reasons); note that the only requirement is that the correct behaviour, f_N or F_M, is generated—any other (smaller) terms may not be correct if these expansions were continued further. The issue now is how we find $\mathfrak{F}$; we present two commonly used constructions that lead to a suitable choice for $\mathfrak{F}$. The simpler uses a straightforward additive rule, and the other a multiplicative rule.

Definition (*composite expansion—additive*)

We write

$$\phi_{NM}(x;\varepsilon) = f_N(x;\varepsilon) + F_M(x/\delta;\varepsilon) - f_{ov}(x;\varepsilon), \tag{1.77}$$

where f_{ov} denotes the 'overlap' terms which are those terms involved in the matching.

The inclusion of f_{ov} becomes obvious when we note that the terms that match must appear in both f_N and F_M, and so are counted twice; f_{ov} then removes one of them. Given the pair f_N and F_M, it is an elementary exercise, in particular cases, to check that the expansion of (1.77), either for $x = \mathrm{O}(1)$ or $x = \mathrm{O}(\delta)$, recovers the appropriate leading-order terms. Further, it is then possible to compare the approximation, ϕ_{NM}, against the original function, $f(x;\varepsilon)$, by estimating $|f - \phi_{NM}|$ for $\forall x \in D$; we will return to this shortly. An example of the additive rule is now presented.

E1.14 A composite expansion (additive)

From E1.8 (see equations (1.52), (1.53) and (1.55)), we have

$$f(x;\varepsilon) = \left(1 + x + \varepsilon + e^{-x/\varepsilon}\right)^{-1}, \quad x \geq 0, \tag{1.78}$$

$$f \sim f_2(x;\varepsilon) = \frac{1}{1+x}\left(1 - \frac{\varepsilon}{1+x}\right), \quad x = O(1),$$

and

$$f \sim F_2(X;\varepsilon) = \frac{1}{1+e^{-X}}\left(1 - \frac{\varepsilon(1+X)}{1+e^{-X}}\right), \quad X = O(1),$$

where $x = \varepsilon X$ and the matching involves (see E1.11)

$$f_{ov}(x;\varepsilon) = 1 - x - \varepsilon.$$

Following (1.77), we define

$$\phi_{22}(x;\varepsilon) = \frac{1}{1+x}\left(1 - \frac{\varepsilon}{1+x}\right) + \frac{1}{1+e^{-x/\varepsilon}}\left(1 - \frac{x+\varepsilon}{1+e^{-x/\varepsilon}}\right) - (1 - x - \varepsilon) \tag{1.79}$$

and then we obtain directly

$$\begin{aligned}\phi_{22} &\sim \frac{1}{1+x}\left(1 - \frac{\varepsilon}{1+x}\right) + (1 - x - \varepsilon) - (1 - x - \varepsilon)\\ &= \frac{1}{1+x}\left(1 - \frac{\varepsilon}{1+x}\right) = f_2 \quad \text{for } x = O(1),\end{aligned}$$

retaining O(1) and O(ε) terms. Correspondingly, for $x = \varepsilon X$, $X = O(1)$, we have

$$\begin{aligned}\phi_{22} &= \frac{1}{1+\varepsilon X}\left(1 - \frac{\varepsilon}{1+\varepsilon X}\right) + \frac{1}{1+e^{-X}}\left(1 - \frac{\varepsilon(1+X)}{1+e^{-X}}\right) - (1 - \varepsilon X - \varepsilon)\\ &\sim 1 - \varepsilon X - \varepsilon + \frac{1}{1+e^{-X}}\left(1 - \frac{\varepsilon(1+X)}{1+e^{-X}}\right) - (1 - \varepsilon X - \varepsilon)\\ &= \frac{1}{1+e^{-X}}\left(1 - \frac{\varepsilon(1+X)}{1+e^{-X}}\right) = F_2 \quad \text{for } X = O(1),\end{aligned}$$

retaining terms O(1) and O(ε). Thus, to this order, (1.79) is a composite expansion for the function (1.78), for $\forall x \geq 0$. To confirm the accuracy of this approximation, we compare (1.78) and (1.79), for $\varepsilon = 0.1$, in figure 3.

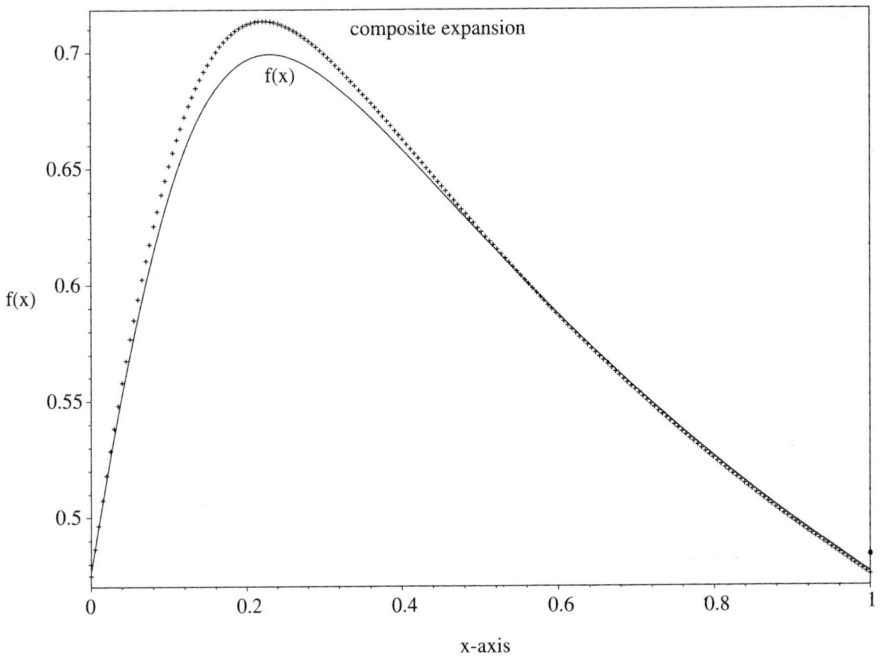

Figure 3. Comparison of the composite expansion, equation (1.79), with the exact function $f = (1 + x + \varepsilon + e^{-x/\varepsilon})^{-1}$ for $0 \leq x \leq 1$, with $\varepsilon = 0.1$.

The agreement between the composite expansion, (1.79), and the original function, as depicted in figure 3, is quite remarkable (and we have not chosen a particularly small value of ε which would make the agreement even more pronounced). This suggests that we should be able to obtain a formal estimate for the error when we use the composite expansion; we will examine this conjecture in a simpler version of E1.14 (so that the manipulation is particularly routine and transparent).

E1.15 Error using a composite expansion

From E1.14, we are given

$$f(x;\varepsilon) = (1 + x + \varepsilon + \mathrm{e}^{-x/\varepsilon})^{-1}, \quad x \geq 0;$$

$$f_1(x;\varepsilon) = \frac{1}{1+x}; \quad F_1(X;\varepsilon) = \frac{1}{1+\mathrm{e}^{-X}}; \quad f_{ov} = 1,$$

and so we form the composite expansion

$$\phi_{11}(x;\varepsilon) = \frac{1}{1+x} + \frac{1}{1+\mathrm{e}^{-x/\varepsilon}} - 1. \tag{1.80}$$

The error in using this composite expansion is, with $x \geq 0$, $\varepsilon \geq 0$,

$$E = |f - \phi_{11}| = \left|(1 + x + \varepsilon + \mathrm{e}^{-x/\varepsilon})^{-1} - \frac{1 - x\mathrm{e}^{-x/\varepsilon}}{(1 + x)(1 + \mathrm{e}^{-x/\varepsilon})}\right|$$

$$= \left|\frac{x\mathrm{e}^{-x/\varepsilon}(2 + x + \varepsilon + \mathrm{e}^{-x/\varepsilon}) - \varepsilon}{(1 + x)(1 + \mathrm{e}^{-x/\varepsilon})(1 + x + \varepsilon + \mathrm{e}^{-x/\varepsilon})}\right|$$

$$= \left|\frac{x\mathrm{e}^{-x/\varepsilon}[1 + (1 + \varepsilon + \mathrm{e}^{-x/\varepsilon})/(1 + x)] - \varepsilon/(1 + x)}{(1 + \mathrm{e}^{-x/\varepsilon})(1 + x + \varepsilon + \mathrm{e}^{-x/\varepsilon})}\right|$$

$$\leq \left|\frac{x\mathrm{e}^{-x/\varepsilon}[1 + (1 + \varepsilon + \mathrm{e}^{-x/\varepsilon})/(1)] + \varepsilon/(1)}{(1)(1)}\right|$$

$$\leq \left|x\mathrm{e}^{-x/\varepsilon}(3 + \varepsilon) + \varepsilon\right|.$$

But the maximum value of $x\mathrm{e}^{-x/\varepsilon}$ occurs at $x = \varepsilon$, for then

$$\frac{\mathrm{d}(x\mathrm{e}^{-x/\varepsilon})}{\mathrm{d}x} = (1 - x/\varepsilon)\mathrm{e}^{-x/\varepsilon} = 0,$$

and so we have the estimate

$$E \leq \varepsilon\mathrm{e}^{-1}(3 + \varepsilon) + \varepsilon = \varepsilon[1 + (3 + \varepsilon)\mathrm{e}^{-1}] \tag{1.81}$$

and hence the error, for $\forall x \geq 0$, is $\mathrm{O}(\varepsilon)$ at most (indeed, this upper bound is approximately 2.1ε for small ε).

This example demonstrates that estimates for the error in using a composite expansion are readily derived, at least in particular cases, when the original function is known. However, corresponding results for the solutions of differential equations are not so easily obtained—and this is the situation of most practical interest. Then we have available only the equation and its asymptotic solution, but not the exact solution of the equation, of course. Thus any estimates will need to be based on an analysis of appropriate differential inequalities; we will touch on these ideas later in this text. (We should add that only rarely is it possible satisfactorily to complete such calculations; in addition, it is often deemed not worthwhile to devote much energy to this exercise, essentially because the results and the context persuade us that we have the appropriate and correct asymptotic form of the solution.)

To conclude this discussion, we briefly describe the multiplicative rule which can be used as an alternative for the construction of a composite expansion. So, with the same notation as introduced earlier:

Definition (*composite expansion—multiplicative*)

Now we write

$$\Phi_{NM}(x;\varepsilon) = \frac{f_N(x;\varepsilon)F_M(x/\delta;\varepsilon)}{f_{ov}(x;\varepsilon)} \tag{1.82}$$

provided that $f_{ov} \neq 0$ for $\forall x \in D$.

Here, the terms involved in the matching (f_{ov}) appear in both f_N and F_M, and so dividing by f_{ov} cancels one of these. It is clear, therefore, that a composite expansion, to a given order, is not unique; indeed, any number of variants exists—we have presented two possible choices only. We will provide an example of the application of (1.82), but we are unable to use the expansions quoted in E1.14 (which would have been useful as a comparison) because, there, $f_{ov} = 1 - x - \varepsilon$ (=0 for $x = 1 - \varepsilon \in D$). However, a slightly simpler version of E1.14 is possible.

E1.16 A composite expansion (multiplicative)

See E1.14; let us be given

$$f \sim f_2 = \frac{1}{1+x}\left(1 - \frac{\varepsilon}{1+x}\right), \quad x = O(1),$$

and

$$f \sim F_1 = \frac{1}{1+e^{-X}}, \quad X = O(1) \quad \text{where } x = \varepsilon X;$$

the matching of these expansions involves simply $f_{ov} = 1$. Thus, (1.82) gives

$$\Phi_{21}(x;\varepsilon) = \frac{1}{1+x}\left(1 - \frac{\varepsilon}{1+x}\right)\left(\frac{1}{1+e^{-x/\varepsilon}}\right) \tag{1.83}$$

and so

$$\Phi_{21} \sim \frac{1}{1+x}\left(1 - \frac{\varepsilon}{1+x}\right) = f_2 \quad \text{for } x = O(1),$$

and

$$\Phi_{21} = \frac{1}{1+\varepsilon X}\left(1 - \frac{\varepsilon}{1+\varepsilon X}\right)\left(\frac{1}{1+e^{-X}}\right) \sim \frac{1}{1+e^{-X}} = F_1 \quad \text{for } X = O(1);$$

thus (1.83) is a (multiplicative) composite expansion for (1.78). (An examination of the accuracy of this expansion can be found in Q1.24, and a few additional examples of composite expansions appear in Q1.22, 1.23.)

This concludes our presentation and discussion of all the elements that constitute classical singular perturbation theory. However, this is far from a complete description of all the techniques that are needed to solve differential equations. These will be introduced as they are required; suffice it to record, at this stage, that one approach that subsumes much that we have done thus far will be carefully developed later—the *method of multiple scales*.

FURTHER READING

In this chapter we have introduced, with carefully chosen explanations and examples, the basic ideas that are at the heart of singular perturbation theory. However, other approaches are possible, emphasising one aspect or another, and a number of good texts are available, which some readers may find both interesting and instructive. We offer the following as some additional (but not essential) reading, with a few observations about each. Of course, such a list is unlikely to be exhaustive, so I have included only my own favourites; I apologise if yours has been omitted.

Any list of texts must include that written by van Dyke (1964), and especially its annotated edition (1975); this provides an excellent introduction to the ideas, together with their applications to many of the classical problems in fluid mechanics. Two other good texts that present the material from a rather elementary stand-point are Hinch (1991) and Bush (1992), although the former is somewhat sophisticated in places; both these cover quite a wide range of applications. A text that also provides an introduction, although perhaps in not so much detail as those already cited, is Kevorkian & Cole (1996), but it offers an excellent introduction to the application of these methods to various types of problem. (This publication is a revised and updated edition of Kevorkian & Cole (1981), which was itself a rewritten and extended version of Cole (1968), both of which are worth some exploration.) A nice introduction to the subject, mainly following the work of Kaplun (see below), is given in Lagerstrom (1988). Nayfeh (1973, 1981) provides many examples discussed in detail (and often these are the same problem tackled in different ways); these two books are useful as references to various applications, but perhaps are less useful as introductions to the relevant underlying ideas.

A carefully presented discussion of how singular perturbation theory manifests itself in ordinary differential equations, and a detailed description of methods of solution with applications, can be found in O'Malley (1991). An outstanding collection of many and varied problems, most of them physically interesting and important, is given in Holmes (1995), but this text is probably best avoided by the novice. Eckhaus (1979) presents a very formal and rigorous approach to the subject, and some interesting applications are included. Smith (1985) uses an instructive mixture of the formal approach combined with examples and applications (although almost all of these relate to ordinary, not partial, differential equations); a large number of references are also included. A discussion of the matching principle, *via* expansion operators, can be found in the papers by Fraenkel (1969).

We now turn to some texts that are more specialised in their content. A collection of the work and ideas of Kaplun, particularly with reference to applications to problems in fluid mechanics, can be found in Kaplun (1967); this is advisable reading only for those with a deep interest in fluids. The construction of asymptotic solutions to ordinary differential equations (that is, in the absence of a small parameter) is described in Wasow (1965) and also in Dingle (1973); this latter book provides a very extensive analysis of asymptotic expansions, their properties and how to construct useful forms of them that provide the basis for numerical estimates.

There are various texts that describe the techniques for finding asymptotic expansions of functions defined, for example, by differential equations or as integrals; excellent examples in this category are Erdelyi (1956), Copson (1967) and Murray (1974). But the outstanding text has got to be Olver (1974), for its depth and breadth; furthermore, this provides an excellent reference for the behaviour of many standard functions (and the results are presented, comprehensively, for functions in the complex plane). Finally, we mention two texts that discuss the properties of divergent series: Hardy (1949) and Ford (1960). The former has become a classical text; it covers a lot of ground but is written in a pleasant and accessible style—it cannot be recommended highly enough.

EXERCISES

Q1.1 *Maclaurin expansions I.* The following real functions are defined for suitable real values of x. Use your knowledge of Maclaurin expansions to find power-series representations of these functions as $x \to 0^+$, giving the first three terms in each case; state those values of x for which the expansions are convergent. (Some of these may not be expressible wholly in terms of integral powers of x.)

(a) $\dfrac{1}{1+\sqrt{1+x}}$; (b) $\dfrac{1}{1+\mathrm{e}^x}$; (c) $\cos(\mathrm{e}^x)$; (d) $\ln(1+x+x^2)$;

(e) $\ln(1+\mathrm{e}^x)$; (f) $\ln(1+x+x^{-1})$; (g) $x^x - 1$.

Q1.2 *Maclaurin expansions II.* Repeat as for Q1.1, but for each of these series also find the general term in the expansion.

(a) $\sqrt{1+x^{-1}}$; (b) $\sin(1+x)$; (c) $\ln(1+x^{-1})$.

Q1.3 *Exact solutions of ODEs.* Find the exact solutions of these ordinary differential equations and explore their properties as $\varepsilon \to 0^+$ (for example, by sketching or plotting the solutions for various ε). You may wish to investigate how an appropriate approximation, for $\varepsilon \to 0^+$, might be obtained directly from the differential equation. (The over-dot denotes the derivative with respect to t, time, for problems in $t \geq 0$; the prime represents the derivative with respect to x.)

(a) $\ddot{x} + 2\varepsilon\dot{x} + (1+\varepsilon^2)x = 0$ with $x(0) = 1,\ \dot{x}(0) = 0$;

(b) $\ddot{x} + x = \sin[(1-\varepsilon)t]$ with $x(0) = \dot{x}(0) = 0$;

(c) $\varepsilon y''' + y'' + \varepsilon y' + y = 0,\ 0 \leq x \leq \pi$, with $y(0) = 2,\ y'(0) = 0,\ y(\pi) = \mathrm{e}^{-\pi/\varepsilon} - 1$;

(d) $2\varepsilon\sqrt{x}y' + y = 1,\ x \geq 0$, with $y(0) = 2$;

(e) $\varepsilon y' = (a - 2x)y,\ x \geq 0$, with $y(0) = 1$ for $a = 1$;
(f) repeat (e) for $a = -1$;
(g) $(\varepsilon + x)y'' + y' = 4x + 1,\ x \geq 0$, with $y(0) = y'(0) = 0$;
(h) $\varepsilon y'' + y' + (y')^2 = 0,\ x \geq 0$, with $y(0) = 0,\ y'(0) = -1/2$ [Hint: seek a solution $y' = f(y)$];
(i) $\frac{1}{2}\varepsilon y'' = yy',\ x \geq 0,\ -1 \leq x \leq 1$, with $y(-1) = 1,\ y(1) = -1$;
(j) repeat (i) for $y(-1) = 0,\ y(1) = \varepsilon\pi/8$;
(k) $\varepsilon y'' = (y')^n,\ 0 \leq x \leq 1$, with $y(0) = 0,\ y(1) = 1$, for $n = 1$;
(l) repeat (k) for $n = 1/2$ and $y(0) = 1,\ y'(0) = 0$;
(m) repeat (k) for $n = 2$ and $y(0) = 1,\ y'(0) = -1$;
(n) repeat (k) for $n = 3$ and $y(0) = 0,\ y'(0) = \varepsilon$.

Q1.4 *Uniform or non-uniform?* Is the behaviour of each of these functions uniformly valid as $x \to 0^+$ and $y \to 0^+$? (To answer this, compare $x \to 0^+$ followed by $y \to 0^+$ with the reversed order of taking the limits.)

(a) $\dfrac{x}{1+y}$; (b) $\dfrac{x}{x+y}$; (c) $\ln\{x + (y/x)\}$; (d) x^y.

Q1.5 *Limits involving exp(x) and ln(x).* (a) For $x > 1$ and $\alpha > 0$, show that $0 < \int_1^x \mathrm{d}t/t < \int_1^x t^{\alpha-1}\mathrm{d}t$, and hence that $0 < \ln x < x^\alpha/\alpha$. (b) Interpret these as inequalities for $x^{-\beta}\ln x$, choose $\beta > \alpha$ and show that $x^{-\beta}\ln x \to 0^+$ as $x \to +\infty$ for $\forall\beta > 0$. (c) Now write $x = 1/y$ and hence obtain $-y^\beta \ln y \to 0^+$ as $y \to 0^+$ for $\forall\beta > 0$. (d) In (b), write $\beta = 1/\gamma$ and show that $x^{-1}(\ln x)^\gamma \to 0^+$ as $x \to +\infty$ for $\forall\gamma > 0$. (e) In (d), write $x = \exp(y)$ and deduce that $\mathrm{e}^{-y}y^\gamma \to 0^+$ as $y \to +\infty$ for $\forall\gamma$.

Q1.6 *Examples of O, o and ~.* Determine which of the following are correct as $x \to 0^+$.

(a) $\tan x = \mathrm{O}(x)$;
(b) $\sin^2(x) = \mathrm{o}(x)$;
(c) $n^x - 1 = \mathrm{O}(x)\ (n > 0,\ n \neq 1)$;
(d) $n^x - 1 \sim x\ (n > 0,\ n \neq 1)$;
(e) $\ln\{(x + x^2)/2\} \sim \frac{1}{2}\ln x$;
(f) $\exp(-1/x) = \mathrm{o}(x^{51})$;
(g) $\sinh(x + x^{-1}) \sim \frac{1}{2}\mathrm{e}^{1/x}$;
(h) $x^2 + x\ln x = \mathrm{o}(x^{1/2})$;
(i) $\ln(\frac{1}{2} + x + 2x^{-1}) \sim -2\ln x$;
(j) $\dfrac{\sin x \tan x}{\cosh(x^{-1})} = \mathrm{O}(x^2 e^{-1/x})$;
(k) $x^x(\ln x)^{-1} = \mathrm{o}(x)$.

Q1.7 *Dominant behaviour.* Find the dominant behaviour, i.e. find $g(x)$ so that $f(x) \sim g(x)$, as $x \to 0^+$, for each of these functions, $f(x)$.

(a) $\dfrac{(1-\cos x)\tan x}{\cosh(x^{-1})}$; (b) $\dfrac{\tanh x}{\ln(\frac{1}{2} + x + x^{-1})}$; (c) $\dfrac{x^x}{\ln(1+\mathrm{e}^{-1/x})}$;

(d) $\dfrac{\ln(1+\mathrm{e}^{-1/x})}{\sinh(1+x^{-1})}$; (e) $\sin(x^x - 1)$; (f) $\dfrac{\ln(x + x^{-1})}{\tanh(x^{-1})}$;

(g) $F[G(x)]$ where $F(x) = G(x) = \ln(1 + \mathrm{e}^{-1/x})$.

Q1.8 *Properties of O and o.* For the limit $x \to a$:

(a) given that $f(x) = O[g(x)]$, $F(x) = O[G(x)]$, show that $fF = O(gG)$;
(b) given that $f(x) = O[g(x)]$, $F(x) = o[G(x)]$, show that $fF = o(gG)$.

Q1.9 *Asymptotic sequences.* Verify that these are asymptotic sequences, where $n = 0, 1, 2, \ldots$.

(a) $\{\ln(1 + x^n)\}$ as $x \to 0$; (b) $\{x^n e^{-nx}\}$ as $x \to +\infty$;

(c) $\{x^n[2 + \sin(x^{-n})]\}$ as $x \to 0$.

Q1.10 *Non-uniqueness of asymptotic expansions.* Find two asymptotic expansions of the function $(1 + x)^{-1}$, valid as $x \to \infty$, based on the asymptotic sequences:

(a) $\{x^{-n}\}$; (b) $\{x^{-2n} p(x)\}$, for a suitable function $p(x)$, which is to be determined.

Q1.11 *Error function.* The error function is defined by

$$\operatorname{erf}(x) = \frac{2}{\sqrt{\pi}} \int_0^x e^{-t^2}\, dt,$$

where $\operatorname{erf}(x) \to 1$ as $x \to +\infty$. Obtain asymptotic expansions of erf(x) for: (a) $x \to 0^+$; (b) $x \to +\infty$, and in each case find the general term and decide if these series are convergent or divergent. Use your result in (b) to give an estimate for the value of erf(2).

Q1.12 *A sine integral.* Obtain an asymptotic expansion of

$$\operatorname{si}(x) = \int_x^\infty \frac{\sin t}{t}\, dt$$

for $x \to +\infty$, and decide if the series is convergent or divergent. Also obtain an expression for the remainder, as an integral, and find an estimate for it.

Q1.13 *Exponential integral.* Find an asymptotic expansion of

$$\operatorname{Ei}(x) = \int_x^\infty \frac{e^{-t}}{t}\, dt,$$

as $x \to 0^+$; this will involve Euler's constant

$$\gamma = \int_0^1 \frac{1}{t}(1 - e^{-t})\, dt - \int_1^\infty \frac{e^{-t}}{t}\, dt.$$

It might be helpful, first, to show that

$$\lim_{x \to 0^+} \left[\int_x^\infty \frac{e^{-t}}{t}\, dt + \ln x \right] = -\gamma.$$

Q1.14 *An integral I.* Find the first four terms in an asymptotic expansion of the integral

$$\int_x^{\infty} \left(\frac{1+t}{t^2}\right) e^{-t^2}\, dt$$

for (a) $x \to +\infty$; (b) $x \to 0^+$.

Q1.15 *An integral II.* Find the first four terms in an asymptotic expansion of the integral

$$\int_0^x \left(\frac{e^t + \ln t}{t + e^t}\right) dt, \qquad x \geq 0,$$

for (a) $x \to 0^+$; (b) $x \to +\infty$ (and in this case, introduce the constants $\alpha = \int_0^{\infty} \frac{\ln t}{t+e^t}\, dt$, $\beta = \int_0^{\infty} \frac{t}{t+e^t}\, dt$).

Q1.16 *An integral III.* Find the first two terms in an asymptotic expansion of the integral

$$I(x;\varepsilon) = \int_0^x \exp\left[-\frac{1}{\varepsilon}(2 - \sin t)\right] dt, \qquad 0 \leq x < \pi/2,$$

as $\varepsilon \to 0^+$. Explain why your result is not valid if $0 \leq x \leq \pi/2$. [Hint : simply use integration by parts.]

Q1.17 *An integral IV.* Use an appropriate integration by parts to find an expansion, valid as $x \to +\infty$, of the integral

$$I(x) = \int_0^{\infty} \frac{e^{-xt}}{1+t}\, dt, \qquad x > 0;$$

in particular, find the general term and also an expression for the remainder. Show that this is an asymptotic expansion and that the series diverges. Find an estimate for the remainder and use this to select the number of terms, for a given x, which will minimise the error when using the series to find the value of $I(x)$.

Q1.18 *Approximation for a Bessel function.* The Bessel function $J_\nu(x)$ is a solution of the equation

$$x^2 y'' + xy' + (x^2 - \nu^2)y = 0;$$

first write $y = x^{-1/2} z(x)$, find the equation for $z(x)$ and show that, for large x, this becomes $z'' + z = 0$, approximately. Hence seek a solution

$$z(x) \sim e^{\omega x} \sum_{n=0}^{\infty} A_n x^{\lambda - n} \qquad \text{as } x \to +\infty,$$

where ω and λ are constants to be determined; also find the recurrence relation for the coefficients A_n. (This is to be compared with the familiar Frobenius method for second order ODEs.) Write down the most general solution available (by using both values of ω), and retain terms as far as x^{-2}. Show that this is consistent with the solution

$$J_\nu(x) \sim \frac{C}{\sqrt{x}} \cos(x + \alpha), \quad x \to +\infty,$$

although it is not possible, here, to find values for C and α.

Q1.19 *Expansion of a function with a parameter.* For these functions (all with the domain $x \geq 0$), expand as $\varepsilon \to 0^+$ for each of $x = \mathrm{O}(1)$, $x = \mathrm{O}(\varepsilon)$ and $x = \mathrm{O}(\varepsilon^{-1})$, and find the first two terms in each asymptotic expansion. Show that your expansions satisfy the matching principle (and you may wish to note wherever any breakdowns are evident in your two-term expansions). Remember that the matching principle applies only to *adjacent* regions.

(a) $\sqrt{\dfrac{x + \varepsilon e^{-x/\varepsilon}}{x + \varepsilon + \varepsilon x^2}}$; (b) $\dfrac{1 + \varepsilon x + \sqrt{x + \varepsilon}}{1 + \sqrt{x + \varepsilon e^{-x/\varepsilon}}}$; (c) $\cosh(\varepsilon x) - 1 + \dfrac{\varepsilon(1+x)}{x^2 + \varepsilon^2}$;

(d) $(1 + x + \varepsilon x^2 \sin(\varepsilon x) + e^{-x/\varepsilon})^{-1}$; (e) $\sqrt{\sinh(\varepsilon x) + \dfrac{\varepsilon(1+x)}{x + \varepsilon}}$;

for this one, also include $x = \mathrm{O}(\varepsilon^2)$: (f) $\dfrac{\sin(\sqrt{\varepsilon x} + \varepsilon e^{-x/\varepsilon})}{\sqrt{1 + \varepsilon + \varepsilon^2 x^2 + \exp(-x/\varepsilon^2)}}$.

Q1.20 *Expansion with exponentially small terms.* For these two functions (both with the domain $x \geq 0$), expand as $\varepsilon \to 0^+$ for $x = \mathrm{O}(1)$ and retain terms $\mathrm{O}(1)$, $\mathrm{O}(\varepsilon)$ and $\mathrm{O}(e^{-x/\varepsilon})$; for $X = x/\varepsilon = \mathrm{O}(1)$ expand and now retain the first two terms only. Show that your expansions satisfy the matching principle.

(a) $\sqrt{\dfrac{1 + x + e^{-x/\varepsilon}}{1 + x + \varepsilon}}$; (b) $\left(1 + \dfrac{x}{1 + x + \varepsilon} + \operatorname{sech}(x/\varepsilon)\right)^{-1}$.

Q1.21 *Matching with logarithms.* The domain of these functions is given as $x > 0$ with $\varepsilon > 0$; in each case, find the first two terms in each of the asymptotic expansions valid for $x = \mathrm{O}(1)$ and for $x = \mathrm{O}(\varepsilon)$ as $\varepsilon \to 0^+$. Show that, with the interpretation '$\ln \varepsilon = \mathrm{O}(1)$', your expansions satisfy the matching principle.

(a) $x - \dfrac{\ln\{1 + (x/\varepsilon)\}}{\ln \varepsilon - \ln(1 + \varepsilon)}$ (this example was introduced by Eckhaus);

(b) $\sqrt{\varepsilon + \ln(1 + x) + x \ln[1 + (x/\varepsilon)]}$;

(c) $(1 + x^2 + \varepsilon x)\ln([1 + (x/\varepsilon)] - \varepsilon \ln[x/(\varepsilon + \varepsilon^2)]$.

Q1.22 *Composite expansions I.* For these functions, given that $x \geq 0$ and $\varepsilon > 0$, find the first two terms in asymptotic expansions valid for $x = \mathrm{O}(1)$ and for $x = \mathrm{O}(\varepsilon)$

as $\varepsilon \to 0^+$. Hence construct additive composite expansions.

(a) $\dfrac{x+\varepsilon+\mathrm{e}^{-x/\varepsilon}}{1+x+\varepsilon}$; (b) $\sqrt{1+\dfrac{\varepsilon(1+x)}{x+\varepsilon}}$.

What do you observe about your composite expansion obtained in (b)?

Q1.23 *Composite expansions II.* See Q1.22; now, when they are defined, find the corresponding multiplicative expansions. (You may wish to compare, by plotting the appropriate functions, the original function and the two composite expansions.)

Q1.24 *Estimate of error.* See E1.16; find an estimate of the error in using this form of the composite expansion.

2. INTRODUCTORY APPLICATIONS

In the previous chapter, we laid the foundations of singular perturbation theory and, although we will need to add some specific techniques for solving certain types of differential equations, we can already tackle simple examples. In addition, we will see that we can apply these ideas directly to other, more routine problems—and this is where we shall begin. Here, we will describe how to approach the problem of finding roots of equations (which contain a small parameter), and how to evaluate integrals of functions which are represented by asymptotic expansions with respect to a parameter. Finally, we begin our study of differential equations by examining a few important, fairly straightforward examples which are, nonetheless, not trivial.

2.1 ROOTS OF EQUATIONS

At some stage in many mathematical problems, it is not unusual to be faced with the need to solve an equation for specific values of an unknown. Such a problem might be as simple as solving a quadratic equation:

$$\varepsilon x^2 + x + 1 = 0,$$

or finding the solution of more complicated equations such as

$$\varepsilon x^3 + \sqrt{x} - 1 = 0 \quad \text{or} \quad \varepsilon \sin x = 1 + \varepsilon - e^{-x/\varepsilon}.$$

In this section, we will describe a technique (for equations which contain a small parameter, as in those above) which is a natural extension of simply obtaining an asymptotic expansion of a function, examining its breakdown, rescaling, and so on.

We will begin by examining the simple quadratic equation

$$f(x;\varepsilon) = \varepsilon x^2 + x + 1 = 0 \tag{2.1}$$

and seek the solutions for $\varepsilon \to 0$. The essential idea is to obtain different asymptotic approximations for $f(x;\varepsilon)$, valid for different sizes of x, and see if these admit (approximate) roots. Given that $-\infty < x < \infty$, we could have roots anywhere on the real line, and so all sizes of x must be examined. (We will consider, first, only the real roots of equations; the extension to complex roots will be discussed in due course.) One further comment is required at this stage: we describe here a technique for finding roots that builds on the ideas of singular perturbation theory. In practice, other approaches are likely to be used in conjunction with ours to solve particular equations e.g. sketching or plotting the function, or using a standard numerical procedure (such as Newton-Raphson). There is no suggestion that this expansion technique should be used in isolation—it is simply one of a number of tools available.

Returning to (2.1), if $x = \mathrm{O}(1)$, then

$$f(x;\varepsilon) \sim 1 + x \quad \text{as } \varepsilon \to 0 \tag{2.2}$$

and so we have a root $x = -1$ (approximately). In order to generate a better approximation, we may use any appropriate method. For example, we could invoke the familiar procedure of iteration, so we may write

$$x_{n+1} = -1 - \varepsilon x_n^2, \quad n = 0, 1, 2, \ldots,$$

with $x_0 = -1$. Then we obtain

$$x_1 = -1 - \varepsilon, \quad x_2 = -1 - \varepsilon - 2\varepsilon^2 - \varepsilon^3,$$

and so on (but note that iteration may not generate a correct asymptotic expansion at a *given* order in ε). It is clear from this approach that a complete representation of the root will be obtained if we use the asymptotic sequence $\{\varepsilon^n\}$, and so an alternative is to seek this form directly—and this is more in keeping with the ideas of perturbation theory. Thus we might seek a root in the form

$$x \sim -1 + \sum_{n=1}^{\infty} \varepsilon^n a_n \tag{2.3}$$

so that (2.1) can be written as

$$\varepsilon(1 - 2\varepsilon a_1 + \cdots) + (-1 + \varepsilon a_1 + \varepsilon^2 a_2 + \cdots) + 1 = 0$$

(where the use of '$= 0$' here is to imply 'equal to zero to all orders in ε'); thus $a_1 = -1$, $a_2 = -2$, and so on. We have one root

$$x \sim -1 - \varepsilon - 2\varepsilon^2 \quad \text{as } \varepsilon \to 0. \tag{2.4}$$

It is clear that (2.2) admits only one root; the other—it is a quadratic equation that we are solving—must appear for a different size of x.

The 'asymptotic expansion' (we treat the function as such)

$$f(x;\varepsilon) \sim 1 + x + \varepsilon x^2 \tag{2.5}$$

remains valid for $x \to 0$, and so there is no new root for $x = o(1)$; however, this expansion does break down where $\varepsilon x^2 = O(x)$ i.e. $x = O(\varepsilon^{-1})$. We define $x = X/\varepsilon$ and write

$$f(X/\varepsilon;\varepsilon) \equiv \varepsilon^{-1} F(X;\varepsilon) = 1 + \frac{X}{\varepsilon} + \frac{X^2}{\varepsilon} (= 0)$$

or

$$\begin{aligned} F(X;\varepsilon) &= X^2 + X + \varepsilon \\ &\sim X^2 + X \quad \text{for } X = O(1) \quad \text{as } \varepsilon \to 0. \end{aligned} \tag{2.6}$$

This approximation admits the roots $X = 0$ and $X = -1$, so now the quadratic equation has a total of three roots! Of course, this cannot be the case; indeed, it is clear that the root $X = 0$ is inadmissible, because the 'asymptotic expansion' $F \sim X^2 + X + \varepsilon$ breaks down where $X = O(\varepsilon)$ (which is $x = O(1)$ and so returns us to (2.5)). The only available root is $X = -1$, and this is the second (approximate) root of the equation (leaving $X = 0$ as no more than a 'ghost' of the root $x \sim -1$). The expansion for F does not further breakdown (as $X \to \infty$) and so there are no other roots—not that we expected any more! We may seek a better approximation, as we did before, in the form

$$X \sim -1 + \sum_{n=1}^{\infty} \varepsilon^n A_n$$

which gives

$$(1 - 2\varepsilon A_1 + \varepsilon^2 A_1^2 - 2\varepsilon^2 A_2 + \cdots) + (-1 + \varepsilon A_1 + \varepsilon^2 A_2 + \cdots) + \varepsilon = 0$$

i.e. $A_1 = 1$, $A_2 = 1$; thus

$$X \sim -1 + \varepsilon + \varepsilon^2 \quad \text{or} \quad x \sim -\frac{1}{\varepsilon} + 1 + \varepsilon \quad \text{as } \varepsilon \to 0.$$

The two roots of the quadratic equation, (2.1), are therefore

$$x \sim -1 - \varepsilon - 2\varepsilon^2, \quad x \sim -\frac{1}{\varepsilon} + 1 + \varepsilon \quad \text{as } \varepsilon \to 0;$$

it is left as an exercise to confirm that these results can be obtained directly from the familiar solution of the quadratic equation, suitably approximated (by using the binomial expansion) for $\varepsilon \to 0$. (Similar problems based on quadratic equations can be found in exercise Q2.1.)

This simple introductory example covers the essentials of the technique: for $-\infty < x < \infty$, find all the different asymptotic forms of $f(x;\varepsilon)$ $(=0)$, and investigate if roots exist for each (dominant) asymptotic representation. Let us now apply this to a slightly more difficult equation which, nevertheless, has a similar structure.

E2.1 A cubic equation

We are to find approximations to all the real roots of the cubic equation

$$f(x;\varepsilon) = \varepsilon x^3 - x^2 + \varepsilon x + 1 = 0,$$

for $\varepsilon \to 0$. First, for $x = \mathrm{O}(1)$, we have

$$f(x;\varepsilon) \sim 1 - x^2 \quad \text{as } \varepsilon \to 0,$$

and this approximation admits the roots $x = \pm 1$; a better approximation is then obtained by writing

$$x \sim \pm 1 + \sum_{n=1}^{\infty} \varepsilon^n a_n$$

so that we obtain

$$\varepsilon(\pm 1 + 3\varepsilon a_1 \cdots) - (1 \pm 2\varepsilon a_1 + \varepsilon^2 a_1^2 \pm 2\varepsilon^2 a_2 \cdots) + \varepsilon(\pm 1 + \varepsilon a_1 \cdots) + 1 = 0.$$

This equation requires that $a_1 = 1$, $a_2 = \pm 3/2$, and so on; two roots are therefore

$$x \sim 1 + \varepsilon + \frac{3}{2}\varepsilon^2, \quad x \sim -1 + \varepsilon - \frac{3}{2}\varepsilon^2 \quad \text{as } \varepsilon \to 0.$$

Now the 'asymptotic expansion'

$$f(x;\varepsilon) \sim 1 - x^2 + \varepsilon(x^3 + x)$$

remains valid as $x \to 0$ but not as $|x| \to \infty$; it breaks down where $\varepsilon x^3 = \mathrm{O}(x^2)$ or $x = \mathrm{O}(\varepsilon^{-1})$. We write $x = X/\varepsilon$ and then

$$f(X/\varepsilon;\varepsilon) \equiv \varepsilon^{-2} F(X;\varepsilon) = \frac{X^3}{\varepsilon^2} - \frac{X^2}{\varepsilon^2} + X + 1 \ (=0)$$

or

$$F(X;\varepsilon) = X^3 - X^2 + \varepsilon^2(1 + X)$$
$$\sim X^3 - X^2 \quad \text{for } X = \mathrm{O}(1) \quad \text{as } \varepsilon \to 0.$$

The only relevant root is $X = 1$ (because the other two are the ghosts of the roots that appear for $x = O(1)$). To improve this approximation, we set

$$X \sim 1 + \sum_{n=1}^{\infty} \varepsilon^n A_n$$

to give

$$1 + 3\varepsilon A_1 + 3\varepsilon^2 A_1^2 + 3\varepsilon^2 A_2 \cdots - \left(1 + 2\varepsilon A_1 + \varepsilon^2 A_1^2 + 2\varepsilon^2 A_2 \cdots\right) + \varepsilon^2(1 + \varepsilon A_1 \cdots + 1) = 0$$

i.e. $A_1 = 0$, $A_2 = -2$, so that a third root is $X \sim 1 - 2\varepsilon^2$ as $\varepsilon \to 0$. Thus the three (real) roots are

$$x \sim \pm 1 + \varepsilon \pm \frac{3}{2}\varepsilon^2, \quad x \sim \frac{1}{\varepsilon} - 2\varepsilon \quad \text{as } \varepsilon \to 0.$$

Our introductory example, and the one above, have been rather conventional polynomial equations, but the technique is particularly powerful when we have to solve, for example, transcendental equations (which contain a small parameter). We will now see how the approach works in a problem of this type.

E2.2 A transcendental equation

We require the approximate (real) roots, as $\varepsilon \to 0^+$, of the equation

$$f(x;\varepsilon) = x^2 - 3\varepsilon x - 1 - \varepsilon + e^{-x/\varepsilon} = 0. \tag{2.7}$$

For $x = O(1)$, we now have two possibilities:

$$f(x;\varepsilon) \sim \begin{cases} x^2 - 1 & \text{for } x > 0 \\ e^{-x/\varepsilon} & \text{for } x < 0 \end{cases}$$

but only the first option admits any roots for real, finite x. Thus $x = \pm 1$ (approximately) and then only the choice $x = +1$ is acceptable (because we require $x > 0$); a better approximation follows directly:

$$x \sim 1 + 2\varepsilon + \varepsilon^2 \quad \text{as } \varepsilon \to 0^+.$$

The 'expansion' is written

$$f(x;\varepsilon) \sim x^2 - 1 - \varepsilon(3x + 1) + e^{-x/\varepsilon} \tag{2.8}$$

where the term $\exp(-x/\varepsilon)$ must be exponentially small for $x = O(1)$ or larger (because no roots exist if this term dominates). Now, for $x > 0$, there is no breakdown as

$x \to \infty$; thus any other roots that might exist must arise as $x \to 0$. Indeed, as $x \to 0$, we see that the expansion (2.8) breaks down where $x = \mathrm{O}(\varepsilon)$, and so we set $x = \varepsilon X$, to give

$$f(\varepsilon X; \varepsilon) \equiv F(X; \varepsilon) = \varepsilon^2 X^2 - 1 - \varepsilon(3\varepsilon X + 1) + \mathrm{e}^{-X} \ (= 0) \tag{2.9}$$
$$\sim \mathrm{e}^{-X} - 1 \quad \text{for } X = \mathrm{O}(1) \quad \text{as } \varepsilon \to 0^+.$$

This approximation has one root at $X = 0$, but this cannot be used directly because the expansion, (2.9), itself breaks down as $X \to 0$. This occurs where $\mathrm{e}^{-X} - 1 = \mathrm{O}(\varepsilon)$ i.e. $X = \mathrm{O}(\varepsilon)$, so a further scaling must be introduced: $X = \varepsilon\chi$, to produce

$$F(\varepsilon\chi; \varepsilon) \equiv \varepsilon\mathfrak{F}(\chi; \varepsilon) = \varepsilon^4\chi^2 - 1 - \varepsilon(3\varepsilon^2\chi + 1) + \mathrm{e}^{-\varepsilon\chi}$$
$$\sim -\varepsilon\chi - \varepsilon + \frac{1}{2}\varepsilon^2\chi^2 - \frac{1}{6}\varepsilon^3\chi^3 - 3\varepsilon^3\chi \quad \text{for } \chi = \mathrm{O}(1) \quad \text{as } \varepsilon \to 0.$$

Since we have
$$\mathfrak{F}(\chi; \varepsilon) \sim -\chi - 1$$

we have a root near $\chi = -1$; to obtain an improved approximation, we write

$$\chi \sim -1 + \sum_{n=1}^{\infty} \varepsilon^n a_n$$

and so obtain

$$1 - \varepsilon a_1 - \varepsilon^2 a_2 \cdots - 1 + \frac{1}{2}\varepsilon\left(1 - 2\varepsilon a_1 + \varepsilon^2 a_1^2 - 2\varepsilon^2 a_2 \cdots\right) - \frac{\varepsilon^2}{6}(-1 \cdots) - 3\varepsilon^2(-1 \cdots) = 0$$

i.e. $a_1 = 1/2, a_2 = 8/3$:
$$\chi \sim -1 + \frac{1}{2}\varepsilon + \frac{8}{3}\varepsilon^2.$$

As $\chi \to 0$, there is no further breakdown, and so we have found two real roots

$$x \sim 1 + 2\varepsilon + \varepsilon^2, \quad x \sim -\varepsilon^2 + \frac{1}{2}\varepsilon^3 + \frac{8}{3}\varepsilon^4 \quad \text{as } \varepsilon \to 0^+.$$

A number of other equations, both polynomial and transcendental, are discussed in the exercises Q2.2, 2.3 and 2.4. However, all these involve the search for *real* roots; we now turn, therefore, to a brief discussion of the corresponding problem of finding *all* roots, whether real or complex. It will soon become clear that we may often adopt precisely the same approach when any roots are being sought (although, sometimes, there may be an advantage in writing $x = \alpha + \mathrm{i}\beta$ and working with two, coupled, real equations). The only small word of warning is that the size of the real and imaginary parts, measured in terms of ε, may be different e.g. $x = \mathrm{O}(1)$ now implies that $|\alpha + \mathrm{i}\beta| = \mathrm{O}(1)$, which can be satisfied if either, but not necessarily both, α and β are O(1). Let us see how this arises in an example.

E2.3 An equation with complex roots

Here, using the more usual notation for a complex number, we consider

$$f(z;\varepsilon) = \varepsilon z^3 + z^2 + 1 = 0$$

and immediately we obtain

$$f(z;\varepsilon) \sim z^2 + 1 \quad \text{for } |z| = \mathrm{O}(1) \quad \text{as } \varepsilon \to 0,$$

and so we have roots $z = \pm\mathrm{i}$, approximately. Thus we write

$$z \sim \pm\mathrm{i} + \sum_{n=1}^{\infty} \varepsilon^n a_n$$

and so the equation becomes

$$\varepsilon(\mp\mathrm{i} - 3\varepsilon a_1 \cdots) + (-1 \pm 2\mathrm{i}\varepsilon a_1 \pm 2\mathrm{i}\varepsilon^2 a_2 + \varepsilon^2 a_1^2 \cdots) + 1 = 0.$$

This is satisfied if $a_1 = 1/2$, $a_2 = \mp 5\mathrm{i}/8$, etc., and thus we have two complex roots

$$z \sim \pm\mathrm{i} + \frac{1}{2}\varepsilon \mp \mathrm{i}\frac{5}{8}\varepsilon^2,$$

and we observe that the imaginary part is O(1), but that the real part is O(ε).

The full 'expansion' is clearly not uniformly valid as $|z| \to \infty$: there is a breakdown where $\varepsilon|z|^3 = \mathrm{O}(|z|^2)$ or $z = \mathrm{O}(\varepsilon^{-1})$. We introduce $z = Z/\varepsilon$ and write

$$f(Z/\varepsilon;\varepsilon) \equiv \varepsilon^{-2}F(Z;\varepsilon) = \frac{Z^3}{\varepsilon^2} + \frac{Z^2}{\varepsilon^2} + 1 (= 0)$$

and so

$$F(Z;\varepsilon) \sim Z^3 + Z^2 \quad \text{for } Z = \mathrm{O}(1) \quad \text{as } \varepsilon \to 0$$

which produces the single, available root near $Z = -1$ and then, more accurately, we have $Z \sim -1 - \varepsilon^2$. The equation has three roots, two of which are complex:

$$z \sim \pm\mathrm{i} + \frac{1}{2}\varepsilon \mp \mathrm{i}\frac{5}{8}\varepsilon^2, \quad z \sim -\frac{1}{\varepsilon} - \varepsilon \quad \text{as } \varepsilon \to 0.$$

Finally, a class of equations for which this direct approach (for complex roots) is not useful is characterised by the appearance of terms such as $\exp(z)$ (or anything equivalent).

In this case, it is almost always convenient to formulate the problem in real and imaginary parts, and the appearance of a small parameter does not affect this approach in any significant way; we present an example of this type.

E2.4 A real-imaginary problem

We seek all the roots of the equation

$$\sin z = 1 + \frac{\varepsilon}{1+|z|} \tag{2.9}$$

as $\varepsilon \to 0^+$; note that

$$\sin z = \frac{1}{2\mathrm{i}}(\mathrm{e}^{\mathrm{i}z} - \mathrm{e}^{-\mathrm{i}z})$$

and so the form of this problem does indeed exhibit this more complicated structure. Let us write $z = x + \mathrm{i}y$, and then (2.9) becomes

$$\sin x \cosh y + \mathrm{i} \cos x \sinh y = 1 + \frac{\varepsilon}{1+\sqrt{x^2+y^2}}$$

or

$$\sin x \cosh y = 1 + \frac{\varepsilon}{1+\sqrt{x^2+y^2}} \quad \text{and} \quad \cos x \sinh y = 0. \tag{2.10a,b}$$

We see immediately that the right-hand sides of these two equations do not presage a breakdown of these contributions, as x or y increases or decreases; thus we proceed with $x = \mathrm{O}(1)$ and $y = \mathrm{O}(1)$. Now equation (2.10b) possesses the solutions

$$x = \pi/2 + n\pi \quad (n = 0, \pm 1, \pm 2, \ldots),$$

and this is the relevant choice (rather than $y = 0$) because we require $\sin x \cosh y > 1$ (from (2.10a)). Then equation (2.10a) gives

$$(-1)^n \cosh y = 1 + \frac{\varepsilon}{1+\sqrt{\left(\frac{1}{2}+n\right)^2\pi^2 + y^2}}$$

and this is consistent only if $n = 2m$ ($m = 0, \pm 1, \pm 2, \ldots$) because $\cosh y > 0$. Finally, the solution arises only for $y \to 0$; since $\cosh y \sim 1 + y^2/2$ as $y \to 0$, we see that a solution exists where $y^2 = \mathrm{O}(\varepsilon)$ and so we introduce $y = \sqrt{\varepsilon}Y$. Thus we obtain

$$\frac{1}{2}Y^2 \sim \frac{1}{1+\left|\frac{1}{2}+2m\right|\pi} \quad \text{or} \quad Y \sim \frac{\pm 2}{\sqrt{2+|1+4m|\pi}}$$

and so we have the set of (approximate) roots

$$z = x + \mathrm{i}y \sim \frac{\pi}{2} + 2m\pi \pm \mathrm{i}\frac{2\sqrt{\varepsilon}}{\sqrt{2 + |1 + 4m|\pi}}, \qquad m = 0, \pm 1, \pm 2, \ldots.$$

A few examples of other equations with complex roots (some of which may be real, of course) are set as exercises in Q2.5.

2.2 INTEGRATION OF FUNCTIONS REPRESENTED BY ASYMPTOTIC EXPANSIONS

Our second direct, and rather routine application of these ideas is to the evaluation of integrals. In particular, we consider integrals of functions that are represented by asymptotic expansions in a small parameter; this may involve one or more expansions, but if it is the latter—and it often is—then the expansions will satisfy the matching principle.

The procedure that we adopt calls upon two general properties: the first is the existence of an intermediate variable (valid in the overlap region; see §1.7), and the second is the familiar device of splitting the range of integration, as appropriate. We then express the integral as a sum of integrals over each of the asymptotic expansions of the integrand, the switch from one to the next being at a point which is in the overlap region. The expansions are then valid for each integration range selected and, furthermore, the value of the original integral (assuming that it exists) is *independent* of how we split the integral. Thus the particular choice of intermediate variable is unimportant; indeed, it may be quite general, satisfying only the necessary conditions for such a variable; see (1.56), for example. Let us apply this technique to a simple example.

E2.5 An elementary integral

We are given

$$f(x;\varepsilon) = \sqrt{x+\varepsilon} + \frac{\varepsilon}{\sqrt{1+x}} + \mathrm{e}^{-x/\varepsilon}, \quad x \geq 0, \quad \varepsilon > 0,$$

and we require the value, as $\varepsilon \to 0^+$, of the integral

$$I(\varepsilon) = \int_0^1 f(x;\varepsilon)\,\mathrm{d}x.$$

(Note that the integral here is elementary, to the extent that it may be evaluated directly, although we will integrate only the relevant asymptotic expansions; this example has been selected so that the interested reader may check the results against the expansion of the exact value.)

First, we expand $f(x;\varepsilon)$ for $x = \mathrm{O}(1)$ and for $x = \varepsilon X = \mathrm{O}(\varepsilon)$, to give

$$f(x;\varepsilon) \sim \sqrt{x} + \varepsilon\left(\frac{1}{\sqrt{1+x}} + \frac{1}{2\sqrt{x}}\right) - \frac{\varepsilon^2}{8x\sqrt{x}}, \qquad x = \mathrm{O}(1),$$

and
$$f(\varepsilon X;\varepsilon) \equiv F(X;\varepsilon) \sim \mathrm{e}^{-X} + \sqrt{\varepsilon}\sqrt{1+X} + \varepsilon - \frac{\varepsilon^2 X}{2}, \qquad X = \mathrm{O}(1),$$

both for $\varepsilon \to 0^+$; we have retained terms as far as $\mathrm{O}(\varepsilon^2)$ in each expansion. (You should confirm that these two expansions satisfy the matching principle.)

Now these two expansions are valid in the overlap region, represented by $x = \mathrm{O}(\delta)$, defined by

$$\delta(\varepsilon) \to 0 \quad \text{and} \quad \varepsilon/\delta(\varepsilon) \to 0 \quad \text{as } \varepsilon \to 0;$$

thus we express the integral as

$$I(\varepsilon) = \int_0^{\delta} F(X;\varepsilon)\,\mathrm{d}x + \int_{\delta}^{1} f(x;\varepsilon)\,\mathrm{d}x \sim \varepsilon \int_0^{\delta/\varepsilon} \left\{\mathrm{e}^{-X} + \sqrt{\varepsilon}\sqrt{1+X} + \varepsilon - \frac{\varepsilon^2 X}{2}\right\} \mathrm{d}X$$
$$+ \int_{\delta}^{1} \left\{\sqrt{x} + \left(\frac{1}{\sqrt{1+x}} + \frac{1}{2\sqrt{x}}\right) - \frac{\varepsilon^2}{8x\sqrt{x}}\right\} \mathrm{d}x.$$

The only requirement, at this stage, is that we are able to perform the integration of the various functions that appear in the asymptotic expansions. Note that the first integral has been expressed as an integration in X—the most natural choice of integration variable in this context. To proceed, we obtain

$$I(\varepsilon) \sim \varepsilon\left[-\mathrm{e}^{-X} + \frac{2\sqrt{\varepsilon}}{3}(1+X)^{3/2} + \varepsilon X - \frac{\varepsilon^2 X^2}{4}\right]_0^{\delta/\varepsilon}$$
$$+ \left[\frac{2}{3}x^{3/2} + \varepsilon\{2(1+x)^{1/2} + x^{1/2}\} + \frac{\varepsilon^2 x^{-1/2}}{4}\right]_{\delta}^{1}$$
$$= \varepsilon\left\{-\mathrm{e}^{-\delta/\varepsilon} + \frac{2\sqrt{\varepsilon}}{3}\left(1 + \frac{\delta}{\varepsilon}\right)^{3/2} + \delta - \frac{\delta^2}{4} + 1 - \frac{2\sqrt{\varepsilon}}{3}\right\}$$
$$+ \left\{\frac{2}{3} + \varepsilon(2\sqrt{2}+1) + \frac{\varepsilon^2}{4} - \frac{2\delta^{3/2}}{3} - \varepsilon[2(1+\delta)^{1/2} + \delta^{1/2}] - \frac{\varepsilon^2}{4\sqrt{\delta}}\right\}$$

and this is to be expanded for $\varepsilon \to 0$, $\delta \to 0$ and $\varepsilon/\delta \to 0$ (note!). Thus we obtain

$$I(\varepsilon) \sim \varepsilon\left\{\frac{2\sqrt{\varepsilon}}{3}\left(\frac{\delta}{\varepsilon}\right)^{3/2}\left(1 + \frac{3}{2}\frac{\delta}{\varepsilon} + \frac{3}{8}\frac{\varepsilon^2}{\delta^2}\cdots\right) + \delta - \frac{\delta^2}{4} + 1 - \frac{2\sqrt{\varepsilon}}{3}\right\}$$
$$+ \left\{\frac{2}{3} + \varepsilon(1+2\sqrt{2}) + \frac{\varepsilon^2}{4} - \frac{2\delta^{3/2}}{3} - 2\varepsilon\left(1 + \frac{1}{2}\delta - \frac{1}{8}\delta^2\cdots\right) - \varepsilon\sqrt{\delta} - \frac{\varepsilon^2}{4\sqrt{\delta}}\right\},$$

where the ellipsis ($\cdots$) indicates further terms in the various binomial expansions; we keep as many as required in order to demonstrate that $\delta(\varepsilon)$ vanishes identically (at this order), to leave

$$\varepsilon - \frac{2}{3}\varepsilon\sqrt{\varepsilon} + \frac{2}{3} + \varepsilon(1 + 2\sqrt{2}) + \frac{\varepsilon^2}{4} - 2\varepsilon \cdots.$$

Thus we have found that

$$I(\varepsilon) \sim \frac{2}{3} + \varepsilon 2\sqrt{2} - \frac{2\varepsilon^{3/2}}{3} + \frac{\varepsilon^2}{4}$$

as $\varepsilon \to 0^+$, as far as terms at $O(\varepsilon^2)$; here we see that the integration over $x = O(1)$ provides the dominant contribution to this value.

This example has presented, *via* a fairly routine calculation, the essential idea that underpins this method for evaluating integrals. Of course, there is no need to exploit this technique if the integral can be evaluated directly (as was the case here); let us therefore examine another problem which is less elementary.

E2.6 Another integral

We wish to evaluate the integral

$$I(\varepsilon) = \int_0^1 \frac{dx}{\sqrt{\dfrac{x+\varepsilon}{(1+x)^2} + \varepsilon^2 \exp(-x/\varepsilon^2)}}$$

as $\varepsilon \to 0^+$; here, the expansion of the integrand requires three different asymptotic expansions (valid for $x = O(1)$, $x = O(\varepsilon)$, $x = O(\varepsilon^2)$). Thus we obtain

$$f(x;\varepsilon) = \left\{\frac{x+\varepsilon}{(1+x)^2} + \varepsilon^2 \exp(-x/\varepsilon^2)\right\}^{-1/2}$$

$$\sim \frac{1+x}{\sqrt{x}}\left(1 - \frac{\varepsilon}{2x} + \frac{3\varepsilon^2}{8x^2}\right), \quad \text{for } x = O(1);$$

$$f(\varepsilon X;\varepsilon) \equiv F(X;\varepsilon) = \left\{\frac{\varepsilon(1+X)}{(1+\varepsilon X)^2} + \varepsilon^2 e^{-X/\varepsilon}\right\}^{-1/2}$$

$$\sim \frac{1}{\sqrt{\varepsilon}}\frac{1}{\sqrt{1+X}}(1+\varepsilon X), \quad \text{for } X = O(1);$$

$$f(\varepsilon^2\chi;\varepsilon) \equiv \mathfrak{F}(\chi;\varepsilon) = \left\{ \frac{\varepsilon(1+\varepsilon\chi)}{(1+\varepsilon^2\chi)^2} + \varepsilon^2 \mathrm{e}^{-\chi} \right\}^{-1/2}$$

$$\sim \frac{1}{\sqrt{\varepsilon}} \left\{ 1 - \frac{1}{2}\varepsilon(\chi + \mathrm{e}^{-\chi}) + \varepsilon^2\chi + \frac{3}{8}\varepsilon^2(\chi + \mathrm{e}^{-\chi})^2 \right\}, \quad \text{for } \chi = \mathrm{O}(1).$$

In this problem, we require two intermediate variables; these are defined by

$$\delta(\varepsilon) \to 0 \quad \text{and} \quad \varepsilon/\delta(\varepsilon) \to 0; \quad \Delta(\varepsilon)/\varepsilon \to 0 \quad \text{and} \quad \varepsilon^2/\Delta(\varepsilon) \to 0,$$

all as $\varepsilon \to 0^+$. The integral is then written as

$$I(\varepsilon) = \int_0^{\Delta} \mathfrak{F}(\chi;\varepsilon)\,\mathrm{d}x + \int_{\Delta}^{\delta} F(X;\varepsilon)\,\mathrm{d}x + \int_{\delta}^{1} f(x;\varepsilon)\,\mathrm{d}x$$

$$\sim \varepsilon^{3/2} \int_0^{\Delta/\varepsilon^2} \left\{ 1 - \frac{\varepsilon}{2}(\chi + \mathrm{e}^{-\chi}) + \varepsilon^2\chi + \frac{3\varepsilon^2}{8}(\chi + \mathrm{e}^{-\chi})^2 \right\} \mathrm{d}\chi$$

$$+ \varepsilon^{1/2} \int_{\Delta/\varepsilon}^{\delta/\varepsilon} \frac{1+\varepsilon X}{\sqrt{1+X}} \mathrm{d}X + \int_{\delta}^{1} \frac{1+x}{\sqrt{x}} \left(1 - \frac{\varepsilon}{2x} + \frac{3\varepsilon^2}{8x^2} \right) \mathrm{d}x$$

and we will now retain terms that will enable us to find an expression for $I(\varepsilon)$ correct at $\mathrm{O}(\varepsilon^2)$. Thus we find that

$$I(\varepsilon) \sim \varepsilon^{3/2} \left[\chi\right]_0^{\Delta/\varepsilon^2} + \varepsilon^{1/2} \left[2\sqrt{1+X} + \frac{2\varepsilon}{3}(1+X)^{3/2} - 2\varepsilon\sqrt{1+X} \right]_{\Delta/\varepsilon}^{\delta/\varepsilon}$$

$$+ \left[2\sqrt{x} + \frac{2}{3}x^{3/2} + \varepsilon\left(\frac{1}{\sqrt{x}} - \sqrt{x} \right) + \frac{3\varepsilon^2}{8}\left(-\frac{2x^{-3/2}}{3} - \frac{2}{\sqrt{x}} \right) \right]_{\delta}^{1}$$

$$= \frac{\Delta}{\sqrt{\varepsilon}} + \sqrt{\varepsilon} \left\{ 2\sqrt{1+\delta/\varepsilon} + \frac{2\varepsilon}{3}(1+\delta/\varepsilon)^{3/2} - 2\varepsilon\sqrt{1+\delta/\varepsilon} \right.$$

$$\left. - 2\sqrt{1+\Delta/\varepsilon} - \frac{2\varepsilon}{3}(1+\Delta/\varepsilon)^{3/2} + 2\varepsilon\sqrt{1+\Delta/\varepsilon} \right\}$$

$$+ \left\{ 2 + \frac{2}{3} - \varepsilon^2 - 2\sqrt{\delta} - \frac{2\delta^{3/2}}{3} - \varepsilon\left(\frac{1}{\sqrt{\delta}} - \sqrt{\delta} \right) + \frac{3\varepsilon^2}{8}\left(\frac{2\delta^{-3/2}}{3} + \frac{2}{\sqrt{\delta}} \right) \right\}$$

$$= \frac{\Delta}{\sqrt{\varepsilon}} + \sqrt{\varepsilon}\left\{2\sqrt{\frac{\delta}{\varepsilon}}\left(1+\frac{\varepsilon}{2\delta}-\frac{\varepsilon^2}{8\delta^2}\cdots\right)+\frac{2\delta^{3/2}}{3\sqrt{\varepsilon}}\left(1+\frac{3\varepsilon}{2\delta}+\frac{3\varepsilon^2}{8\delta^2}\cdots\right)\right.$$

$$\left.-2\sqrt{\varepsilon\delta}\left(1+\frac{\varepsilon}{2\delta}\cdots\right)-2\left(1+\frac{\Delta}{2\varepsilon}\cdots\right)-\frac{2\varepsilon}{3}\left(1+\frac{3\Delta}{2\varepsilon}\cdots\right)+2\varepsilon\left(1+\frac{\Delta}{2\varepsilon}\cdots\right)\right\}$$

$$+2+\frac{2}{3}-\varepsilon^2-2\sqrt{\delta}-\frac{2\delta^{3/2}}{3}-\frac{\varepsilon}{\sqrt{\delta}}+\varepsilon\sqrt{\delta}+\frac{\varepsilon^2}{4\delta^{3/2}}+\frac{3\varepsilon^2}{4\sqrt{\delta}}$$

$$\sim \frac{8}{3}-2\sqrt{\varepsilon}+\frac{4}{3}\varepsilon\sqrt{\varepsilon}-\varepsilon^2 \quad \text{as } \varepsilon \to 0^+.$$

(You should confirm that, in the above, both δ and Δ cancel identically, to this order.)

Further examples that make use of these ideas can be found in exercises Q2.6, 2.7 and 2.8. With a little experience, it should not be too difficult to recognise how many terms need to be retained in each expansion in order to produce $I(\varepsilon)$ to a desired accuracy. The region that gives the dominant contribution is usually self-evident, and quite often this alone will provide an acceptable approximation to the value of the integral. Furthermore, terms that contain the overlap variables can be ignored altogether, because they must cancel (although there is a case for their retention—which was our approach above—as a check on the correctness of the details).

2.3 ORDINARY DIFFERENTIAL EQUATIONS: REGULAR PROBLEMS

We now turn to an initial discussion of how the techniques of singular perturbation theory can be applied to the problem of finding solutions of differential equations—unquestionably the most significant and far-reaching application that we encounter. The relevant ideas will be developed, first, for problems that turn out to be *regular* (but we will indicate how singular versions of these problems might arise, and we will discuss some simple examples of these later in this chapter). Clearly, we need to lay down the basic procedure that must be followed when we seek solutions of differential equations. However, these techniques are many and varied, and so we cannot hope to present, at this stage, an all-encompassing recipe. Nevertheless, the fundamental principles can be developed quite readily; to aid us in this, we consider the differential equation

$$\frac{dy}{dx}+y+\varepsilon y^2 = x, \quad 0 \le x \le 1, \quad \text{with} \quad y(1;\varepsilon)=1, \qquad (2.11)$$

for $\varepsilon \to 0$. This problem, we observe, is not trivial; it is an equation which, although first order, is nonlinear and with a forcing term on the right-hand side.

The first stage is to decide on a suitable asymptotic sequence for the representation of $y(x;\varepsilon)$. Here, we note that the process of iteration on the equation, which can be written for this purpose (with a prime for the derivative) as

$$y'_{n+1} + y_{n+1} - x = -\varepsilon y_n^2, \quad n \geq 0, \quad \text{with} \quad y_0 = 0,$$

gives

$$y'_1 + y_1 - x = 0; \quad y'_2 + y_2 - x = -\varepsilon y_1^2,$$

and so on, so that y_2 takes the form $y_2 = u_2(x) + \varepsilon v_2(x)$ (for appropriate functions u_2, v_2). When this solution is used to generate y_3, it is clear that we will produce terms in ε, ε^2 and ε^3, and so this pattern will continue: the equation implies the 'natural' asymptotic sequence $\{\varepsilon^n\}$, so this is what we will assume to initiate the solution method. (It should be noted that the boundary condition is consistent with this assumption, as is the alternative condition $y(1;\varepsilon) = 1 + \varepsilon$. On the other hand, a boundary value $y(1;\varepsilon) = 1 + \sqrt{\varepsilon}$ would force the asymptotic sequence to be adjusted to accommodate this i.e. $\{\varepsilon^{n/2}\}$, $n = 0, 1, 2, \ldots$.)

Thus we seek a solution of the problem (2.11) in the form

$$y(x;\varepsilon) \sim \sum_{n=0}^{N} \varepsilon^n y_n(x), \tag{2.12}$$

for some $x \in D$ (and we do not know which xs will be allowed, at this stage). The expansion (2.12) is used in the differential equation to give

$$(y'_0 + \varepsilon y'_1 + \varepsilon^2 y'_2 \cdots) + (y_0 + \varepsilon y_1 + \varepsilon^2 y_2 \cdots) + \varepsilon(y_0 + \varepsilon y_1 \cdots)^2 - x = 0,$$

where '= 0' means zero to all orders in ε; thus we require

$$y'_0 + y_0 - x = 0; \quad y'_1 + y_1 + y_0^2 = 0; \quad y'_2 + y_2 + 2y_0 y_1 = 0, \tag{2.13a,b,c}$$

and so on. Similarly, the boundary condition gives

$$y_0(1) + \varepsilon y_1(1) + \varepsilon^2 y_2(1) \cdots - 1 = 0$$

so that

$$y_0(1) = 1, \quad y_n(1) = 0 \quad \text{for } n \geq 1; \tag{2.14a,b}$$

of course, to evaluate on $x = 1$ implies that the asymptotic expansion, (2.12), is valid here—but we do not know this yet. This is written down because, if the problem turns out to be well-behaved i.e. regular, then we will have this ready for use; essentially, all we are doing is noting (2.14)—we can reject it if the expansion will not permit evaluation on $x = 1$.

The next step is simply to solve each equation (for $y_0, y_1, \ldots$) in turn; we see directly (from (2.13a)) that the general solution for y_0 is

$$y_0(x) = x - 1 + A_0 e^{-x}, \tag{2.15}$$

where A_0 is an arbitrary constant. Then, from (2.13b), we have

$$y_1' + y_1 = -y_0^2 = -(x-1)^2 - 2A_0(x-1)\mathrm{e}^{-x} - A_0^2\mathrm{e}^{-2x}$$

which can be written (on using the integrating factor e^x) as

$$(\mathrm{e}^x y_1)' = -(x-1)^2\mathrm{e}^x - 2A_0(x-1) - A_0^2\mathrm{e}^{-x}.$$

This produces the general solution

$$y_1(x) = -(x-1)^2 + 2x - 4 - A_0(x-1)^2\mathrm{e}^{-x} + A_0^2\mathrm{e}^{-2x} + A_1\mathrm{e}^{-x}, \qquad (2.16)$$

where A_1 is a second arbitrary constant. It is immediately clear that these first two terms in the expansion are defined (and well-behaved i.e. no hint of a non-uniformity) for $0 \le x \le 1$, so we may impose the boundary conditions, (2.14a,b); these produce

$$y_0(x) = x - 1 + \mathrm{e}^{1-x}; \quad y_1(x) = -x^2 + 4x - 5 - (x^2 - 2x)\mathrm{e}^{1-x} + \mathrm{e}^{2(1-x)}.$$

Thus our asymptotic expansion, so far, is

$$y(x;\varepsilon) \sim x - 1 + \mathrm{e}^{1-x} + \varepsilon\left\{-x^2 + 4x - 5 - (x^2 - 2x)\mathrm{e}^{1-x} + \mathrm{e}^{2(1-x)}\right\}, \qquad (2.17)$$

and this is certainly uniformly valid for $0 \le x \le 1$: we have a 2-term expansion of the solution. (Note that the specification of the domain is critical here; if, for example, we were seeking the solution with the same boundary condition, but in $x \ge 1$, then (2.17) would *not* be uniformly valid: there is a breakdown where $x = \mathrm{O}(\varepsilon x^2)$ i.e. $x = \mathrm{O}(\varepsilon^{-1})$; see the problem in (2.34), below.) The evidence in (2.17) suggests that we have the beginning of a uniformly valid asymptotic expansion i.e. (2.12) is valid for $\forall N \ge 0$ and for $\forall x \in D$ (and it is left as an exercise to find $y_2(x)$ and to check that the inclusion of this term does not alter this proposition).

In order to investigate the uniform validity, or otherwise, of (2.12), one approach is to examine the general term in the expansion; this is the solution of

$$y_{n+1}' + y_{n+1} = f_n \quad \text{with} \quad y_{n+1}(1) = 0 \quad \text{for } n \ge 0, \qquad (2.18)$$

where $f_n = f_n(y_0, y_1, \ldots, y_n)$ with $f_0 = -y_0^2$, $f_1 = -2y_0y_1$. The solution to (2.18) is

$$y_{n+1}(x) = -\mathrm{e}^{-x}\int_x^1 f_n\mathrm{e}^x\,\mathrm{d}x;$$

but $y_0(x)$ and $y_1(x)$ are bounded functions for $x \in [0,\ 1]$, and hence so is $y_2(x)$, and then so is $y_3(x)$ and hence all the y_ns. In particular, $y_n(x) \to 0$ as $x \to 1$ $(n \ge 1)$, and $y_n(x) \to c_n$ (c_n constants) as $x \to 0$ $(n \ge 0)$: there is no breakdown of the asymptotic expansion. The problem posed in (2.11) is therefore *regular*, resulting in a uniformly valid asymptotic expansion.

More complete, formal and rigorous discussions of uniform validity, in the context of differential equations, can be found in other texts, such as Smith (1985), O'Malley (1991) and Eckhaus (1979). Typically, these arguments involve writing

$$y(x;\varepsilon) = Y_N(x;\varepsilon) + \varepsilon^{N+1} R_{N+1}(x;\varepsilon),$$

where $Y_N(x;\varepsilon) = \sum_{n=0}^{N} \varepsilon^n y_n(x)$, and then showing that R_{N+1} remains bounded for $\forall N \geq 0$ and for $\forall x \in D$. We will outline how this can be applied to our problem, (2.11); first, we obtain

$$Y_N' + \varepsilon^{N+1} R_{N+1}' + Y_N + \varepsilon^{N+1} R_{N+1} + \varepsilon(Y_N + \varepsilon^{N+1} R_{N+1})^2 = x$$

with $R_{N+1}(1;\varepsilon) = 0$ for $N \geq 0$. Since each $y_n(x)$ satisfies an appropriate differential equation and boundary condition, this gives

$$R_{N+1}' + R_{N+1} + F_N(y_0, y_1, \ldots, y_N; \varepsilon) + 2\varepsilon Y_N R_{N+1} + \varepsilon^{N+2} R_{N+1}^2 = 0, \tag{2.19}$$

where F_N comprises the $\mathrm{O}(\varepsilon^N)$ terms, and smaller, from the expansion of Y_N^2 (after division by ε^{N+1}). A uniform asymptotic expansion requires that R_{N+1} is bounded as $\varepsilon \to 0$, for $\forall x \in D$ and $\forall N \geq 0$. To prove such a result is rarely an elementary exercise in general, and it is not trivial here, although a number of approaches are possible. One method is based on Picard's iterative scheme (which is a standard technique for proving the existence of solutions of first order ordinary differential equations in some appropriate region of (x, y)-space); this will be described in any good basic text on ordinary differential equations (e.g. Boyce & DiPrima, 2001). Another possibility, closely related to Picard's method, is formally to integrate the equation for R_{N+1}, thereby obtaining an integral equation, and then to derive estimates for the integral term (and hence for R_{N+1}). We will outline a third technique, which involves the construction of estimates directly for the differential equation, and then integrating a reduced version of the equation for R_{N+1}.

At this stage we do not know if R_{N+1} is of one sign, for $0 \leq x \leq 1$, or if it changes sign on this interval; however, we may proceed without specifying or assuming the nature of this property, but it will affect the details; first we write

$$R_{N+1}' + R_{N+1} + 2\varepsilon Y_N R_{N+1} + \varepsilon^{N+2} R_{N+1}^2 = -F_N.$$

But we do know that each y_n is bounded (for $x \in [0, 1]$) and hence so is F_N, which we will express in the form $|F_N| \leq k$ (a constant independent of ε), and so

$$-k \leq R_{N+1}' + R_{N+1} + 2\varepsilon Y_N R_{N+1} + \varepsilon^{N+2} R_{N+1}^2 \leq k.$$

This same property of the functions y_n leads to a corresponding statement for Y_N: $|Y_N| \leq c$ (again, independent of ε), which now gives

$$-k \mp 2\varepsilon c\, R_{N+1} \leq R'_{N+1} + R_{N+1} + \varepsilon^{N+2} R^2_{N+1} \leq k \pm 2\varepsilon c\, R_{N+1}, \tag{2.20}$$

where the upper sign applies if $R_{N+1} \geq 0$, and the lower if $R_{N+1} \leq 0$. (Often in arguments of this type, we cannot incorporate both signs, and we are reduced to working with the modulus of the function; we will see here that we can allow the form given in (2.20).) Between the two inequalities, we have an expression associated with a constant coefficient Riccati equation; let us therefore consider

$$R'_{N+1} + R_{N+1} + \varepsilon^{N+2} R^2_{N+1} = \lambda + 2\varepsilon\mu\, R_{N+1}, \tag{2.21}$$

where $-k \leq \lambda(x) \leq k$ and $-c \leq \mu(x) \leq c$ for arbitrary (bounded) functions λ and μ. If an appropriate unique solution of (2.21), satisfying $R_{N+1}(1;\varepsilon) = 0$, exists for all λ and all μ as specified, then we will certainly have satisfied (2.20). However, we will, in this text, give only the flavour of how the development proceeds, by considering a restricted version of the problem with the special choice: λ and μ constant (but satisfying the given bounds).

To solve (2.21), we introduce $R_{N+1} = \varepsilon^{-N-2}\phi'/\phi$ to obtain

$$\phi'' + (1 - 2\varepsilon\mu)\phi' - \varepsilon^{N+2}\lambda\phi = 0$$

which has, in our special case, the general solution

$$A\exp[-(1 - \varepsilon\alpha_1)x] + B\exp\left[\varepsilon^{N+2}\alpha_2 x\right]$$

where the arbitrary constants are A and B, and the auxiliary equation for the exponents is

$$\alpha^2 + (1 - 2\varepsilon\mu)\alpha - \varepsilon^{N+2}\lambda = 0.$$

The roots of this equation have been written as

$$\alpha = -1 + \varepsilon\alpha_1, \quad \varepsilon^{N+2}\alpha_2,$$

where $\alpha_i(\varepsilon) = \mathrm{O}(1)$, for $i = 1,\ 2$, as $\varepsilon \to 0$. Finally, the solution which satisfies the condition on $x = 1$ is

$$R_{N+1}(x;\varepsilon) = -\alpha_2 \frac{\left\{\exp\left[(1 - \varepsilon\alpha_1 + \varepsilon^{N+2}\alpha_2)(1 - x)\right] - 1\right\}}{1 + \varepsilon^{N+2}\alpha_2 \exp\left[(1 - \varepsilon\alpha_1 + \varepsilon^{N+2}\alpha_2)(1 - x)\right]},$$

which is bounded for $\forall\lambda \in [-k,\ k]$, $\forall\mu \in [-c,\ c]$, $\forall x \in [0,\ 1]$ as $\varepsilon \to 0$. Thus the error is $\varepsilon^{N+1} R_{N+1} = \mathrm{O}(\varepsilon^{N+1})$ for $\forall N \geq 0$, as required. (A comprehensive discussion requires the analysis of the equation for ϕ with general, bounded $\lambda = \lambda(x)$ and $\mu = \mu(x)$, which is possible, but beyond the aims of this text.)

As should be clear from this calculation, it is to be anticipated that special properties, relevant to a particular problem, may have to be invoked. Here, for example, we took advantage of the underlying Riccati equation; other problems may require quite different approaches. However, we must also emphasise that, for many practical and important problems encountered in applied mathematics, these calculations are often too difficult to succumb to such a general analysis. Indeed, the conventional wisdom is that, if breakdowns have been identified, rescaling employed and asymptotic solutions found (and matched, as required), then we have produced a sufficiently robust description. It should be noted that the process of rescaling might involve a consideration of all possible scalings allowed by the governing equation, which will then greatly strengthen our trust in the results obtained. Those readers who prefer the more rigorous approach that such discussions afford are encouraged to study the texts previously mentioned. In this text, however, we shall proceed without much further consideration of these more formal aspects of the asymptotic solution of differential equations.

Now that we have presented the salient features of the method of constructing solutions, we apply it to another example.

E2.7 A regular second-order problem

We seek an asymptotic solution, as $\varepsilon \to 0$, of

$$y'' + y' + \varepsilon y^2 = 0, \quad 0 \le x \le 1, \tag{2.22}$$

with $y(0;\varepsilon) = 1$ and $y(1;\varepsilon) = \mathrm{e}^{-1}$; the primes here denote derivatives. First, we assume that there is a solution, for some $x \in D$, of the form

$$y(x;\varepsilon) \sim \sum_{n=0}^{N} \varepsilon^n y_n(x).$$

Thus we obtain

$$y_0'' + y_0' = 0; \quad y_1'' + y_1' + y_0^2 = 0, \tag{2.23a,b}$$

and so on, with

$$y_0(0) = 1, \quad y_0(1) = \mathrm{e}^{-1}; \quad y_n(0) = y_n(1) = 0 \quad \text{for } n \ge 1 \tag{2.24}$$

(if the expansion is valid at the end-points). The general solution of (2.23a) is

$$y_0(x) = A_0 + B_0\mathrm{e}^{-x} \quad \text{(where } A_0,\ B_0 \text{ are the arbitrary constants)}$$

and then (2.23b) becomes

$$y_1'' + y_1' = -y_0^2 = -A_0^2 - 2A_0B_0\mathrm{e}^{-x} - B_0^2\mathrm{e}^{-2x}$$

which, in turn, has the general solution

$$y_1(x) = A_1 + B_1\mathrm{e}^{-x} - A_0^2(x-1) + 2A_0B_0x\mathrm{e}^{-x} - \tfrac{1}{2}B_0^2\mathrm{e}^{-2x}$$

where A_1, B_1 are the new arbitrary constants. The functions $y_0(x)$ and $y_1(x)$ are clearly defined for $0 \le x \le 1$, and there is no suggestion of a breakdown, so we impose the boundary conditions (2.24) to give

$$y_0(x) = \mathrm{e}^{-x}; \quad y_1(x) = \frac{1}{2}(\mathrm{e}^{-x} - \mathrm{e}^{-1} + \mathrm{e}^{-1-x} - \mathrm{e}^{-2x})$$

and then our asymptotic expansion (to this order) is

$$y(x;\varepsilon) \sim \mathrm{e}^{-x} + \frac{\varepsilon}{2}\{\mathrm{e}^{-x} - \mathrm{e}^{-1} + \mathrm{e}^{-1-x} - \mathrm{e}^{-2x}\}. \tag{2.25}$$

Now that we have obtained the expansion, (2.25), we are able to confirm that we have a 2-term uniformly valid representation of the solution. In order to examine the general term in this asymptotic expansion, if this is deemed necessary, we can follow the method described earlier. Thus we may write

$$y_n'' + y_n' = f_{n-1}(y_0, \ldots, y_{n-1}), \quad n \ge 1,$$

where, in particular, we have $f_0 = -y_0^2$, and $y_n(0) = y_n(1) = 0$ for $n \ge 1$; the general solution for y_n is

$$y_n(x) = \mathrm{e}^{-x}\int_0^x \mathrm{e}^x\left(\int_0^x f_{n-1}\,\mathrm{d}x\right)\mathrm{d}x + A_n + B_n\mathrm{e}^{-x},$$

where A_n and B_n are determined to satisfy the two boundary conditions. The essentials of the argument are then as we have already outlined in our first, simple, presentation: y_0 is bounded (on [0, 1]), so is y_1, and hence so is y_2, etc., for all $y_n(x)$. Further, $y_n(x) \to 0$ as $x \to 0$ and as $x \to 1$, for $\forall n \ge 1$: the asymptotic expansion is uniformly valid.

Some further examples of regular expansions can be found in Q2.9 and 2.10, and an interesting variant of E2.7 is discussed in Q2.15.

2.4 ORDINARY DIFFERENTIAL EQUATIONS: SIMPLE SINGULAR PROBLEMS

Now that we have introduced the simplest ideas that enable solutions of differential equations to be constructed, we must extend our horizons. The first point to record is that, only quite rarely, do we encounter problems that can be represented by uniformly valid expansions (although, somewhat after the event, we can often construct such expansions—in the form of a composite expansion, for example; see §1.10). The more common equations exhibit singular behaviour, in one form or another; the simplest situation, we suggest, is when the techniques used above (§2.3) produce asymptotic expansions that break down, resulting in the need to rescale, expand again and (probably) invoke the matching principle. (Other types of singularity can arise, and these will be described in due course.) To see how this approach is a natural extension of what we have done thus far, we will present a problem based on the equation given in (2.11).

We consider

$$\frac{dy}{dx} + \left(1 + \frac{\varepsilon^2}{x^2 + \varepsilon^2}\right) y + \varepsilon y^2 = 0, \quad 0 \le x \le 1, \quad \text{with} \quad y(1; \varepsilon) = 1, \tag{2.26}$$

for $\varepsilon \to 0$; the important new ingredient here is the variable coefficient (which, we note, is $1 + O(\varepsilon^2)$ for $x = O(1)$). We seek a solution in the form

$$y(x; \varepsilon) \sim \sum_{n=0}^{N} \varepsilon^n y_n(x)$$

and we will need to find the terms y_0, y_1 and y_2 (at least) in order to include a contribution from the new part of the coefficient. The equations for the $y_n(x)$ are

$$y_0' + y_0 = 0; \quad y_1' + y_1 + y_0^2 = 0; \quad y_2' + y_2 + \frac{y_0}{x^2} + 2y_0 y_1 = 0,$$

and so on; the boundary condition requires that

$$y_0(1) = 1, \quad y_n(1) = 0 \quad \text{for } n \ge 1.$$

In this problem, we should expect that evaluation of the expansion on $x = 1$ is allowed—all terms are defined for $x = O(1)$—but we must anticipate difficulties as $x \to 0$.

The solutions for the functions $y_0(x)$ and $y_1(x)$ follow from the results given in (2.15) and (2.16), respectively, but with the particular integral omitted; thus

$$y_0(x) = e^{1-x}; \quad y_1(x) = e^{2-2x} - e^{1-x}.$$

The 2-term asymptotic expansion, $y \sim y_0 + \varepsilon y_1$ is uniformly valid for $\forall x \in [0,\ 1]$. Let us find the next term in the expansion; this is the solution of

$$y_2' + y_2 + \frac{e^{1-x}}{x^2} + 2e^{1-x}(e^{2-2x} - e^{1-x}) = 0 \quad \text{with} \quad y_2(1) = 0.$$

Thus, introducing the integrating factor e^x, we have

$$(e^x y_2)' + \frac{e}{x^2} + 2e(e^{2-2x} - e^{1-x}) = 0$$

and so
$$y_2(x) = \frac{e^{1-x}}{x} + e^{1-x}(e^{2-2x} - 2e^{1-x}) + Ae^{-x}$$

where the arbitrary constant must be $A = 0$ (to satisfy $y_2(1) = 0$). This third term in the asymptotic expansion is very different from the first two: it is not defined on $x = 0$, so we must expect a breakdown. The expansion, to this order, is now

$$y(x;\varepsilon) \sim e^{1-x} + \varepsilon\left\{e^{2-2x} - e^{1-x}\right\} + \varepsilon^2\left\{\frac{e^{1-x}}{x} + e^{3-3x} - 2e^{2-2x}\right\} \tag{2.27}$$

as $\varepsilon \to 0$ for $x = O(1)$; as $x \to 0$, we clearly have a breakdown where the second and third terms in the expansion become the same size i.e. $\varepsilon = O(\varepsilon^2/x)$ or $x = O(\varepsilon)$. Note that this breakdown occurs for a larger size of x (as x is decreased from $O(1)$) than the breakdown associated with the first and third terms, so we must consider $x = O(\varepsilon)$.

The problem for $x = O(\varepsilon)$ is formulated by writing

$$x = \varepsilon X \quad \text{and} \quad y(\varepsilon X;\varepsilon) \equiv Y(X;\varepsilon),$$

where the relabelling of y is an obvious convenience (and we note that $y = O(1)$ for $x = O(\varepsilon)$). The original equation, in (2.26), expressed in terms of X and Y, requires the identity

$$\frac{dy}{dx} = \frac{dY}{dx} = \frac{d}{dx}Y(x/\varepsilon;\varepsilon) = \varepsilon^{-1}\frac{dY}{dX}$$

and then we obtain

$$\varepsilon^{-1}\frac{dY}{dX} + \left(1 + \frac{\varepsilon^2}{\varepsilon^2 X^2 + \varepsilon^2}\right)Y + \varepsilon Y^2 = 0 \quad \text{or} \quad \frac{dY}{dX} + \varepsilon\left(1 + \frac{1}{1 + X^2}\right)Y + \varepsilon^2 Y^2 = 0, \tag{2.28}$$

but the boundary condition is not available, because this is specified where $x = \mathrm{O}(1)$. Equation (2.28) suggests that we seek a solution in the form

$$Y(X;\varepsilon) \sim \sum_{n=0}^{N} \varepsilon^n Y_n(X),$$

which gives

$$Y_0' = 0; \quad Y_1' + \left(1 + \frac{1}{1+X^2}\right) Y_0 = 0, \quad \text{and so on.} \tag{2.29a,b}$$

Immediately we obtain $Y_0 = A_0$ (an arbitrary constant), and then equation (2.29b) becomes

$$Y_1' = -\left(1 + \frac{1}{1+X^2}\right) A_0$$

which integrates to give

$$Y_1(X) = -A_0(X + \arctan X) + A_1,$$

where A_1 is a second arbitrary constant.

The resulting 2-term expansion is therefore

$$Y(X;\varepsilon) \sim A_0 + \varepsilon\left\{A_1 - A_0(X + \arctan X)\right\}, \quad X = \mathrm{O}(1); \tag{2.30}$$

the two arbitrary constants are determined by invoking the matching principle: (2.30) and (2.27) are to match. Thus we write the terms in (2.27) as functions of X, let $\varepsilon \to 0$ (for $X = \mathrm{O}(1)$) and retain terms $\mathrm{O}(1)$ and $\mathrm{O}(\varepsilon)$ (which are used in (2.30)); conversely, write (2.30) as a function of x, expand and retain terms $\mathrm{O}(1)$, $\mathrm{O}(\varepsilon)$ and $\mathrm{O}(\varepsilon^2)$. From (2.27) we construct

$$\begin{aligned} &\mathrm{e}^{1-\varepsilon X} + \varepsilon\left\{\mathrm{e}^{2-2\varepsilon X} - \mathrm{e}^{1-\varepsilon X}\right\} + \varepsilon^2\left\{\frac{\mathrm{e}^{1-\varepsilon X}}{\varepsilon X} + \mathrm{e}^{3-3\varepsilon X} - 2\mathrm{e}^{2-2\varepsilon X}\right\} \\ &\quad \sim \mathrm{e} + \varepsilon\left(\frac{\mathrm{e}}{X} + \mathrm{e}^2 - \mathrm{e} - \mathrm{e}X\right) \quad \text{for } X = \mathrm{O}(1); \end{aligned} \tag{2.31}$$

and from (2.30) we write

$$\begin{aligned} &A_0 + \varepsilon\left\{A_1 - A_0\left(\frac{x}{\varepsilon} + \arctan\frac{x}{\varepsilon}\right)\right\} \\ &\quad \sim A_0 - A_0 x + \varepsilon\left(A_1 - A_0\frac{\pi}{2}\right) + \frac{\varepsilon^2 A_0}{x} \quad \text{for } x = \mathrm{O}(1). \end{aligned} \tag{2.32}$$

(This expansion requires the standard result:

$$\arctan X \sim \frac{\pi}{2} - \frac{1}{X} + \frac{1}{3X^3} \quad \text{as } X \to +\infty,$$

which the interested reader may wish to derive.) The two 'expansions of expansions', (2.31) and (2.32), match when we choose $A_0 = \mathrm{e}$ and $A_1 = \mathrm{e}^2 - \mathrm{e} + \mathrm{e}\pi/2$; the asymptotic expansion for $X = \mathrm{O}(1)$ is therefore

$$Y(X;\varepsilon) \sim \mathrm{e} + \varepsilon \left\{ \frac{\mathrm{e}\pi}{2} + \mathrm{e}^2 - \mathrm{e} - \mathrm{e}(X + \arctan X) \right\}. \tag{2.33}$$

We now observe that, although the expansion (2.27) is not defined on $x = 0$, the expansion valid for $x = \mathrm{O}(\varepsilon)$ does allow evaluation on $x = 0$ i.e. $X = 0$; indeed, from (2.33), we see that

$$y(0;\varepsilon) = Y(0;\varepsilon) \sim \mathrm{e}\left\{1 + \varepsilon\left(\tfrac{1}{2}\pi + \mathrm{e} - 1\right)\right\} \quad \text{as } \varepsilon \to 0.$$

In summary, the procedure involves the construction of an asymptotic expansion valid for $x = \mathrm{O}(1)$ and applying the boundary condition(s) if the expansion remains valid here. The expansion is then examined for $\forall x \in D$, seeking any breakdowns, rescaling and hence rewriting the equation in terms of the new, scaled variable; this problem is then solved as another asymptotic expansion, matching as necessary. A couple of general observations are prompted by this example. First, the matching principle has been used to determine the arbitrary constants of integration because the boundary condition does not sit where $x = \mathrm{O}(\varepsilon)$; thus the process of matching is equivalent, here, to imposing boundary conditions (and thereby obtaining unique solutions for $\forall x \in D$). In the context of differential equations, this is the usual rôle of the matching principle, and it is fundamental in seeking complete solutions.

The second issue is rather more general. In this example, the expansion for $x = \mathrm{O}(1)$, (2.27), had to be taken to the term at $\mathrm{O}(\varepsilon^2)$ before the non-uniformity (as $x \to 0$) became evident. This prompts the obvious question: how many terms should be determined so that we can be (reasonably) sure that all possible contributions to a breakdown have been identified? A very good rule of thumb is to ensure that the asymptotic expansion contains information generated by every term in the differential equation. Thus our recent example, (2.26),

$$y' + \left(1 + \frac{\varepsilon^2}{x^2 + \varepsilon^2}\right) y + \varepsilon y^2 = 0$$

requires $\mathrm{O}(\varepsilon)$ terms to include the nonlinearity (εy^2) and $\mathrm{O}(\varepsilon^2)$ terms for the dominant representation of the varying part of the variable coefficient. In a physically based problem, the interpretation of this rule is simply to ensure that every different physical effect is included at some stage in the expansion. As an example of this idea, consider the nonlinear, damped oscillator with variable frequency described by the equation:

$$\ddot{x} + \varepsilon^3 \dot{x} + \left(1 + \frac{\varepsilon}{t + \varepsilon}\right) x + \varepsilon^2 x^2 = 0, \quad t \geq 0.$$

At O(1) we have the basic oscillator; at $O(\varepsilon)$ the variable frequency; at $O(\varepsilon^2)$ the non-linearity; at $O(\varepsilon^3)$ the damping. Thus, in order to investigate the leading contributions (at least) to each of these properties of the oscillation, the asymptotic expansion must be taken as far as the inclusion of terms $O(\varepsilon^3)$. (There is no suggestion that each will necessarily lead to a breakdown, and an associated scaling, but each needs to be examined.) One further important observation will be discussed in the next section; we conclude this section with two examples that exploit all these ideas.

E2.8 Problem (2.11) extended

We consider the problem

$$\frac{dy}{dx} + y + \varepsilon y^2 = x, \quad x \geq 1, \quad \text{with} \quad y(1;\varepsilon) = 1, \tag{2.34}$$

which is the same equation and boundary condition as we introduced in (2.11), but now the domain is $x \geq 1$. The asymptotic solution for $x = O(1)$, which satisfies the boundary condition, is (2.17) i.e.

$$y(x;\varepsilon) \sim x - 1 + e^{1-x} + \varepsilon\left\{-x^2 + 4x - 5 - (x^2 - 2x)e^{1-x} + e^{2(1-x)}\right\} \tag{2.35}$$

and this breaks down where $x = O(\varepsilon x^2)$ or $x = O(\varepsilon^{-1})$. Thus we introduce $x = X/\varepsilon$; but for this size of x, we observe that $y = O(\varepsilon^{-1})$ and so this also must be scaled: we write

$$y(X/\varepsilon;\varepsilon) \equiv \varepsilon^{-1} Y(X;\varepsilon).$$

Equation (2.34) becomes

$$\frac{\varepsilon}{\varepsilon}\frac{dY}{dX} + \frac{1}{\varepsilon}Y + \frac{\varepsilon}{\varepsilon^2}Y^2 = \frac{X}{\varepsilon} \quad \text{or} \quad \varepsilon\frac{dY}{dX} + Y + Y^2 = X,$$

and we seek a solution

$$Y(X;\varepsilon) \sim \sum_{n=0}^{N} \varepsilon^n Y_n(X),$$

which gives

$$Y_0 + Y_0^2 = X; \quad Y_1 + 2Y_0 Y_1 + Y_0' = 0, \tag{2.36a,b}$$

and so on. This result may cause some surprise: this sequence of problems is purely *algebraic*—there is no integration of differential equations required at any stage.

Equation (2.36a) has the solution

$$Y_0(X) = \frac{1}{2}\left(-1 \pm \sqrt{1+4X}\right), \tag{2.37}$$

and we will need to invoke the matching principle to decide which sign is appropriate. Thus from the first term in (2.35), we see that

$$\frac{Y}{\varepsilon} \sim \frac{X}{\varepsilon} - 1 + e^{1-X/\varepsilon} \quad \text{and so} \quad Y \sim X; \tag{2.38}$$

from (2.37) we obtain

$$\varepsilon y \sim \frac{1}{2}\left(-1 \pm \sqrt{1+4\varepsilon x}\right) \quad \text{so} \quad y \sim \frac{1}{2\varepsilon}(-1 \pm 1) \pm x \tag{2.39}$$

and then (2.38) and (2.39) match only for the positive sign. (Note that, because y has also been scaled, this must be included in the construction which enables the matching to be completed.) The solution for the first term is therefore

$$Y_0(X) = \frac{1}{2}\left(\sqrt{1+4X} - 1\right),$$

and then the second term is obtained directly as

$$Y_1(X) = -\frac{1}{1+4X};$$

the resulting 2-term asymptotic expansion is

$$Y(X;\varepsilon) \sim \frac{1}{2}\left(\sqrt{1+4X} - 1\right) - \frac{\varepsilon}{1+4X} \quad \text{for } X = O(1). \tag{2.40}$$

We have found that this problem, (2.34), requires an asymptotic expansion for $x = O(1)$ and another for $x = O(\varepsilon^{-1})$. In addition, it is clear that (2.40) does not further break down as $X \to \infty$ (and it is fairly easy to see that no later terms in the expansion will alter this observation): two asymptotic expansions are sufficient. The appearance of an algebraic problem implies that all solutions are the same—any variation by virtue of different boundary values $y(1;\varepsilon)$ is lost for $x = O(\varepsilon^{-1})$; how is this possible? The explanation becomes clear when (2.35) is examined more closely; the terms associated with the arbitrary constants (at each order) are exponential functions, and for $x = X/\varepsilon$ these are all proportional to $\exp(-nX/\varepsilon)$: they are exponentially small. Such terms have been omitted from the asymptotic expansion for $Y(X;\varepsilon)$; if they had been included, then the matching of these terms would have ensured that information about the boundary values would have been transmitted to the solution valid for $x = O(\varepsilon^{-1})$, albeit in exponentially small terms.

E2.9 An equation with an interesting behaviour near $x = 0$

We consider the problem

$$(x + \varepsilon y)\frac{dy}{dx} + y + \varepsilon y^2 = 0, \quad 0 \le x \le 1, \quad \text{with} \quad y(1;\varepsilon) = 1, \tag{2.41}$$

as $\varepsilon \to 0^+$; we assume that

$$y(x;\varepsilon) \sim \sum_{n=0}^{N} \varepsilon^n y_n(x)$$

for some $x = O(1)$. We obtain the sequence of equations

$$xy_0' + y_0 = 0; \quad xy_1' + y_1 + y_0 y_0' + y_0^2 = 0, \tag{2.42a,b}$$

and so on; the boundary condition (if available here) gives

$$y_0(1) = 1; \quad y_n(1) = 0 \quad \text{for } n \ge 1. \tag{2.43}$$

The general solution of (2.42a) is simply

$$y_0 = \frac{A_0}{x}$$

(which is not defined on $x = 0$, so we anticipate the need for a scaling as $x \to 0$, if a bounded solution exists), and then (2.42b) gives

$$xy_1' + y_1 = -y_0 y_0' - y_0^2 = \frac{A_0^2}{x^3} - \frac{A_0^2}{x^2}$$

or

$$xy_1 = -\frac{1}{2}\frac{A_0^2}{x^2} + \frac{A_0^2}{x} + A_1.$$

The asymptotic expansion is therefore

$$y(x;\varepsilon) \sim \frac{A_0}{x} + \varepsilon\left\{\frac{A_1}{x} - \frac{1}{2}\frac{A_0^2}{x^3} + \frac{A_0^2}{x^2}\right\}$$

and this is defined for $x = O(1)$, including $x = 1$, but not as $x \to 0$. So the boundary condition, (2.43), can be applied, requiring the arbitrary constants to be $A_0 = 1$ and $A_1 = -1/2$ i.e.

$$y(x;\varepsilon) \sim \frac{1}{x} + \varepsilon\left\{\frac{1}{x^2} - \frac{1}{2x} - \frac{1}{2x^3}\right\} \quad \text{for } x = O(1). \tag{2.44}$$

As $x \to 0$, expansion (2.44) breaks down where $x^{-1} = \mathrm{O}(\varepsilon x^{-3})$ or $x = \mathrm{O}(\varepsilon^{1/2})$, and then $y = \mathrm{O}(\varepsilon^{-1/2})$; we introduce the scaled variables

$$x = \sqrt{\varepsilon} X, \quad y(\sqrt{\varepsilon} X; \varepsilon) \equiv \frac{1}{\sqrt{\varepsilon}} Y(X; \varepsilon)$$

and then the equation in (2.41) becomes

$$\left(\sqrt{\varepsilon} X + \varepsilon \frac{Y}{\sqrt{\varepsilon}}\right) \frac{1}{\varepsilon} \frac{\mathrm{d}Y}{\mathrm{d}X} + \frac{Y}{\sqrt{\varepsilon}} + \varepsilon \frac{Y^2}{\varepsilon} = 0 \quad \text{or} \quad (X + Y)\frac{\mathrm{d}Y}{\mathrm{d}X} + Y + \sqrt{\varepsilon}\, Y^2 = 0. \qquad (2.45)$$

For this equation, it is clear that we must seek a solution in the form

$$Y(X; \varepsilon) \sim \sum_{n=0}^{N} \varepsilon^{n/2} Y_n(X)$$

and then (2.45) yields

$$(X + Y_0) Y_0' + Y_0 = 0; \quad (X + Y_0) Y_1' + Y_0 Y_1' + Y_1 + Y_0^2 = 0, \qquad (2.46\text{a,b})$$

and so on. The first equation here can be written as

$$\frac{\mathrm{d}}{\mathrm{d}X}\left(XY_0 + \frac{1}{2} Y_0^2\right) = 0 \quad \text{and so} \quad XY_0 + \frac{1}{2} Y_0^2 = \frac{1}{2} B_0,$$

where B_0 is an arbitrary constant; thus

$$Y_0(X) = -X \pm \sqrt{X^2 + B_0}. \qquad (2.47)$$

and both B_0 and the choice of sign are to be determined by matching.

From the first term in (2.44) we obtain

$$\frac{Y}{\sqrt{\varepsilon}} \sim \frac{1}{\sqrt{\varepsilon} X} \quad \text{or} \quad Y \sim \frac{1}{X}, \qquad (2.48)$$

and from (2.47) we have

$$\sqrt{\varepsilon} y \sim -\frac{x}{\sqrt{\varepsilon}} \pm \sqrt{\frac{x^2}{\varepsilon} + B_0} \quad \text{so} \quad y \sim \frac{1}{\varepsilon}\left[-x \pm x\left(1 + \frac{1}{2}\frac{\varepsilon B_0}{x^2}\right)\right]$$

which matches with (2.48) only for the positive sign and then with $B_0 = 2$. Thus (2.47) becomes

$$Y_0(X) = -X + \sqrt{2 + X^2},$$

and then equation (2.46b) is

$$(XY_1 + Y_0Y_1)' = -Y_0^2 = -\left(X^2 - 2X\sqrt{2+X^2} + 2 + X^2\right)$$

or

$$XY_1 + Y_0Y_1 = -\frac{2}{3}X^3 + \frac{2}{3}(2+X^2)^{3/2} - 2X + B_1.$$

The general expression for Y_1 is therefore

$$Y_1(X) = \frac{2}{3}(2+X^2) - \frac{2X}{3}\frac{(3+X^2)}{\sqrt{2+X^2}} + \frac{B_1}{\sqrt{2+X^2}}$$

and so we have the expansion

$$Y(X;\varepsilon) \sim -X + \sqrt{2+X^2} + \sqrt{\varepsilon}\left\{\frac{2}{3}(2+X^2) - \frac{2X}{3}\frac{(3+X^2)}{\sqrt{2+X^2}} + \frac{B_1}{\sqrt{2+X^2}}\right\} \tag{2.49}$$

and this is to be matched with (2.44). From (2.44) we obtain

$$\frac{Y}{\sqrt{\varepsilon}} \sim \frac{1}{\sqrt{\varepsilon}X} + \varepsilon\left\{\frac{1}{\varepsilon X^2} - \frac{1}{2\varepsilon\sqrt{\varepsilon}X^3} - \frac{1}{2\sqrt{\varepsilon}X}\right\} \quad \text{or} \quad Y \sim \frac{1}{X} - \frac{1}{2X^3} + \frac{\sqrt{\varepsilon}}{X^2} \tag{2.50}$$

and, correspondingly from (2.49), we have

$$\sqrt{\varepsilon}y \sim -\frac{x}{\sqrt{\varepsilon}} + \sqrt{2+\frac{x^2}{\varepsilon}} + \sqrt{\varepsilon}\left\{\frac{2}{3}\left(2+\frac{x^2}{\varepsilon}\right) - \frac{2x}{3\sqrt{\varepsilon}}\frac{(3+x^2/\varepsilon)}{\sqrt{2+x^2/\varepsilon}} + \frac{B_1}{\sqrt{2+x^2/\varepsilon}}\right\}$$

or

$$y \sim \frac{1}{x} - \frac{\varepsilon}{2x^3} + \frac{\varepsilon}{x^2} + \frac{B_1\sqrt{\varepsilon}}{x}$$

which matches only if $B_1 = 0$ (because the term $O(\sqrt{\varepsilon})$ in B_1 must be eliminated). The solution valid for $x = O(\sqrt{\varepsilon})$ is therefore

$$Y(X;\varepsilon) \sim -X + \sqrt{2+X^2} + \sqrt{\varepsilon}\left\{\frac{2}{3}(2+X^2) - \frac{2X}{3}\frac{(3+X^2)}{\sqrt{2+X^2}}\right\} \tag{2.51}$$

and this expansion is defined on $X = 0$, yielding

$$Y(0;\varepsilon) \sim \sqrt{2} + \frac{4}{3}\sqrt{\varepsilon} \quad \text{or} \quad y(0;\varepsilon) \sim \sqrt{\frac{2}{\varepsilon}} + \frac{4}{3} \quad \text{as } \varepsilon \to 0^+.$$

We observe, in this example, that the value of the function on $x = 0$ is well-defined from (2.51), but that it diverges as $\varepsilon \to 0^+$. This demonstrates the important property that we require, for a solution to exist, that the asymptotic representation be defined for $\forall x \in D$, but that solutions obtained from these expansions may diverge as $\varepsilon \to 0$: we may have $x = 0$ in the domain, but $\varepsilon \neq 0$.

The examples that we have presented thus far (and others can be found in Q2.11–2.15), and particularly those that involve a rescaling after a breakdown, possess an important but rather less obvious property. This relates to the existence of general scalings of the differential equation, and the resulting 'balance' of (dominant) terms in the equation; this leads us to the introduction of an additional fundamental tool. This idea will now be explored in some detail, and use made of it in some further examples.

2.5 SCALING OF DIFFERENTIAL EQUATIONS

Let us first return to our most recent example

$$(x + \varepsilon y)\frac{dy}{dx} + y + \varepsilon y^2 = 0 \tag{2.52}$$

given in (2.41). We may, if it is convenient or expedient, choose to use new variables defined by

$$x = \delta X, \quad y = \Delta Y,$$

where δ and Δ are arbitrary positive constants; X and Y are now scaled versions of x and y, respectively. Thus, with $dy/dx = (\Delta/\delta)dY/dX$, equation (2.52) becomes

$$\left(X + \frac{\varepsilon\Delta}{\delta}Y\right)\frac{dY}{dX} + Y + \varepsilon\Delta Y^2 = 0, \tag{2.53}$$

and then a choice for δ and Δ might be driven by the requirement to find a new asymptotic expansion valid in an appropriate region of the domain. In this example, the first term of the asymptotic expansion valid for $x = O(1)$ is $y \sim 1/x$ (see (2.44)) and so any scaling that is to produce a solution which matches to this must satisfy $Y \sim 1/X$ i.e. $\Delta = \delta^{-1}$. With this choice, equation (2.53) becomes

$$\left(X + \frac{\varepsilon}{\delta^2}Y\right)\frac{dY}{dX} + Y + \frac{\varepsilon}{\delta}Y^2 = 0, \tag{2.54}$$

and from our previous analysis of this problem, we know that the breakdown of the asymptotic expansion valid for $x = O(1)$ occurs for $x = O(\sqrt{\varepsilon})$ i.e. $\delta = \sqrt{\varepsilon}$; the issue here is whether this can be deduced directly from the (scaled) equation.

The clue to the way forward can be found when we examine the terms, in the differential equation, that produce the leading terms in the two asymptotic expansions, one valid for $x = O(1)$ and the other for $x = O(\sqrt{\varepsilon})$. From (2.41) and (2.45), these are

$$(\hat{\underline{x}} + \varepsilon\hat{y})\frac{\mathrm{d}y}{\mathrm{d}x} + \hat{\underline{y}} + \varepsilon y^2 = 0,$$

where '_' denotes terms used, in the first approximation, with $x = \mathrm{O}(1)$, and '^' denotes, correspondingly, the terms used where $x = \mathrm{O}(\sqrt{\varepsilon})$. (The derivative has not been labelled, but it will automatically be retained, by virtue of the multiplication of terms, when labelled terms are used in an approximation.) The important interpretation is that some terms—here all—used where $x = \mathrm{O}(1)$ are balanced against some terms *not* used previously (in the first approximation), but now required where $x = \mathrm{O}(\sqrt{\varepsilon})$. When we impose this requirement on (2.54), and note that the breakdown is as $x \to 0$ i.e. $\delta \to 0$, then the only balance occurs when we choose $\varepsilon/\delta^2 = \mathrm{O}(1)$, or, because we may define δ in any appropriate way, simply $\delta = \sqrt{\varepsilon}$. It is impossible to balance the term in ε/δ against the O(1) terms here (to give different leading terms) because, when we do this, the dominant term then becomes $\varepsilon/\delta^2 = 1/\varepsilon$, which is plainly inconsistent. Note that ε/δ^2 and ε/δ, as $\delta \to 0$, can never be balanced.

Thus, armed only with the general scaling property, the behaviour $y \sim 1/x$ as $x \to 0$ and the requirement to balance terms, we are led to the choice $\delta = \sqrt{\varepsilon}$ $(= \Delta^{-1})$; this does not involve any discussion of the *nature* of the breakdown of the asymptotic expansion. This new procedure is very easily applied, is very powerful and is the most immediate and natural method for finding the relevant scaled regions for the solution of a differential equation. We use this technique to explore two examples that we have previously discussed, and then we apply it to a new problem.

E2.10 Scaling for problem (2.11) (see also (2.34))

Consider the equation

$$y' + y + \varepsilon y^2 = x \tag{2.55}$$

with a boundary condition given at $x = x_0 > 0$, where $x_0 = \mathrm{O}(1)$ as $\varepsilon \to 0$; the domain is either $0 \le x \le x_0$ or $x \ge x_0$. The solution of the first term in the asymptotic expansion valid for $x = \mathrm{O}(1)$ is (see (2.15))

$$y(x;\varepsilon) \sim x - 1 + A_0\mathrm{e}^{-x}. \tag{2.56}$$

The general scaling, $x = \delta X$, $y = \Delta Y$, in (2.55) gives

$$\frac{1}{\delta}\frac{\mathrm{d}Y}{\mathrm{d}X} + Y + \varepsilon\Delta Y^2 = \frac{\delta}{\Delta}X; \tag{2.57}$$

f the domain is $0 \geq x \geq x_0$ then from (2.56) we see that $y = O(1)$ as $x \to 0$ (unless $A_0 = 1$; see below). So we select $\Delta = 1$, (2.57) becomes

$$\frac{1}{\delta}Y' + Y + \varepsilon Y^2 = \delta X,$$

and there is no choice of scaling, as $\delta \to 0$, which balances the term δ^{-1} against εY^2. We conclude that a second asymptotic region does not exist, and hence that the expansion for $x = O(1)$ is uniformly valid on this (bounded) domain, which agrees with the discussion following (2.17). (In the special case $A_0 = 1$, $y = O(x^2)$ as $x \to 0$, and so a matched solution now requires $\Delta = \delta^2$, producing

$$\frac{1}{\delta}Y' + Y + \varepsilon\delta^2 Y^2 = \frac{1}{\delta}X;$$

again, there is no choice of δ, as $\delta \to 0$, which balances δ^{-1} against $\varepsilon\delta^2$.)

On the other hand, if the domain is $x \geq x_0$, then $y = O(x)$ as $x \to \infty$, and we require $\Delta = \delta$ for a matched solution to exist; equation (2.57) now becomes

$$\frac{1}{\delta}Y' + Y + \varepsilon\delta^2 Y^2 = X.$$

This time, with $\delta \to \infty$ (because $x \to \infty$), the O(1) terms balance $\varepsilon\delta^2 Y^2$ if $\varepsilon\delta = O(1)$ e.g. $\varepsilon\delta = 1$ or $\delta = \varepsilon^{-1}$, which recovers the scaling used to give (2.40) in E2.8.

E2.11 Scaling for problem (2.22)

Consider the equation

$$y'' + y' + \varepsilon y^2 = 0 \tag{2.58}$$

with $0 \leq x \leq 1$ and suitable boundary conditions (which may both be at one end, or one at each end). The general solution of the dominant terms from (2.58), with $x = O(1)$ as $\varepsilon \to 0$, is

$$y(x;\varepsilon) \sim A + Be^{-x} \tag{2.59}$$

(as used to generate (2.25)). In this example, the asymptotic expansion, of which (2.59) is the first term, may break down as $x \to 0$ or as $x \to 1$ or, just possibly, as $x \to x_0$ with $0 < x_0 < 1$. All these may be subsumed into one calculation by introducing a simple extension of our method of scaling: let $x = x_0 + \delta X$, where we may allow $0 \leq x_0 \leq 1$ in this formulation. Then with the usual $y = \Delta Y$, (2.58) becomes

$$\frac{\Delta}{\delta^2}Y'' + \frac{\Delta}{\delta}Y' + \varepsilon\Delta^2 Y^2 = 0 \quad \text{or} \quad Y'' + \delta Y' + \varepsilon\Delta\delta^2 Y^2 = 0.$$

(Note that we have used the identity

$$\frac{dy}{dx} = \Delta \frac{d}{dx} Y(X) = \Delta \frac{d}{dx} Y\left(\frac{x - x_0}{\delta}\right) = \frac{\Delta}{\delta} \frac{dY}{dX},$$

and similarly for the second derivative; this is valid for $\forall x_0$.) For general A and B (2.59) implies that $y = O(1)$ as $x \to x_0$, so we select $\Delta = 1$:

$$Y'' + \delta Y' + \varepsilon \delta^2 Y^2 = 0,$$

and no balance exists (that is, between Y'' and $\varepsilon\delta^2 Y^2$) as $\delta \to 0$. Again we deduce, on the basis of this scaling argument, that the asymptotic expansion is uniformly valid for $0 \le x \le 1$ (exactly as we found was the case for expansion (2.25)). The special case in which $A + Be^{-x_0} = 0$ produces $y = O(x - x_0)$ as $x \to x_0$, and so we now require $\Delta = \delta$, but any balance is still impossible.

E2.12 Scaling procedure applied to a new equation

For our final example, we consider the equation

$$\varepsilon y'' + y' + y = r(x), \quad 0 \le x \le 1, \tag{2.60}$$

where $r(x)$ is either zero or $r(x) = x$; the two boundary conditions are either one at each end of the domain, or both at one end—it is immaterial in this discussion. The general solution of the dominant terms in (2.60) as $\varepsilon \to 0$, for $x = O(1)$, is

$$y = Ae^{-x} \quad \text{or} \quad y = Ae^{-x} + x - 1, \tag{2.61a,b}$$

the latter applying when $r(x) = x$. The general scaling in the neighbourhood of any x_0, $0 \ge x_0 \ge 1$, is $x = x_0 + \delta X$ and $y = \Delta Y$, which gives

$$\frac{\varepsilon\Delta}{\delta^2} Y'' + \frac{\Delta}{\delta} Y' + \Delta Y = 0 \quad \text{or} \quad \frac{\varepsilon}{\delta^2} Y'' + \frac{1}{\delta} Y' + Y = 0, \tag{2.62}$$

for the equation with $r(x) \equiv 0$. Because this equation is linear and homogeneous, the scaling in y is redundant: it cancels identically. (We may still require Δ to measure the size of y, but it can play no rôle in the determination, from the equation, of any appropriate scaling near $x = x_0$.) The balance of terms, as $\delta \to 0$, requires either that $\varepsilon/\delta^2 = O(1)$ or that $\varepsilon/\delta^2 = O(\delta^{-1})$, and only the latter is consistent, so $\delta = \varepsilon$; the former balances the terms Y'' and Y, but then Y' is the dominant contributor! Thus any scaled region must be described by $x = x_0 + \varepsilon X$, although this analysis cannot help us decide if an x_0 exists, or what x_0 might be; we will examine this issue in the next section.

The case of $r(x) = x$ is slightly different, because the equation is no longer homogeneous: there is a right-hand side. The same scaling now produces

$$\frac{\varepsilon}{\delta^2}Y'' + \frac{1}{\delta}Y' + Y = \frac{x_0}{\Delta} + \frac{\delta}{\Delta}X,$$

but Δ can be found by the condition that any solution we seek for Y must match to (2.61b). If $y = O(1)$ as $x \to x_0$, then $\Delta = 1$ and the result is as before: the only balance is provided by $\delta = \varepsilon$. On the other hand, if $Ae^{-x_0} + x_0 - 1 = 0$, then $y = O[(x - x_0)]$ as $x \to x_0$ $(x_0 \neq 0)$, so $\Delta = \delta$ and then we have

$$\frac{\varepsilon}{\delta^2}Y'' + \frac{1}{\delta}Y' + Y = \frac{x_0}{\delta} + X.$$

But the new term is the same size as an existing term, and so the same result follows yet again: $\delta = \varepsilon$ is the only available choice. (Taking $x_0 = 0$ gives the same result.)

The technique of scaling differential equations, coupled with the required behaviour necessary if matching is to be possible, is simple but powerful (as the above examples demonstrate). It is often incorporated at an early stage in most calculations, and that is how we will view it in the final introductory examples that we present; additional examples are available in Q2.16. We will shortly turn to a discussion of a classical type of problem: those that exhibit a boundary-layer behaviour (a phenomenon that we have already met in E1.4; see (1.16)). However, before we start this, a few comments of a rather more formal mathematical nature are in order, and may be of interest to some readers.

Let us suppose that we have scaled an equation according to $x = \delta X$ and $y = \Delta Y$, and that we have chosen $\Delta(\delta)$ to satisfy matching requirements; we will express this as $\Delta = \delta^n$, for some known n. The scaling, or *transformation*,

$$x = \delta X, \quad y = \delta^n Y \quad (\text{for real } \delta \neq 0)$$

will be represented by T_δ; this transformation of variables belongs to a *continuous group* or *Lie group*. (Note that this discussion has not invoked the balance of terms, which then leads to a choice $\delta(\varepsilon)$; this would constitute a selection of one member of the group.) We now explore the properties of this transformation. First we apply, successively, the transformation T_δ and then T_γ; this is equivalent to the single transformation $T_{\delta\gamma}$, and so we have the multiplication rule:

$$T_\delta T_\gamma = T_{\delta\gamma}.$$

But we also have $T_\gamma T_\delta = T_{\gamma\delta} = T_{\delta\gamma}$, so this law is *commutative*. Furthermore, the *associative* law is satisfied:

$$T_\delta(T_\gamma T_\beta) = T_\delta T_{\gamma\beta} = T_{\delta\gamma\beta} = T_{\delta\gamma} T_\beta = (T_\delta T_\gamma) T_\beta;$$

in addition, we have the identity transformation T_1 i.e. $T_1 T_\delta = T_{\delta 1} = T_\delta$ for all $\delta \neq 0$. Finally, we form

$$T_{1/\delta} T_\delta = T_1 \quad \text{and} \quad T_\delta T_{1/\delta} = T_1,$$

and so $T_{1/\delta}$ is both the left and right inverse of T_δ. Thus the elements of T_δ, for all real $\delta \neq 0$, form an *infinite group*, where δ is the *parameter* of this continuous group. (If n is fractional, then we may have to restrict the parameter to $\delta > 0$.)

Although we have not used the full power of this continuous group—we eventually select only one member for a given ε—there are other significant applications of this fundamental property in the theory of differential equations. For example, if a particular scaling transformation leaves the equation unchanged (except for a change of the symbols!) i.e. the equation is *invariant*, then we may seek solutions which satisfy the same invariance. Such solutions are, typically, *similarity solutions* (if they exist) of the equation; this aspect of differential equations is generally outside the considerations of singular perturbation theory (although these solutions may be the relevant ones in certain regions of the domain, in particular problems).

2.6 EQUATIONS WHICH EXHIBIT A BOUNDARY-LAYER BEHAVIOUR

There are many problems, posed in terms of either ordinary or partial differential equations, that have solutions which include a thin region near a boundary of the domain which is required to accommodate the boundary value there. Such regions are thin by virtue of a scaling of the variables in the appropriate parameter and, typically, this involves large values of the derivatives near the boundary. The terminology—boundary layer—is rather self-evident, although it was first associated with the viscous boundary layer in fluid mechanics (which we will describe in Chapter 5). Here, we will introduce the essential idea *via* some appropriate ordinary differential equations, and make use of the relevant scaling property of the equation.

The nature of this problem is best described, first, by an analysis of equation (1.16):

$$\varepsilon \frac{d^2 y}{dx^2} + (1+\varepsilon)\frac{dy}{dx} + y = 0, \quad 0 \leq x \leq 1, \tag{2.63}$$

with $y(0;\varepsilon) = \alpha$ and $y(1;\varepsilon) = \beta$ (and we will assume that α and β are not functions of ε, and that $\alpha \neq \beta\mathrm{e}$). In this presentation of the construction of the asymptotic solution, valid as $\varepsilon \to 0^+$, we will work directly from (2.63) (although the exact solution is available in (1.22), which may be used as a check, if so desired). Because we wish to incorporate an application of the scaling property, we need to know *where* the boundary layer (interpreted as a scaled region) is situated: is it near $x = 0$ or near

$x = 1$ (or, possibly, both)? Here, we will assume that there is a single boundary layer near $x = 0$; the problem of finding the position of a boundary layer will be addressed in the next section, at least for a particular class of ordinary differential equations. Let us now return to equation (2.63).

As should be evident from example E1.4, and will become very clear in what follows, it is the appearance of the small parameter multiplying the highest derivative that is critical here. The presence of ε in another coefficient is altogether irrelevant to the general development; it is retained only to allow direct comparison with E1.4. We seek a solution of (2.63) in the form

$$y(x;\varepsilon) \sim \sum_{n=0}^{N} \varepsilon^n y_n(x), \quad \text{for } x = \mathrm{O}(1),$$

and then we obtain

$$y_0' + y_0 = 0; \quad y_1' + y_1 + y_0'' + y_0' = 0, \tag{2.64a,b}$$

and so on. The only boundary condition available to us (because of the assumption about the position of the boundary layer) is

$$y(1;\varepsilon) = \beta \quad \text{i.e.} \quad y_0(1) = \beta,\ y_n(1) = 0 \quad \text{for } n \geq 1.$$

Thus

$$y_0(x) = \beta \mathrm{e}^{1-x}, \quad y_1(x) \equiv 0,$$

and indeed, in this problem, we then have $y_n(x) \equiv 0$, $n \geq 1$, although exponentially small terms would be required for a more complete description of the asymptotic solution valid for $x = \mathrm{O}(1)$; so we have

$$y(x;\varepsilon) \sim \beta \mathrm{e}^{1-x}. \tag{2.65}$$

The scaled version of (2.63) is obtained by writing $x = \delta X$, for $\delta \to 0$, and $y \equiv Y(X;\varepsilon)$ (because $y = \mathrm{O}(1)$ as $x \to 0$, although any scaling on Y will vanish identically from the equation); thus

$$\frac{\varepsilon}{\delta^2} Y'' + (1+\varepsilon)\frac{1}{\delta} Y' + Y = 0.$$

The relevant balance, as we have already seen in (2.62), is $\varepsilon/\delta^2 = \mathrm{O}(\delta^{-1})$ or $\delta = \varepsilon$, giving

$$Y'' + (1+\varepsilon)Y' + \varepsilon Y = 0. \tag{2.66}$$

We seek a solution of this equation in the usual form:

$$Y(X;\varepsilon) \sim \sum_{n=0}^{N} \varepsilon^n Y_n(X)$$

which gives

$$Y_0'' + Y_0' = 0; \quad Y_1'' + Y_1' + Y_0' + Y_0 = 0, \tag{2.67a,b}$$

and so on. We have available the one boundary condition prescribed at $x = 0$ (which is in the region of the boundary layer) i.e. at $X = 0$, so

$$Y(0;\varepsilon) = \alpha \quad \text{or} \quad Y_0(0) = \alpha,\ Y_n(0) = 0 \quad \text{for } n \geq 1. \tag{2.68}$$

From (2.67a) and (2.68) we obtain

$$Y_0(X) = A_0 + (\alpha - A_0)\mathrm{e}^{-X} \tag{2.69}$$

where A_0 is an arbitrary constant, and then (2.67b) becomes

$$Y_1'' + Y_1' - (\alpha - A_0)\mathrm{e}^{-X} + A_0 + (\alpha - A_0)\mathrm{e}^{-X} = 0.$$

Thus

$$Y_1'' + Y_1' = -A_0 \quad \text{and so} \quad Y_1(X) = A_1 + B_1\mathrm{e}^{-X} - A_0 X,$$

where A_1 and B_1 are the arbitrary constants, which must satisfy $A_1 + B_1 = 0$ from $Y_1(0) = 0$; this gives the solution

$$Y_1(X) = A_1(1 - \mathrm{e}^{-X}) - A_0 X.$$

The 2-term asymptotic expansion valid for $x = \mathrm{O}(\varepsilon)$ is therefore

$$Y(X;\varepsilon) \sim A_0 + (\alpha - A_0)\mathrm{e}^{-X} + \varepsilon\left\{A_1(1 - \mathrm{e}^{-X}) - A_0 X\right\}, \quad X = \mathrm{O}(1), \tag{2.70}$$

which is to be matched to (2.65); this should uniquely determine A_0 and A_1. We write (2.65) as

$$\beta\mathrm{e}^{1-\varepsilon X} \sim \beta\mathrm{e}(1 - \varepsilon X), \quad X = \mathrm{O}(1), \tag{2.71}$$

retaining terms O(1) and O(ε) (as used in (2.70)); correspondingly, we write (2.70) as

$$A_0 + (\alpha - A_0)e^{-x/\varepsilon} + \varepsilon\left\{A_1(1 - e^{-x/\varepsilon}) - A_0\frac{x}{\varepsilon}\right\} \sim A_0 + \varepsilon A_1 - A_0 x, \quad x = \mathrm{O}(1), \tag{2.72}$$

and the $O(\varepsilon)$ term is included because that was obtained for (2.65)—although it had a zero coefficient. In order to match (2.71) and (2.72), we require $A_0 = \beta e$ and $A_1 = 0$ i.e.

$$Y(X; \varepsilon) \sim \beta e + (\alpha - \beta e)e^{-X} - \varepsilon \beta e X. \tag{2.73}$$

(It is left as an exercise to show that (2.65) and (2.73) are recovered from suitable expansions of (1.22).) We should note that (2.65) exhibits no breakdown—there is only one term here, after all—but (2.73) does break down where $\varepsilon X = O(1)$, i.e. $x = O(1)$, as we would expect. Any indication of a breakdown in the asymptotic expansion valid for $x = O(1)$ will come from the exponentially small terms; let us briefly address this aspect of the problem.

The first point to note is that, from (2.70) and (2.73), we would require not only $O(1)$ and $O(\varepsilon)$, but also $O(e^{-x/\varepsilon})$ terms (and others), in order to complete the matching procedure; this is simply because we need, at least in principle, to match to all the terms (cf. (2.72))

$$A_0 - A_0 x + \varepsilon A_1 + (\alpha - A_0)e^{-x/\varepsilon} \ldots.$$

Thus the expansion valid for $x = O(1)$ must include a term $e^{-x/\varepsilon}$ to allow matching to this order (and then it is not too difficult to see that a *complete* asymptotic expansion requires all the terms in the sequence $\{\varepsilon^n e^{-mx/\varepsilon}\}$, $n = 0, 1, 2, \ldots, m = 0, 1, 2, \ldots$). In passing, we observe that this use of the matching principle is new in the context of our presentation here. We are using it, first, in a general sense, to determine the *type(s)* of term(s) required in the expansion valid in an adjacent region in order to allow matching. Then, with these terms included, the 'full' matching procedure may be employed to check the details and fix the values of any arbitrary constants left undetermined. We will find the first of the exponentially small terms here.

For $x = O(1)$, we seek a solution

$$y(x; \varepsilon) \sim \tilde{y}_N(x; \varepsilon) + \hat{Y}(x; \varepsilon)e^{-x/\varepsilon} \tag{2.74}$$

where $\tilde{y}_N(x; \varepsilon) \equiv \sum_{n=0}^{N} \varepsilon^n y_n(x)$ ($\sim \beta e^{1-x}$ in this example); thus equation (2.63) becomes

$$\varepsilon \tilde{y}_N'' + (1 + \varepsilon)\tilde{y}_N' + \tilde{y}_N + \{\varepsilon \hat{Y}'' - (1 - \varepsilon)\hat{Y}'\}\, e^{-x/\varepsilon} = 0.$$

(Again '= 0' means zero to all orders in ε.) We already have that $\tilde{y}_N(x; \varepsilon)$ satisfies the original equation, and so $\hat{Y}(x; \varepsilon)$ must satisfy

$$\varepsilon \hat{Y}'' - (1 - \varepsilon)\hat{Y}' = 0;$$

with $\hat{Y} \sim \hat{Y}_M \equiv \sum_{n=0}^{M} \varepsilon^n \hat{y}_n(x)$ we obtain $\hat{y}_0' = 0$ or $\hat{y}_0 = C_0$ (constant). Thus the expansion (2.74) becomes

$$y(x;\varepsilon) \sim \beta e^{1-x} + C_0 e^{-x/\varepsilon} \tag{2.75}$$

and this is to be matched to the asymptotic expansion (2.73); from (2.75) we obtain

$$\beta e(1 - \varepsilon X) + C_0 e^{-X} \quad \text{for } X = O(1)$$

and from (2.73) we have

$$\beta e(1 - x) + (\alpha - \beta e)e^{-x/\varepsilon} \quad \text{for } x = O(1),$$

which match if $C_0 = \alpha - \beta e$. Thus the solution valid for $x = O(1)$, i.e. away from the boundary layer near $x = 0$, incorporating the first exponentially small term, is

$$y(x;\varepsilon) \sim \beta e^{1-x} + (\alpha - \beta e)e^{-x/\varepsilon}. \tag{2.76}$$

One final comment: this solution, (2.76), produces the value $y(1;\varepsilon) \sim \beta + (\alpha - \beta e)e^{-1/\varepsilon}$, so now the boundary value is in error by $O(e^{-1/\varepsilon})$. To correct this, a further term is required; we must write (at the order of all the first exponentially small terms)

$$y(x;\varepsilon) \sim \tilde{y}_N(x;\varepsilon) + \hat{Y}_M(x;\varepsilon)e^{-x/\varepsilon} + \overline{Y}_M(x;\varepsilon)e^{-1/\varepsilon}$$

where $\overline{Y}_M(x;\varepsilon) \sim \sum_{n=0}^{M} \varepsilon^n \overline{y}_n(x)$. It is left as an exercise to show that $\overline{y}_0' + \overline{y}_0 = 0$, with the solution $\overline{y}_0(x) = -(\alpha - \beta e)e^{1-x}$, ensures that the boundary condition on $x = 1$ is correct at this order. The inclusion of the term $O(e^{-1/\varepsilon})$ in the expansion valid for $x = O(1)$ forces, *via* the matching principle, a term of this same order in the expansion for $x = O(\varepsilon)$, and so the pattern continues. (The appearance of all these terms, in both expansions, can be seen by expanding the exact solution, (1.22), appropriately.)

E2.13 A nonlinear boundary-layer problem

We consider the problem

$$\varepsilon y'' - y' + \varepsilon x y^2 = 2x, \quad 0 \le x \le 1, \tag{2.77}$$

with $y(0;\varepsilon) = 2$, $y(1;\varepsilon) = 2+\varepsilon$, for $\varepsilon \to 0^+$; we are given that the boundary layer is near $x = 1$. We seek a solution, for $1 - x = O(1)$, in the form

$$y(x;\varepsilon) \sim \sum_{n=0}^{N} \varepsilon^n y_n(x)$$

and so
$$y_0' = -2x; \quad y_1' = y_0'' + xy_0^2,$$

etc., with $y_0(0) = 2$, $y_n(0) = 0$ for $n \geq 1$. Thus we obtain

$$y_0(x) = 2 - x^2, \quad y_1(x) = \tfrac{1}{6}x^6 - x^4 + 2x^2 - 2x$$

which leads to

$$y(x;\varepsilon) \sim 2 - x^2 + \varepsilon\left\{\tfrac{1}{6}x^6 - x^4 + 2x^2 - 2x\right\}, \tag{2.78}$$

but this solution does not satisfy the boundary condition on $x = 1$.

For the solution near $x = 1$, we introduce $x = 1 - \delta X$ (with $X \geq 0$) and $y \equiv Y(X;\varepsilon)$, so that $\delta = \varepsilon$ i.e.

$$Y'' + Y' + \varepsilon^2(1 - \varepsilon X)Y^2 = 2\varepsilon(1 - \varepsilon X).$$

The solution in the form

$$Y(X;\varepsilon) \sim \sum_{n=0}^{N} \varepsilon^n Y_n(X)$$

gives
$$Y_0'' + Y_0' = 0, \quad Y_1'' + Y_1' = 2,$$

and so on; the available boundary condition requires that

$$Y_0(0) = 2; \quad Y_1(0) = 1; \quad Y_n(0) = 0, \quad n \geq 2.$$

Thus

$$Y_0(X) = A_0 + (2 - A_0)e^{-X}$$

and
$$Y_1(X) = 2X + A_1 + (1 - A_1)e^{-X},$$

where A_0 and A_1 are the arbitrary constants to be determined by matching; that is, we must match

$$Y(X;\varepsilon) \sim A_0 + (2 - A_0)e^{-X} + \varepsilon\left\{2X + A_1 + (1 - A_1)e^{-X}\right\} \tag{2.79}$$

with (2.78). From (2.79) we obtain

$$y \sim A_0 + 2(1-x) + \varepsilon A_1 \quad \text{for } x = \mathrm{O}(1);$$

from (2.78) we have

$$Y \sim 1 + 2\varepsilon X - \tfrac{5}{6}\varepsilon \quad \text{for } X = \mathrm{O}(1),$$

and these match if we select $A_0 = 1$ and $A_1 = -5/6$; thus (2.79) becomes

$$Y(X;\varepsilon) \sim 1 + \mathrm{e}^{-X} + \varepsilon\left\{2X - \tfrac{5}{6} + \tfrac{11}{6}\mathrm{e}^{-X}\right\}.$$

The fundamental issue relating to boundary-layer-type problems, which we have avoided thus far, addresses the question of where the boundary layer might be located. In the examples discussed above, we allowed ourselves the advantage of knowing where this layer was situated—and the consistency of the resulting asymptotic solution confirmed that unique, well-defined solutions existed, so presumably we started with the correct information. We now examine this important aspect of boundary-layer problems.

2.7 WHERE IS THE BOUNDARY LAYER?

For this discussion, we consider the general second-order ordinary differential equation in the form

$$\varepsilon y'' + a(x)y' + f(x, y; \varepsilon) = 0, \qquad x_0 \le x \le x_1, \tag{2.80}$$

with suitable boundary conditions and for $\varepsilon \to 0^+$; note that the coefficient of y'' must be ε. The coefficient $a(x)$ will satisfy either $a(x) > 0$ or $a(x) < 0$, for $\forall x \in D$; the term $f(x, y; \varepsilon) = \mathrm{O}(1)$ (or smaller) as $\varepsilon \to 0^+$, for $\forall x \in D$ and for all solutions $y(x;\varepsilon)$ that may be of interest. Of course, this describes only one class of such boundary-layer problems, but this does cover by far the most common ones encountered in mathematical modelling. (Some of these conditions, both explicitly written and implied, can be relaxed; we will offer a few generalisations later.) The guiding principle that we will adopt is to seek solutions of (2.80) which remain bounded as $\varepsilon \to 0^+$, for $\forall x \in D$.

The starting point is the construction of the asymptotic solution valid for suitable $x = \mathrm{O}(1)$, directly from the equation as written in (2.80)—but this will necessarily generate a sequence of first-order equations. It is therefore impossible to impose the *two* boundary conditions (as we have already demonstrated in our examples above); the inclusion of a boundary layer remedies this deficiency in the solution. We introduce the boundary layer in the most general way possible: define $X = g(x)/\delta$ for some $g(x)$

and some $\delta(\varepsilon)$, and write $y \equiv \Delta Y(X;\varepsilon)$. Then we have

$$\frac{\mathrm{d}y}{\mathrm{d}x} = \frac{\Delta}{\delta} g'(x) \frac{\mathrm{d}Y}{\mathrm{d}X}, \quad \frac{\mathrm{d}^2 y}{\mathrm{d}x^2} = \frac{\Delta}{\delta} g''(x) \frac{\mathrm{d}Y}{\mathrm{d}X} + \Delta \left(\frac{g'}{\delta}\right)^2 \frac{\mathrm{d}^2 Y}{\mathrm{d}X^2}$$

and so equation (2.80) becomes

$$\frac{\varepsilon}{\delta^2} \left\{ (g')^2 \frac{\mathrm{d}^2 Y}{\mathrm{d}X^2} + \delta g'' \frac{\mathrm{d}Y}{\mathrm{d}X} \right\} + \frac{a(x)}{\delta} g' \frac{\mathrm{d}Y}{\mathrm{d}X} + F(X, Y; \varepsilon, \Delta) = 0 \tag{2.81}$$

where $F = f/\Delta$. We will further assume, whatever the choice of Δ (and in almost all problems that we encounter, $\Delta = 1$), that the terms in δ^{-1} and $\varepsilon\delta^{-2}$ dominate as $\delta \to 0$. Thus the 'classical' choice for the balance of terms, $\delta = \varepsilon$, applies here for general $g(x)$ (which we have yet to determine).

The differential equation valid in the boundary layer can now be written

$$(g')^2 \frac{\mathrm{d}^2 Y}{\mathrm{d}X^2} + ag' \frac{\mathrm{d}Y}{\mathrm{d}X} + \varepsilon \left\{ g'' \frac{\mathrm{d}Y}{\mathrm{d}X} + F \right\} = 0, \tag{2.82}$$

which has the first term in an asymptotic expansion, $Y \sim Y_0(X)$, satisfying

$$(g')^2 \frac{\mathrm{d}^2 Y_0}{\mathrm{d}X^2} + ag' \frac{\mathrm{d}Y_0}{\mathrm{d}X} = 0. \tag{2.83}$$

At this stage, $g'(x)$ has yet to be determined; let us choose $g'(x) = a(x)$ (we may not choose $g' = 0$), then we are left with the simple, generic problem for $Y_0(X)$:

$$Y_0'' + Y_0' = 0, \tag{2.84}$$

which also solves the difficulty over the mixing of the x and X notations in (2.83). Thus all boundary-layer problems in this class have the same general solution, from (2.84),

$$Y_0(X) = A_0 + B_0 \mathrm{e}^{-X}. \tag{2.85}$$

However, we are no nearer finding the position of the boundary layer itself; this we now do by examining the available solutions for $g(x)$.

The general form for $g(x)$ is

$$g(x) = \int_C^x a(x')\mathrm{d}x', \tag{2.86}$$

where C is an arbitrary constant (and this proves to be the most convenient way of including the constant of integration). First, we suppose that $a(x) > 0$, and examine

(2.85) when expressed in terms of x (as will be necessary for any matching); this gives

$$Y \sim A_0 + B_0 \exp\left\{-\frac{1}{\varepsilon}\int_C^x a(x')\,dx'\right\}. \tag{2.87}$$

But we are seeking solutions that remain bounded as $\varepsilon \to 0^+$, and this is possible only if the exponent in (2.87) is non-positive for $\forall x \in D$. Thus we require

$$\int_C^x a(x')\,dx' \geq 0 \quad \text{for } x_0 \leq x \leq x_1,$$

and this means that $x \geq C$ for the same xs i.e. $C = x_0$ (for otherwise $x < C$ can occur in the domain, and the integral will change sign). Then, for x sufficiently close to x_0, we have

$$\frac{1}{\varepsilon}\int_{x_0}^x a(x')\,dx' = X = O(1)$$

and we have a choice for the boundary-layer variable. Hence, for $a(x) > 0$, the boundary layer must sit near the left-hand edge of the domain. Conversely, the same argument in the case $a(x) < 0$ requires that the boundary layer be situated near $x = x_1$ (the right-hand edge of the domain). When we apply this rule to equation (2.63): $\varepsilon y'' + (1+\varepsilon)y' + y = 0,\ 0 \leq x \leq 1$, we see that $a(x) = 1 + \varepsilon > 0$ and so the boundary layer is in the neighbourhood of $x = 0$ (and we introduced $x = \varepsilon X$ for this example). Similarly, equation (2.77): $\varepsilon y'' - y' + \varepsilon x y^2 = 2x,\ 0 \leq x \leq 1$, has $a(x) = -1 < 0$, and so the boundary layer is now near $x = 1$ (and we used $x = 1 - \varepsilon X$).

It is rarely necessary to incorporate the formal definition of $g(x)$ to generate the appropriate variable that is to be used to represent the boundary layer (although it will always produce the simplest form of the solution). For example, the equation of this class: $\varepsilon y'' - y'/(2+x) + \cdots = 0,\ -1 \leq x \leq 2$, has a boundary layer near $x = 2$ (because $a(x) = -1/(2+x) < 0$ for $\forall x \in D$). Now an appropriate scaled variable is simply $x = 2 - \varepsilon X$, giving

$$\frac{1}{\varepsilon}\frac{d^2Y}{dX^2} + \frac{1}{\varepsilon}(4 - \varepsilon X)^{-1}\frac{dY}{dX} + \cdots = 0,$$

and this choice will suffice, even though higher-order terms will require the expansion of $(4 - \varepsilon X)^{-1}$ (but this is usually a small price to pay—and we already know that this asymptotic expansion will breakdown for $\varepsilon X = O(1)$, so retaining εX in the coefficient has no unforeseen complications). It should also be noted that, exceptionally, a boundary-layer-type problem may not require a boundary layer at all, in order to accommodate the given boundary value (to leading order or, possibly, to all orders). This is evident for the equation given in (2.63) (and see also the exact solution, (1.22)); in this example, if the boundary values satisfy the special condition $\alpha = \beta e$, then no boundary layer whatsoever is required. Note, however, that if $\alpha - \beta e = O(\varepsilon)$, then the boundary layer is present, but only to correct the boundary value at $O(\varepsilon)$—the leading term (for $x = O(1)$) is uniformly valid. We call upon all these ideas in the next example.

E2.14 A nonlinear, variable coefficient boundary-layer problem

We consider

$$\varepsilon y'' - \frac{y'}{1+2x} - \frac{1}{y} = 0, \quad 0 \le x \le 1, \tag{2.88}$$

with $y(0;\varepsilon) = y(1;\varepsilon) = 3$, for $\varepsilon \to 0^+$. Because the coefficient of y' is negative for $\forall x \in D$, the boundary layer will be situated at the right-hand edge of the domain i.e. near $x = 1$. Away from $x = 1$, we seek a solution

$$y(x;\varepsilon) \sim \sum_{n=0}^{N} \varepsilon^n y_n(x)$$

and so

$$-\frac{y_0'}{1+2x} - \frac{1}{y_0} = 0; \quad y_0'' - \frac{y_1'}{1+2x} + \frac{y_1}{y_0^2} = 0, \tag{2.89}$$

etc., and we may use the boundary condition on $x = 0$:

$$y_0(0) = 3, \quad y_n(0) = 0 \quad \text{for } n \ge 1.$$

Thus we obtain

$$y_0^2(x) = A_0 - 2(x + x^2)$$

and then the application of the boundary condition yields the solution

$$y_0(x) = \sqrt{9 - 2(x + x^2)}\,; \tag{2.90}$$

we note that y_0 remains real and positive as $x \to 1$. (The second term can also be found, but it is a slightly tiresome exercise and its inclusion teaches us little about the solution.) Clearly, $y_0 \to \sqrt{5}$ as the right-hand boundary is approached, which does not satisfy the given boundary condition $y(1;\varepsilon) = 3$, and so a boundary layer is required. (Of course, if $y(1;\varepsilon) = \sqrt{5}$ then (2.90) would be a uniformly valid 1-term asymptotic solution.)

We introduce $x = 1 - \varepsilon X$ (so that $X \ge 0$) and write $y \equiv Y(X;\varepsilon) = O(1)$; equation (2.88) then becomes

$$Y'' + \frac{Y'}{3 - 2\varepsilon X} - \frac{\varepsilon}{Y} = 0 \tag{2.91}$$

which gives, with $Y(X;\varepsilon) \sim Y_0(X)$,

$$Y_0'' + \tfrac{1}{3}Y_0' = 0. \tag{2.92}$$

The solution of (2.92), which satisfies the boundary condition $Y_0(0) = 3$, is

$$Y_0(X) = B_0 + (3 - B_0)\mathrm{e}^{-X/3} \tag{2.93}$$

and B_0 is to be determined by matching. From (2.90) we have directly that $Y \sim \sqrt{5}$, and from (2.93) we see that $y \sim B_0$; thus matching requires $B_0 = \sqrt{5}$ and the first term in the boundary-layer solution is

$$Y(X;\varepsilon) \sim \sqrt{5} + (3 - \sqrt{5})\mathrm{e}^{-X/3}$$

and then a composite expansion can be written down, if that is required.

The fundamental ideas that underpin the notion of boundary-layer-type solutions, in second-order ordinary differential equations, have been developed, but many variants of this simple idea exist; see also Q2.17–2.20. These lead to adjustments in the formulation, or to generalisations, or to a rather different structure (with corresponding interpretation). We now describe a few of these possibilities, but what we present is far from providing a comprehensive list; rather, we present some examples which emphasise the application of the basic technique of scaling to find thin layers where rapid changes occur. In the next section, we briefly describe a number of different scenarios, and present an example of each type.

2.8 BOUNDARY LAYERS AND TRANSITION LAYERS

Our first development from the simple notion of a boundary layer is afforded by an extension of our discussion of the position of this layer, *via* equation (2.80); here, we consider the case where $a(\alpha) = 0$, $x_0 < \alpha < x_1$. Such a point is analogous to a *turning point* (see Q2.24) and the solution valid near $x = \alpha$ takes the form of a *transition layer*. (The terminology 'turning point' is used to denote where the character of the solution changes or 'turns', typically from oscillatory to exponential, in equations such as $y'' + a(x)y = 0$.) The general approach is to seek a scaling–just as for a boundary layer–but now at this interior point. Let us suppose that $a(x) = \lambda(x - \alpha)^n$, for given constants λ and n, then equation (2.80) becomes

$$\varepsilon y'' + \lambda(x - \alpha)^n y' + f(x, y; \varepsilon) = 0$$

and we introduce

$$x = \alpha + \delta X, \quad y \equiv \Delta Y(X;\varepsilon)$$

to give

$$\frac{\varepsilon}{\delta^2}Y'' + \lambda\delta^{n-1}X^n Y' + F = 0,$$

where we have used the same notation as in (2.81). The balance that we seek is given by the choice $\delta^{n+1} = \varepsilon$ or $\delta = \varepsilon^{1/(1+n)}$, provided that $n < 1$ (in order that this balance does indeed produce the dominant terms, with $F = O(1)$ or smaller). The procedure then unfolds as for the boundary-layer problems, although we will now have solutions in $x_0 \leq x < x_2$ and in $x_2 < x \leq x_1$, where $(x_2 - \alpha) = O(1)$, together with a matched solution where $(x - \alpha) = O(\varepsilon^{1/(1+n)})$. If $n \geq 1$, then the balance of terms requires $\delta = \sqrt{\varepsilon}$, and then either ($n = 1$) all *three* terms in the equation contribute to leading order, or ($n > 1$) the balance is between Y'' and F. (This description assumes that $F = O(1)$; note also that the chosen behaviour of $a(x)$ used here need only apply *near* $x = \alpha$ for this approach to be relevant.)

E2.15 An equation with a transition layer

Consider the equation

$$\varepsilon y'' + x^{1/3}y' + y^2 = 0, \quad -1 \leq x \leq 1, \tag{2.94}$$

for $\varepsilon \to 0^+$, where, for real solutions, we interpret $(-x)^{1/3} = -x^{1/3}$; the boundary conditions are

$$y(-1;\varepsilon) = 2/9 \quad \text{and} \quad y(1;\varepsilon) = 1/3. \tag{2.95a,b}$$

We will find the first terms only in the asymptotic expansions valid away from $x = 0$ (where the coefficient $x^{1/3}$ is zero), in $-1 \leq x < 0$, and then in $0 < x \leq 1$, and finally valid near $x = 0$. For $x = O(1)$, we write $y(x;\varepsilon) \sim y_0(x)$ and so, from (2.94), we obtain

$$x^{1/3}y_0' + y_0^2 = 0 \quad \text{or} \quad y_0(x) = \frac{2}{3}(A_0 + x^{2/3})^{-1};$$

we determine the arbitrary constant by imposing the boundary conditions (2.95a,b), thereby producing solutions valid either on one side, or the other, of $x = 0$:

$$y_0(x) = \frac{2}{3\,(2 + x^{2/3})}, \quad -1 \leq x < -|x_1|; \quad y_0(x) = \frac{2}{3\,(1 + x^{2/3})}, \quad |x_1| < x \leq 1. \tag{2.96a,b}$$

Note that these solutions do not hold in the neighbourhood of $x = 0$ i.e. we may allow $x_1 = o(1)$ but with $\varepsilon/x_1 = o(1)$; this solution *would* be valid for $-1 \leq x \leq 1$, to leading order, if the function given in (2.96) were continuous at $x = 0$, and then no transition layer would be needed (to this order, at least).

Near $x = 0$, we write $x = \delta X$ and, in order to match, we require $y = \mathrm{O}(1)$, so we write $y \equiv Y(X; \varepsilon)$; equation (2.94) then gives

$$\frac{\varepsilon}{\delta^2} Y'' + \frac{X^{1/3}}{\delta^{2/3}} Y' + Y^2 = 0$$

and hence we select $\delta = \varepsilon^{3/4}$. Thus in the transition layer we have the equation

$$Y'' + X^{1/3} Y' + \sqrt{\varepsilon}\, Y^2 = 0,$$

and we seek a solution $Y(X; \varepsilon) \sim Y_0(X)$, where

$$Y_0'' + X^{1/3} Y_0' = 0 \quad \text{or} \quad Y_0(X) = B_0 + C_0 \int_0^X \exp\left(-\frac{3}{4}\hat{X}^{4/3}\right) \mathrm{d}\hat{X}. \tag{2.97}$$

(The lower limit in the integral here is simply a convenience; with this choice, the arbitrary constant following the second integration is B_0. Note that this integral exists for $-\infty < X < \infty$.) These arbitrary constants are determined by matching; from (2.96a,b) we obtain directly that

$$Y \sim \frac{1}{3} \quad \text{as } x \to 0^-; \qquad Y \sim \frac{2}{3} \quad \text{as } x \to 0^+. \tag{2.98}$$

From (2.97), we first write

$$Y_0 = B_0 + C_0 \int_0^{\varepsilon^{-3/4} x} \exp\left(-\frac{3}{4}\hat{X}^{4/3}\right) \mathrm{d}\hat{X}; \tag{2.99}$$

now let us introduce the constant

$$k = \int_0^{\infty} \exp\left(-\frac{3}{4}X^{4/3}\right) \mathrm{d}X \quad \left[= \left(\frac{3}{4}\right)^{1/4} \Gamma(3/4)\right]$$

(where $\Gamma(z)$ is the *gamma function*), then

$$y \sim B_0 + C_0 k, \quad x > 0; \quad y \sim B_0 - C_0 k, \quad x < 0, \quad \text{as } \varepsilon \to 0^+ \tag{2.100}$$

and matching (2.98) and (2.100) requires $B_0 = 1/2$ and $C_0 = 1/(6k)$. The first term of the asymptotic expansion valid in the transition layer (around $x = 0$) is therefore

$$Y(X; \varepsilon) \sim \frac{1}{2} + \frac{1}{6k} \int_0^X \exp\left(-\frac{3}{4}\hat{X}^{4/3}\right) \mathrm{d}\hat{X}.$$

This example has demonstrated how boundary-layer techniques are equally applicable to interior (transition) layers and, further, they need not be restricted to single layers. Some problems exhibit multiple layers; for example

$$\varepsilon y'' + \left(x^2 - \tfrac{1}{4}\right) y' + f = 0, \quad -1 \le x \le 1,$$

has transition layers both near $x = 1/2$ and near $x = -1/2$, and the solution away from these layers is now in three parts.

A type of problem which contains elements of both boundary and transition layers occurs if the coefficient, $a(x)$, of y', is zero at one (or both) boundaries. Because $a(x) \neq 0$ at an *internal* point, it is not a transition layer, but the fact that $a(x) \to 0$ at the end-point affects the scaling—it is no longer $O(\varepsilon)$ in general—and this must be determined directly (as we did for the transition layer).

E2.16 A boundary-layer problem with a new scaling

We consider the equation

$$\varepsilon y'' + \sqrt{x}\, y' + y^2 = 0, \quad 0 \le x \le 1, \tag{2.101}$$

for $\varepsilon \to 0^+$, with $y(0;\varepsilon) = 2$ and $y(1;\varepsilon) = 1/3$; note that $\sqrt{x} \ge 0$ and so we must expect a boundary layer near $x = 0$. Away from $x = 0$, we seek a solution

$$y(x;\varepsilon) \sim \sum_{n=0}^{N} \varepsilon^n y_n(x)$$

which gives

$$\sqrt{x}\, y_0' + y_0^2 = 0, \quad \sqrt{x}\, y_1' + 2y_0 y_1 + y_0'' = 0,$$

and so on, with $y_0(1) = 1/3$ and $y_n(1) = 0$ for $n \ge 1$. Thus we obtain

$$y_0 = \frac{1}{A_0 + 2\sqrt{x}}, \quad \text{and then} \quad y_0(x) = \frac{1}{1 + 2\sqrt{x}} \tag{2.102}$$

in order to satisfy the boundary condition (and as $x \to 0$ this does not approach $2 = y(0;\varepsilon)$, and so a boundary layer is certainly required). The equation for $y_1(x)$ is therefore

$$\sqrt{x}\, y_1' + \left(\frac{2}{1 + 2\sqrt{x}}\right) y_1 = \frac{1 + 6\sqrt{x}}{2x\sqrt{x}\left(1 + 2\sqrt{x}\right)^3}$$

or

$$\left[\left(1 + 2\sqrt{x}\right)^2 y_1\right]' = \frac{1 + 6\sqrt{x}}{2x^2\left(1 + 2\sqrt{x}\right)}$$

which has the general solution

$$y_1(x) = \frac{8\ln\left[\left(1+2\sqrt{x}\right)/\sqrt{x}\right] - 4x^{-1/2} - \frac{1}{2}x^{-1} + A_1}{\left(1+2\sqrt{x}\right)^2}$$

and $y_1(1) = 0$ requires $A_1 = \frac{9}{2} - 8\ln 3$. (It should be noted that $y_1(x) \sim -1/(2x)$ as $x \to 0$ which, alone, indicates a breakdown where $x = O(\varepsilon)$—but this, as we shall see below, is irrelevant).

For the boundary layer, we scale $x = \delta X$ with $y \equiv Y(X;\varepsilon)$, then the equation, (2.101), becomes

$$\frac{\varepsilon}{\delta^2}Y'' + \delta^{-1/2}\sqrt{X}Y' + Y^2 = 0$$

and so we must select $\delta = \varepsilon^{2/3}$, which leads to

$$Y'' + \sqrt{X}Y' + \varepsilon^{1/3}Y^2 = 0.$$

(The apparent scaling, $x = \varepsilon X$, is therefore redundant—it is smaller than that required in the boundary layer.) We seek a solution $Y(X;\varepsilon) \sim Y_0(X)$, with

$$Y_0'' + \sqrt{X}Y_0' = 0 \quad \text{and so} \quad Y_0(X) = B_0 + C_0\int_0^X \exp\left(-\frac{2}{3}\hat{X}^{3/2}\right)\mathrm{d}\hat{X}$$

and the available boundary condition, $Y_0(0) = 2$, then gives

$$Y_0(X) = 2 + C_0\int_0^X \exp\left(-\frac{2}{3}\hat{X}^{3/2}\right)\mathrm{d}\hat{X}, \quad X \geq 0. \tag{2.103}$$

Finally, we determine C_0 by invoking the matching principle; from (2.102) we obtain $Y \sim 1$, and from (2.103) we see that

$$y \sim 2 + C_0 k \quad \text{where } k = \int_0^\infty \exp\left(-\frac{2}{3}X^{3/2}\right)\mathrm{d}X \quad \left[= \left(\frac{2}{3}\right)^{1/3}\Gamma(2/3)\right], \tag{2.104}$$

which match if $C_0 = -1/k$; the boundary-layer solution is therefore

$$Y(X;\varepsilon) \sim 2 - \frac{1}{k}\int_0^X \exp\left(-\frac{2}{3}\hat{X}^{3/2}\right)\mathrm{d}\hat{X},$$

where k is given in (2.104).

Boundary layers also arise even in the absence of the first-derivative term; indeed, equations of the form

$$\varepsilon y'' - a(x)y = f(x), \quad \varepsilon \to 0^+,$$

with $a(x) \geq 0$ for $\forall x \in D$, can have a boundary layer at each end of the domain. (If $a(x) \leq 0$, then the relevant part of the solution is oscillatory and boundary layers are not present. If $a(x) = 0$ at an interior point, then we have a classical turning-point problem and near this point we will require a transition layer.) The solution away from these layers is simply given, to leading order, by $y(x) = -f(x)/a(x)$. To see the nature of this problem, consider the case $a(x) = 1$. The equation that controls the solution in the boundary layers is then $Y'' - Y = 0$ and so $Y = Ae^X + Be^{-X}$, and exponentially decaying solutions—ensuring bounded solutions as $\varepsilon \to 0^+$—arise for $A = 0$ or $B = 0$, appropriately chosen, either on the left boundary or on the right boundary.

E2.17 Two boundary layers

Consider the equation

$$\varepsilon y'' - (1 + 3x^2)y = x, \quad 0 \leq x \leq 1, \tag{2.105}$$

for $\varepsilon \to 0^+$, with $y(0; \varepsilon) = y(1; \varepsilon) = 1$; the solution for suitable $x = O(1)$ is written as $y(x; \varepsilon) \sim y_0(x)$, to leading order, where

$$y_0(x) = -x/(1 + 3x^2). \tag{2.106}$$

This solution, (2.106), clearly does not satisfy the boundary conditions as $x \to 0$ nor as $x \to 1$. The boundary layer near $x = 1$ is expressed in terms of $x = 1 - \delta X$ $(X \geq 0)$ with $y \equiv Y(X; \varepsilon)$ $(= O(1))$; equation (2.105) then becomes

$$\frac{\varepsilon}{\delta^2} Y'' - (4 - 6\delta X + 3\delta^2 X^2)Y = 1 - \delta X$$

which leads to the choice $\delta = \sqrt{\varepsilon}$. Seeking a solution $Y(X; \varepsilon) \sim Y_0(X)$, then

$$Y_0'' - 4Y_0 = 1 \quad \text{or} \quad Y_0 = A_0 e^{2X} + B_0 e^{-2X} - \frac{1}{4}$$

and for bounded (i.e. matchable) solutions as $X \to \infty$ we must have $A_0 = 0$. The boundary condition, $Y_0(0) = 1$, gives $B_0 = 5/4$ and so

$$Y(X; \varepsilon) \sim \frac{1}{4}\left(5e^{-2X} - 1\right)$$

and this is completely determined, as is (2.106), so matching is used only to confirm the correctness of these results (and this is left as an exercise).

The boundary layer near $x = 0$ is written in terms of $x = \delta\chi$ $(\chi \geq 0)$ with $y \equiv z(\chi;\varepsilon)$ $(= \mathrm{O}(1)$ again); equation (2.105) now becomes

$$\frac{\varepsilon}{\delta^2} z'' - (1 + 3\delta^2\chi^2)z = \delta\chi.$$

The boundary layer is, not surprisingly, the same size at this end of the domain: $\delta = \sqrt{\varepsilon}$, and then with $z(\chi;\varepsilon) \sim z_0(\chi)$ we obtain

$$z_0'' - z_0 = 0 \quad \text{or} \quad z_0 = C_0 \mathrm{e}^{\chi} + D_0 \mathrm{e}^{-\chi}.$$

A bounded solution, as $\chi \to \infty$, requires the choice $C_0 = 0$, and $D_0 = 1$ will ensure that the boundary condition ($z_0(0) = 1$) is satisfied i.e. $z_0(\chi) = \mathrm{e}^{-\chi}$. The next term in the asymptotic expansion $z(\chi;\varepsilon) \sim \sum_{n=0}^{N} \varepsilon^{n/2} z_n(\chi)$ satisfies

$$z_1'' - z_1 = \chi \quad \text{which gives} \quad z_1 = C_1 \mathrm{e}^{\chi} + D_1 \mathrm{e}^{-\chi} - \chi$$

with $C_1 = 0$ (for boundedness) and $D_1 = 0$ (because $z_1(0) = 0$) i.e.

$$z(\chi;\varepsilon) \sim \mathrm{e}^{-\chi} - \sqrt{\varepsilon}\chi;$$

it is left as another exercise to confirm that this matches with (2.106).

In our examples so far (and see also Q2.21, 2.22), the character and position of the boundary layer (or its interior counterpart, the transition layer) have been controlled by the known function $a(x)$, as in $a(x)y'$ (or $a(x)y$ in our most recent example). We will now investigate how the same approach can be adopted when the relevant coefficient is a function of y. In this situation, we do not know, *a priori*, the sign of y—and this is usually critical. Typically, we make an appropriate assumption, seek a solution and then test the assumption. For example, if the two boundary values for y—we are thinking here of two-point boundary value problems—have the same sign, then we may reasonably suppose that y retains this sign throughout the domain. On the other hand, if the boundary values are of opposite sign, then the solution must have at least one zero somewhere on the domain (and this indicates the existence of a transition layer).

E2.18 A problem which exhibits either a boundary layer or a transition layer: I

(An example similar to this one is discussed in great detail in Kevorkian & Cole, 1981 & 1996; see Q2.23.)

We consider the equation

$$\tfrac{1}{2}\varepsilon y'' + yy' - 2xy = 0, \quad 0 \le x \le 1, \tag{2.107}$$

for $\varepsilon \to 0^+$, with $y(0;\varepsilon) = \alpha$ and $y(1;\varepsilon) = \beta$ (where α and β are independent of ε); for suitable $x = O(1)$, we seek a solution in the usual form

$$y(x;\varepsilon) \sim \sum_{n=0}^{N} \varepsilon^n y_n(x),$$

which gives

$$y_0 y_0' - 2xy_0 = 0; \quad (y_0 y_1)' - 2xy_1 + \tfrac{1}{2} y_0'' = 0, \tag{2.108a,b}$$

and so on. The general solution to equation (2.108a) is

$$y_0(x) = A_0 + x^2; \tag{2.109}$$

we exclude the solution $y_0(x) \equiv 0$ because we will consider problems for which $\alpha \neq 0$ and $\beta \neq 0$. (Any special solutions which may need to make use of the zero solution are easily incorporated if required.) The next term in this asymptotic expansion is obtained from (2.108b) i.e.

$$[(A_0 + x^2)y_1]' - 2xy_1 + 1 = 0$$

which yields

$$y_1(x) = A_1 - \frac{1}{\sqrt{A_0}} \arctan\left(\frac{x}{\sqrt{A_0}}\right), \tag{2.110}$$

where A_1 is the second arbitrary constant (and we have taken $A_0 > 0$).

It is clear, however, that it is impossible to proceed without more information about the boundary values, α and β. Let us examine, first, the problem for which both values are positive; we therefore assume that a solution, $y > 0$, exists and hence that any boundary layer must be situated in the neighbourhood of $x = 0$ (indicated by the term yy' with $y > 0$). With this in mind, we may use the one available boundary condition away from $x = 0$, i.e. $y(1;\varepsilon) = \beta$; thus we obtain

$$y(x;\varepsilon) \sim \beta - 1 + x^2. \simeq Y_0 \tag{2.111}$$

Correspondingly, with $y_1(1) = 0$, we see that $A_1 = 0$; we will assume hereafter that $\beta - 1 > 0$ (but we clearly have an interesting case if $-1 < \beta - 1 \le 0$, for then the solution *does* have a zero near $x = \sqrt{1-\beta}$, even with $\alpha > 0$–a possibility not pursued here).

Now, as $x \to 0$, we see that $y_0(x) \to \beta - 1$ (> 0), and if this does not equal α then a boundary layer is required near $x = 0$. For this layer, we write $x = \delta X$, with $y \equiv Y(X; \varepsilon)$ ($= \mathrm{O}(1)$), and then (2.107) becomes

$$\frac{\varepsilon}{2\delta^2} Y'' + \frac{1}{\delta} YY' - 2\delta XY = 0 \quad \text{or} \quad \tfrac{1}{2} Y'' + YY' - 2\varepsilon^2 XY = 0 \tag{2.112}$$

with the choice $\delta = \varepsilon$. We seek a solution $Y(X; \varepsilon) \sim Y_0(X)$, where

$$\tfrac{1}{2} Y_0'' + Y_0 Y_0' = 0 \quad \text{or} \quad Y_0' = B_0^2 - Y_0^2 \tag{2.113}$$

and we have chosen to write the arbitrary constant of integration as B_0^2 (> 0); any other choice produces a solution which cannot be matched—an investigation that is left as an exercise. The next integral of (2.113) gives the general solution

$$Y_0(X) = B_0 \left(\frac{C_0 e^{2B_0 X} - 1}{C_0 e^{2B_0 X} + 1} \right), \qquad B_0 > 0,$$

and to satisfy $Y_0(0) = \alpha$ this can be written as

$$Y_0(X) = B_0 \left\{ \frac{\alpha + B_0 + (\alpha - B_0) e^{-2B_0 X}}{\alpha + B_0 - (\alpha - B_0) e^{-2B_0 X}} \right\}. \tag{2.114}$$

The value of the remaining arbitrary constant, B_0, is now determined by matching (2.114) and (2.111); from (2.111) we obtain

$$Y \sim \beta - 1 \quad \text{for } X = \mathrm{O}(1)$$

and from (2.114) we see that

$$y \sim B_0 \quad \text{for } x = \mathrm{O}(1),$$

which requires that $B_0 = \beta - 1$. Thus we have successfully completed the initial calculations in the construction of asymptotic expansions valid for $x = \mathrm{O}(1)$ and for $x = \mathrm{O}(\varepsilon)$; these demonstrate that, in the case $\alpha > 0$ and $\beta > 1$, we have a solution which satisfies $y > 0$ for $\forall x \in D$.

E2.19 A problem which exhibits either a boundary layer or a transition layer: II

We repeat example E2.18, but now with the boundary values $\beta > 1$ (chosen to avoid the difficulties already noted) and $\alpha < 0$, and thus the solution must change sign (at least once) in order to accommodate these boundary values. This indicates the need for a transition layer at some $x = x_0$, $0 < x_0 < 1$, and determining x_0 becomes an essential element in the construction of the solution.

For $x = \mathrm{O}(1)$ we have, as before (see (2.111)),

$$y(x;\varepsilon) \sim \beta - 1 + x^2 \quad \text{as } \varepsilon \to 0^+, \tag{2.115}$$

but this can hold only for $x_0 + a < x \le 1$ ($a \to 0$ but $\delta(\varepsilon)/a \to 0$, where $\delta(\varepsilon)$ is introduced below); for $0 \le x < x_0 - a$ we have the corresponding solution

$$y(x;\varepsilon) \sim \alpha + x^2 \quad \text{as} \quad \varepsilon \to 0^+. \tag{2.116}$$

Near $x = x_0$ we write $x = x_0 + \delta X$, with $y \equiv Y(X;\varepsilon)$ $(= \mathrm{O}(1))$, which essentially repeats (2.112) i.e. $\delta = \varepsilon$ and so

$$\tfrac{1}{2}Y'' + YY' - 2\varepsilon(x_0 + \varepsilon X)Y = 0,$$

and this gives the same general solution, to leading order, as before (see (2.113), *et seq.*):

$$Y_0(X) = B_0\left(\frac{C_0 \mathrm{e}^{2B_0X} - 1}{C_0 \mathrm{e}^{2B_0X} + 1}\right). \tag{2.117}$$

However, for a transition layer, we do not have *any* boundary conditions; here, we must match (2.117) to both (2.115) and (2.116).

From (2.115) and (2.116) we obtain

$$\beta - 1 + x^2 \sim \beta - 1 + x_0^2 \quad \text{and} \quad \alpha + x^2 \sim \alpha + x_0^2, \tag{2.118}$$

respectively, both for $X = \mathrm{O}(1)$; from (2.117), with $X = (x - x_0)/\varepsilon$, we have

$$B_0\left(\frac{C_0 e^{2B_0X} - 1}{C_0 e^{2B_0X} + 1}\right) \sim \begin{cases} B_0 & \text{for } x > x_0 \\ -B_0 & \text{for } x < x_0 \end{cases} \tag{2.119}$$

as $\varepsilon \to 0^+$. We observe, immediately, that a property of this transition layer is to admit only a change in value across it from $-B_0$ to $+B_0$ (which will fix the value of x_0) and that the matching excludes C_0 (so this cannot be determined at this stage). Now (2.119) does match with (2.118) when we choose

$$B_0 = \beta - 1 + x_0^2 \quad \text{and} \quad -B_0 = \alpha + x_0^2$$

which requires that

$$x_0^2 = \tfrac{1}{2}(1 - \beta - \alpha)$$

and hence a transition layer exists at $x = x_0$ $(0 < x_0 < 1)$ provided

$$0 < 1 - \beta - \alpha < 2. \qquad (2.120)$$

If this condition is not satisfied, for given α and β, then the adjustment to the given boundary value must be through a boundary layer near $x = 0$. Thus, for example, with $\beta = 2$ (> 1) and $\alpha = -2$ (< 0), there is a transition layer at $x_0 = 1/\sqrt{2}$ and the jump across the layer is between $\pm 3/2$ (to leading order). On the other hand, the problem with $\beta = 2$ and $\alpha = -4$ does not admit a transition layer; the boundary layer near $x = 0$ is used to accommodate the change in value from $\beta - 1 = 1$ (to leading order) to $\alpha = -4$. The dominant solution in the transition layer is given by (2.117) with $B_0 = \frac{1}{2}(\beta - 1 - \alpha) > 0$, although C_0 is unknown at this stage. (The rôle of C_0 is to determine the position of the transition layer, correct at $O(\varepsilon)$.)

These two examples, E2.18 and E2.19 (and see also Q2.23), demonstrate the complexity and richness of solutions that are available for this type of problem, depending on the particular boundary values that are prescribed. All this can be traced to the nonlinearity associated with the y' term; if this term were simply $a(x)y'$, then we would have a fixed boundary layer, or fixed transition layers, independent of the specific boundary values (as we have seen in our earlier examples). We conclude this section with an example which shows how these ideas can be extended, fairly straightforwardly, to higher-order equations. (The following example is based on the type of problem that can arise when examining the displacement of a loaded beam.)

E2.20 A problem with two boundary layers

We consider

$$\varepsilon^2 y'''' - \frac{y''}{1 - \frac{3}{4}x} = 8, \quad 0 \le x \le 1, \qquad (2.121)$$

for $\varepsilon \to 0^+$ (and the use of ε^2 here is merely an algebraic convenience), with

$$y(0;\varepsilon) = y'(0;\varepsilon) = y(1;\varepsilon) = y'(1;\varepsilon) = 0. \qquad (2.122)$$

Before we begin the detailed analysis of this problem, a couple of points should be noted. First, the variable coefficient, $1/(1 - \frac{3}{4}x)$, in (1.121), is positive for $\forall x \in D$ and so, second, this implies that we have available (locally) two exponential solutions. These arise from, approximately,

$$\varepsilon^2 y'''' - \lambda^2 y'' = 0 \quad \text{i.e.} \quad y \propto \exp(\pm\lambda x/\varepsilon)$$

and so we may select one exponential near $x = 0$ and the other near $x = 1$, ensuring exponential decay as we move away from the boundaries. Thus we must anticipate boundary layers near both $x = 0$ and $x = 1$; we will assume that they exist—we can always ignore them if they are not required (because the boundary values are automatically satisfied).

Hence, for $x = O(1)$ away from $x = 0$ and $x = 1$, we seek a solution with $y(x;\varepsilon) \sim y_0(x)$, where

$$-\frac{y_0''}{1-\frac{3}{4}x} = 8 \quad \text{or} \quad y_0(x) = A_0 + B_0 + x^3 - 4x^2, \tag{2.123}$$

but no boundary conditions are available (by virtue of the assumed existence of boundary layers). For the boundary layer near $x = 0$, it is clear that we require $x = \varepsilon X$, $y \equiv \Delta Y(X;\varepsilon)$ (but we do not know Δ at this stage; since $y(0;\varepsilon) = 0$, *presumably* $y_0(x) \to 0$ as $x \to 0$ although the *size* of y_0 in this limit is unknown). Thus equation (2.121) becomes

$$Y'''' - \frac{Y''}{1-\frac{3}{4}\varepsilon X} = \frac{8\varepsilon^2}{\Delta}.$$

We seek a solution $Y(X;\varepsilon) \sim Y_0(X)$, where $Y_0(X)$ satisfies

$$Y_0'''' - Y_0'' = 0 \quad \text{or} \quad Y_0(X) = C_0 + D_0 X + E_0 e^{X} + F_0 e^{-X}$$

provided $\varepsilon^2/\Delta = o(1)$ as $\varepsilon \to 0$; we will assume that this condition applies, and we will check it shortly. The boundary conditions are $Y_0(0) = Y_0'(0) = 0$ and we must not allow the term which grows exponentially away from $X = 0$, so $E_0 = 0$, and then

$$Y_0(X) = C_0(1 - X - e^{-X}).$$

Immediately we observe that the term $C_0 X = C_0 x/\varepsilon$ is unmatchable to $y_0(x)$ $(= O(1))$ unless we select $\Delta = \varepsilon$ (and then $\varepsilon^2/\Delta = \varepsilon$); this we do, so that

$$Y(X;\varepsilon) \sim \varepsilon C_0(1 - X - e^{-X}). \tag{2.124}$$

Correspondingly, for the boundary layer near $x = 1$, we write $x = 1 - \varepsilon\chi$ $(\chi \geq 0)$ and now choose $y \equiv \varepsilon \hat{Y}(\chi;\varepsilon)$ to give

$$\hat{Y}'''' - \frac{\hat{Y}''}{\frac{1}{4}+\frac{3}{4}\varepsilon\chi} = 8\varepsilon.$$

Thus the first term in the expansion, $\hat{Y}(\chi;\varepsilon) \sim \hat{Y}_0(\chi)$, satisfies

$$\hat{Y}_0'''' - 4\hat{Y}_0'' = 0 \quad \text{or} \quad \hat{Y}_0(\chi) = G_0 + H_0\chi + J_0 e^{2\chi} + K_0 e^{-2\chi}$$

and then (exactly as described for Y_0) we obtain

$$\hat{Y}(\chi;\varepsilon) \sim \varepsilon G_0(1 - 2\chi - e^{-2\chi}). \tag{2.125}$$

We now determine the constants A_0, B_0, C_0 and G_0 by matching (2.123) with, in turn, (2.124) and (2.125). First, from (2.123) with $x = \varepsilon X$ and $y = \varepsilon Y$, we obtain

$$\varepsilon Y \sim \varepsilon A_0 X + B_0 \quad \text{for } X = O(1);$$

from (2.124) we have

$$y \sim -\varepsilon C_0 x \quad \text{for } x = O(1)$$

which requires $B_0 = 0$ and then $C_0 = -A_0$. Again, from (2.123), but now with $x = 1 - \varepsilon\chi$ and $y = \varepsilon\hat{Y}$, we obtain

$$\varepsilon\hat{Y} \sim A_0 - 3 + (5 - A_0)\varepsilon\chi \quad \text{for } \chi = O(1),$$

and, finally, (2.125) gives

$$y \sim -2\varepsilon G_0(1 - x) \quad \text{for } x = O(1).$$

Now we require $A_0 = 3$ and $5 - A_0 = -2G_0$; thus, collecting all these results, we see that

$$A_0 = 3, \quad B_0 = 0, \quad C_0 = -3, \quad G_0 = -1$$

and hence, to leading order, we have

$$y(x;\varepsilon) \sim 3x + x^3 - 4x^2$$

with $$Y(X;\varepsilon) \sim 3\varepsilon(X - 1 + e^{-X}); \quad \hat{Y}(\chi;\varepsilon) \sim \varepsilon(2\chi - 1 + e^{-2\chi}).$$

So, indeed, boundary layers are required at each end in order to accommodate the boundary conditions there (although we may note that the solution for $y(x;\varepsilon)$ does satisfy $y(0;\varepsilon) = y(1;\varepsilon) = 0$, but not the derivative conditions).

Some further examples of higher-order equations that exhibit boundary-layer behaviour are offered in Q2.26.

This chapter has been devoted to a presentation of some of the fairly routine applications of singular perturbation theory to various types of mathematical problem.

Although we have touched on methods for finding roots of equations, and on integration, the main thrust has been to develop basic techniques for solving differential equations—the most important use, by far, of these methods. We shall devote the rest of the text to extending and developing the methods for solving differential equations, both ordinary and partial, and their applications to many practical problems that are encountered in various branches of mathematical modelling. Many of the examples and exercises in this chapter are, perforce, invented to make a point or to test ideas; however, a few of the later exercises that are included at the end of this chapter (see Q2.27–2.35) begin to employ the techniques in physically relevant problems. In the next chapter, we will show how these ideas can be applied to a broader class of problems and, in particular, begin our discussion of partial differential equations. This will allow us, in turn, to begin to extend the applications of singular perturbation theory to more problems which arise within a physically relevant context.

FURTHER READING

A few of the existing texts include a discussion of the methods for finding roots of equations, and for evaluating integrals of functions which contain a small parameter; in particular, the interested reader is directed to Holmes (1995) and Hinch (1991). Differential equations that give rise to regular problems are given little consideration—they are quite rare, after all—but some can be found in Holmes (1995) and in Georgescu (1995). We have already mentioned those texts that present a more formal approach to perturbation theory (Eckhaus, 1979; Smith, 1985; O'Malley, 1991), but some further developments along these lines are also given in Chang & Howes (1984).

The whole arena of scaling with respect to a parameter, and we should include here the construction of non-dimensional variables, is fairly routine but very powerful. These ideas play a rôle, not only in the identification of asymptotic regions (as we have seen), but also in providing more general pointers to the construction of solutions. A very thorough introduction to these ideas, and their connection with asymptotics, can be found in Barenblatt (1996). A discussion of the applications of group theory to the study of differential equations is likely to be available in any good, relevant text; one such, which emphasises precisely the application to differential equations, is Dresner (1999).

The nature of a boundary layer (which is, for our current interest, limited to a property of certain ordinary differential equations) is described at length, and carefully, in most available texts on singular perturbation theory. We can mention, as examples of the extent and depth of what is discussed, the excellent presentations on this subject given by Smith (1985) and Holmes (1995). The determination of the position of a boundary layer is also covered in most existing texts, although O'Malley (1991) probably provides the most detailed analysis. (This work also includes a number of relevant references which the interested reader may wish to investigate.) An excellent discussion of the interplay between boundary layers and transition layers (for nonlinear equations) is given in Kevorkian & Cole (1981, 1996). (Those readers who wish to examine techniques applicable to turning points, at this stage, are encouraged to study Wasow (1965) and Holmes (1995); we will touch on these ideas in Chapter 4.)

Finally, some examples of higher-order equations, which exhibit boundary-layer-type solutions, are discussed in Smith (1985) and O'Malley (1991).

EXERCISES

Q2.1 *Quadratic equations.* Write down the exact roots of these quadratic equations, where ε is a positive parameter.

(a) $x^2 + 2\varepsilon x - 1 = 0$; (b) $\varepsilon x^2 + 4x - 4 = 0$; (c) $x^2 - (2+\varepsilon)x + 1 = 0$.

Now, in each case, use the binomial expansion to obtain power-series representations of these roots, valid for $\varepsilon \to 0^+$, writing down the first three terms for each root. (You may wish to investigate how these same expansions can be derived directly from the original quadratic equations.)

Finally, obtain the corresponding power series which are valid for $\varepsilon \to +\infty$.

Q2.2 *Equations I.* Find the first two terms in the asymptotic expansions of all the real roots of these equations, for $\varepsilon \to 0^+$.

(a) $\varepsilon x^3 - x^2 + x + \frac{3}{4} = 0$; (b) $\varepsilon x^3 - x^2 + 2x - 1 = 0$; (c) $\varepsilon x^3 + \sqrt{x} - 1 = 0$;
(d) $\varepsilon x^3 - x + 2\sqrt{x} - 1 = 0$; (e) $\varepsilon x + \varepsilon^2 - \tanh x = 0$;
(f) $\varepsilon x^3 - x^2 + 1 + 2\varepsilon - \mathrm{e}^{-x/\varepsilon} = 0$; (g) $\varepsilon^5 x^3 - \varepsilon x - \varepsilon + 1 - \mathrm{e}^{-x} = 0$;
(h) $3\varepsilon x^4 - x^3 + 6\sqrt{x} - 5 = 0$; (i) $1 + \varepsilon - \mathrm{e}^{-x/\varepsilon} - \varepsilon \sin x = 0$;
(j) $\varepsilon x^3 + x^2 - 1 - \varepsilon \sinh(\varepsilon x) = 0$.

Q2.3 *Equations II.* Repeat Q2.2 for these slightly more involved equations.

(a) $1 + \varepsilon - 2\sqrt{\varepsilon} + (x\sqrt{\varepsilon} - 1)^3 - \sqrt{\varepsilon}(x + \mathrm{e}^{-x/\varepsilon}) = 0$;

(b) $2\varepsilon x + \dfrac{2(1-\varepsilon)}{2+\varepsilon x} - e^{-|x|/\varepsilon} = 0$; (c) $\varepsilon x^2 - x + \varepsilon(\ln \varepsilon)^2 - \dfrac{x^2}{x^2 + \varepsilon^2 \mathrm{e}^{-x/\varepsilon}} = 0$;

(d) $\mathrm{e}^x - 2 + \dfrac{\varepsilon}{1+x^2} = 0$; (e) $\varepsilon^3 x^3 - \varepsilon^4 x^2 - \varepsilon - \tanh x = 0$.

Q2.4 *Kepler's equation.* A routine problem in celestial mechanics is to find the *eccentric anomaly*, u, given both the *eccentricity* e $(0 \le e < 1)$ and the *mean anomaly* nt (where t is time measured from where $u = 0$, and $n = 2\pi/P$, where P is the period); u is then the solution of *Kepler's equation*

$$nt = u - e \sin u$$

(see e.g. Boccaletti & Pucacco, 1996). For many orbits (for example, most planets in our solar system), the eccentricity is very small; find the first three terms in the asymptotic solution for u as $e \to 0^+$. Confirm that your 3-term expansion is uniformly valid for all nt.

Q2.5 *Complex roots.* Find the first two terms in the asymptotic expansions of all the roots of these equations, for $\varepsilon \to 0^+$.

(a) $\varepsilon z^3 + z^2 + z + 1 = 0$; (b) $\varepsilon z^4 - z^2 + 2z - 2 = 0$; (c) $\varepsilon z^4 + z^2 - 2z + 2 = 0$;

(d) $z^4 + z^3 + z^2 + \varepsilon = 0$; (e) $\mathrm{e}^z = 1 + \varepsilon |z|^2$; (f) $\mathrm{e}^{z/\varepsilon} = \mathrm{e}^2 + \dfrac{\varepsilon}{1+|z|}$.

Q2.6 *Simple integrals.* Obtain estimates for these integrals, for $\varepsilon \to 0^+$, by first finding asymptotic expansions of the integrand for each relevant size of x, retaining the first two terms in each case. (These integrals can be evaluated exactly, so you may wish to check your results against the expansions of the exact values.)

(a) $\displaystyle\int_0^1 \left(x^2 + \frac{\varepsilon}{\sqrt{x+\varepsilon}} + \mathrm{e}^{-x/\varepsilon}\right)\mathrm{d}x$; (b) $\displaystyle\int_0^\infty \frac{\mathrm{d}x}{(1+x)(x+\varepsilon)}$;

(c) $\displaystyle\int_0^1 \left(2x + \varepsilon + \frac{\varepsilon}{\sqrt{x}} + 3\varepsilon^2 x^2 + \mathrm{e}^{-x/\varepsilon}\right)\mathrm{d}x$; (d) $\displaystyle\int_0^\infty \frac{(2\varepsilon - 2\varepsilon^2 - 1)x - (1+3\varepsilon)}{\varepsilon x^2(x+1-\varepsilon) + (x+1)^2 - \varepsilon^2}\,\mathrm{d}x$.

Q2.7 *More integrals.* See Q2.6; repeat for these integrals (but here you are not expected to have available the exact values).

(a) $\displaystyle\int_0^1 \frac{1+\varepsilon x + \mathrm{e}^{-x/\varepsilon}}{1+x^2}\,\mathrm{d}x$; (b) $\displaystyle\int_0^1 \frac{x\,\mathrm{d}x}{\sqrt{(1-x)(x^2+\varepsilon^2)}}$;

(c) $\displaystyle\int_0^\infty \frac{\mathrm{d}x}{(1+x)(x+\varepsilon)^n}$ for $n = \frac{1}{2}$; (d) repeat (c) for $n = 2$;

(e) $\displaystyle\int_0^\infty \frac{x\,\mathrm{d}x}{(1+x)(1+\varepsilon x)\sqrt{x^2+\varepsilon^2}}$; (f) $\displaystyle\int_0^1 \frac{x + \varepsilon + \sqrt{1+\varepsilon \mathrm{e}^{-x/\varepsilon}}}{\sqrt{1+x-\varepsilon+2\varepsilon \mathrm{e}^{-x/\varepsilon}}}\,\mathrm{d}x$.

Q2.8 *An integral from thin aerofoil theory.* An integral of the type that can appear in the study of thin aerofoil theory (for the velocity components in the flow field) is

$$I(x;\varepsilon) = \varepsilon \int_{-1}^{1} \frac{\xi + \varepsilon/\sqrt{1+\xi}}{(\xi - x)^2 + \varepsilon^2}\,\mathrm{d}\xi, \quad -1 \le x \le 1;$$

obtain the first terms in the asymptotic expansions of the integrand (for $\varepsilon \to 0^+$), with x away from the end-points, for each of: (a) ξ away from x and away from $\xi = -1$; (b) $\xi = x + \varepsilon\varsigma$; (c) $\xi = -1 + \varepsilon^2 z$. Hence find an estimate for $I(x;\varepsilon)$. Repeat the calculations with $x = 1 - \varepsilon X$, and then with $x = -1 + \varepsilon X$, for ξ away from the end-points, and then with $\xi = 1 - \varepsilon\varsigma$, $\xi = -1 + \varepsilon^2 z$; $\xi = 1 - \varepsilon\varsigma$, $\xi = -1 + \varepsilon\varsigma$, $\xi = -1 + \varepsilon^2 z$, respectively. Again, find estimates for $I(1 - \varepsilon X;\varepsilon)$ and for $I(-1 + \varepsilon X;\varepsilon)$; show that your asymptotic approximations for $I(x;\varepsilon)$ satisfy the matching principle.

Q2.9 *Regular expansions for differential equations.* Find the first two terms in the asymptotic expansions of the solutions of these equations, satisfying the given boundary conditions. In each case you should use the asymptotic sequence $\{\varepsilon^n\}$, and you should confirm that your 2-term expansions are uniformly valid. (You may wish to examine the nature of the general term, and hence produce an argument that shows the uniform validity of the expansion to all orders in ε.)

(a) $y' = 2x + \varepsilon y^2$, $0 \le x \le 1$, $y(0;\varepsilon) = 0$;

(b) $y'' + y' + \varepsilon y^3 = 0$, $0 \le x \le 1$, $y(0;\varepsilon) = 1$, $y(1;\varepsilon) = \mathrm{e}^{-1}$;

(c) $y'' + y + \varepsilon y^3 = 0$, $0 \le x \le \pi/2$, $y(0;\varepsilon) = 1$, $y(\pi/2;\varepsilon) = 0$;

(d) $x^2 y'' - (4x + \varepsilon x^3 \mathrm{e}^x)y' + 6y = 0$, $-1 \le x \le 1$, $y(-1;\varepsilon) = 2$, $y(1;\varepsilon) = 0$;

(e) $\ddot{x} + (1 - \varepsilon t \mathrm{e}^{-t})x = 0$, $t \ge 0$, $x(0;\varepsilon) = 0$, $\dot{x}(0;\varepsilon) = 1$.

Q2.10 *Eigenvalue problems*. A standard problem in many branches of applied mathematics and physics is to find the eigenvalues (and eigenfunctions) of appropriate problems based on ordinary differential equations. In these examples, find the first two terms in the asymptotic expansions of both the eigenvalues (λ) and the eigenfunctions; for each use the asymptotic sequence $\{\varepsilon^n\}$.

(a) $y'' + \lambda(1 + \varepsilon x)y = 0$, $0 \le x \le 1$, $y(0; \varepsilon) = y(1; \varepsilon) = 0$;

(b) $y'' + \lambda(1 + \varepsilon \sin x)y = 0$, $0 \le x \le 1$, $y(0; \varepsilon) = y(1; \varepsilon) = 0$;

(c) $y'' + \lambda y = \varepsilon y^2$, $0 \le x \le 1$, $y(0; \varepsilon) = y(1; \varepsilon) = 0$.

Q2.11 *Breakdown of asymptotic solutions of differential equations*. These ordinary differential equations define solutions on the domain $0 \le x \le 1$, with conditions given on $x = 1$. In each case, find the first two terms in an asymptotic solution valid for $x = \mathrm{O}(1)$ as $\varepsilon \to 0^+$, which allows the application of the given boundary condition(s). Show, in each case, that the resulting expansion is not uniformly valid as $x \to 0$; find the breakdown, rescale and hence find the first term in an asymptotic expansion valid near $x = 0$, matching as necessary. Finally find, for each problem, the dominant asymptotic behaviour of $y(0; \varepsilon)$ as $\varepsilon \to 0^+$.

(a) $(x^2 + \varepsilon y)y' + 2xy = \dfrac{3\varepsilon}{2y}$, $y(1; \varepsilon) = 1$;

(b) $(x^2 + \varepsilon y^2)y' + xy = \varepsilon y^2$, $y(1; \varepsilon) = 1$;

(c) $yy'' - \frac{1}{2}(y')^2 - 2\varepsilon = 2\varepsilon y^{3/2}$, $y(1; \varepsilon) = 1 + 2\varepsilon$, $y'(1; \varepsilon) = 2 + 3\varepsilon$;

(d) $(x^2 + \varepsilon y)y' + \dfrac{2xy}{1+x} = 3x^2(1+x)^2$, $y(1; \varepsilon) = 8$;

(e) $yy'' - 2(y')^2 = \varepsilon\left[2(y')^3 + 6y'\right]$, $y(1; \varepsilon) = 1 - \varepsilon$, $y'(1; \varepsilon) = -1 + \varepsilon$.

Q2.12 *Another breakdown problem*. See Q2.11; repeat for the problem

$$(x^3 + \varepsilon y)y' - \frac{x^4}{y} = \varepsilon\left(3x^3 y + \frac{x}{y}\right), \quad y(1; \varepsilon) = 1 + \varepsilon, \quad x \le 1,$$

but show that, for a real solution to exist, the domain is $x_0(\varepsilon) \le x \le 1$, where $x_0(\varepsilon) \sim (2\varepsilon)^{1/3}$, and then find the dominant asymptotic behaviour of $y(2\varepsilon^{1/3}; \varepsilon)$ as $\varepsilon \to 0^+$.

Q2.13 *Breakdown as $x \to \infty$: I*. Find the first two terms in an asymptotic solution, valid for $x = \mathrm{O}(1)$ as $\varepsilon \to 0^+$, of

$$y''' + y'' + \varepsilon[(y')^2 + y] = \varepsilon e^{-2x}, \quad x \ge 0,$$

with $y(0; \varepsilon) = 2$, $y'(0; \varepsilon) = -1$, $y''(0; \varepsilon) = 1$. Now show that this expansion is not uniformly valid as $x \to \infty$, find the breakdown, rescale and find the first two terms in an expansion valid for large x, matching as necessary. Show, also, that this 2-term expansion breaks down for even larger x, but do not take the analysis further.

Q2.14 *Breakdown as $x \to \infty$: II.* See Q2.11 (a) and (c); for these equations and boundary conditions, and the asymptotic solutions already found for $x = O(1)$, take the domain now to be $x \geq 1$. Hence show that the expansions are not uniformly valid as $x \to \infty$, find the breakdown, rescale and then find the first terms in the expansions valid for large x, matching as necessary.

Q2.15 *Problem E 2.7 reconsidered.* Find the first two terms in an asymptotic expansion, valid for $x = O(1)$ as $\varepsilon \to 0$, of

$$y'' + y' + \varepsilon y^2 = 0, \quad 0 \leq x \leq 1,$$

with $y(0;\varepsilon) = \varepsilon$, $y(1;\varepsilon) = 1 - e^{-1} - \frac{1}{2}\varepsilon e^{-2}$. Show that, formally, this requires two matched expansions, but that the asymptotic solution obtained for $x = O(1)$ correctly recovers the solution for $\forall x \in D$ i.e. it is uniformly valid. (Note the *balance* of terms, when scaled near $x = 0$!)

Q2.16 *Scaling of equations.* See Q2.11 and Q2.14; use the dominant terms only, valid for $x = O(1)$, together with appropriate scalings associated with the relevant balance of terms, to analyse these equations. Compare your results with the scalings obtained from the breakdown of the asymptotic expansions.

Q2.17 *Boundary-layer problems I.* Find the first two terms in asymptotic expansions, valid for $x = O(1)$ (away from the boundary layer) as $\varepsilon \to 0^+$, for each of these equations, with the given boundary conditions. Then, for each, find the first term in the boundary-layer solution, matching as necessary. (You may wish to use your expansions to construct composite expansions valid for $\forall x \in D$, to this order.)

(a) $\varepsilon y'' + (1+x)^2 y' - y^2 = 0$, $0 \leq x \leq 1$, $y(0;\varepsilon) = 2$, $y(1;\varepsilon) = 2 + 4\varepsilon$;

(b) $\varepsilon y'' + (1+\varepsilon x) y' - 2\sqrt{y} = -2\varepsilon x$, $0 \leq x \leq 1$, $y(0;\varepsilon) = 2$, $y(1;\varepsilon) = 4$;

(c) $\varepsilon y'' + (1+\varepsilon x) y' + (1+2x)e^y = 0$, $0 \leq x \leq 1$, $y(0;\varepsilon) = 1$, $y(1;\varepsilon) = 2\ln(2/3)$;

(d) $\varepsilon y' = \dfrac{1+y}{(1+x)^2}\left(\dfrac{1+y}{2+x} - y\right)$, $0 \leq x < \infty$, $y(0;\varepsilon) = \frac{1}{2}$;

(e) $\varepsilon y'' + \sqrt{x}(y' - y) = 0$, $0 \leq x \leq 1$, $y(0;\varepsilon) = 0$, $y(1;\varepsilon) = 1$.

Q2.18 *Boundary-layer problems II.* See Q2.17; repeat for these more involved equations.

(a) $\varepsilon[y'' + (y')^2] + y' + y^2 = \dfrac{2\varepsilon}{1+x}$, $0 \leq x \leq 1$, $y(0;\varepsilon) = 1 + \ln 2$, $y(1;\varepsilon) = \frac{1}{2}$;

(b) $\varepsilon y y'' + y^2 y' - y = \varepsilon x$, $0 \leq x \leq 1$, $y(0;\varepsilon) = 2\sqrt{2}$, $y(1;\varepsilon) = 2$;

(c) $\varepsilon y'' - y y' - y^2 = \varepsilon(e^{-x} + e^{-2x})$, $0 \leq x \leq 1$, $y(0;\varepsilon) = 1$, $y(1;\varepsilon) = \frac{1}{2}e^{-1}$;

(d) $\varepsilon y'' + (x+y)[y' - x - \varepsilon(x+y)] = \varepsilon$, $0 \leq x \leq 1$, $y(0;\varepsilon) = 0$, $y'(0;\varepsilon) = 2/\varepsilon$;

(e) $\varepsilon y' y'' + y(y')^2 - 4y^2 = 4\varepsilon y^2\left[(1+x)^{-3} + 2(1+x)\right]$, $0 \leq x \leq 1$, $y(0;\varepsilon) = \frac{1}{3}$, $y(1;\varepsilon) = 4(1+2\varepsilon)$;

(f) $\varepsilon y''' + y'' + \varepsilon y'(y'+2) + y = 1$, $0 \leq x \leq \pi/2$, $y(0;\varepsilon) = 0$, $y(\pi/2;\varepsilon) = 1 - \frac{1}{3}\varepsilon$, $y'(\pi/2;\varepsilon) = -1 + (\varepsilon\pi)/4$.

Q2.19 *Two boundary layers.* The function $y(x;\varepsilon)$ is defined by the problem

$$\varepsilon^2 y'' + 4\varepsilon x^2 y' - 5y = x^2 - 2, \quad 0 \le x \le 1,$$

with $y(0;\varepsilon) = 1$, $y(1;\varepsilon) = 2$. Find the first two terms in an asymptotic expansion valid for $x = \mathrm{O}(1)$, as $\varepsilon \to 0^+$, away from $x = 0$ and $x = 1$. Hence show that, in this problem, boundary layers exist both near $x = 0$ and near $x = 1$, and find the first term in each boundary-layer solution, matching as necessary.

Q2.20 *A boundary layer within a thin layer.* Consider the equation

$$\varepsilon^2 y'' + y' - \frac{\varepsilon y^2}{x^2 + \varepsilon^2} = 2x, \quad 0 \le x \le 1,$$

for $\varepsilon \to 0^+$, with $y(0;\varepsilon) = 1$, $y(1;\varepsilon) = 2$. Find the first terms in each of three regions, two of which are near $x = 0$, matching as necessary. (Here, only the inner-most region is a true boundary layer; the other is simply a scaled-thin-connecting region.)

Q2.21 *Boundary layers or transition layers?* Decide if these equations, on the given domains, possess solutions which may include boundary layers or transition layers; give reasons for your conclusions.

(a) $\varepsilon y'' + (x-2)y' + y = 0, 0 \le x \le 1$;

(b) $\varepsilon y'' + (x - x^2)y' + y^2 = 0, 0 \le x \le 1$;

(c) $\varepsilon y'' + (2x^2 - x)y' + y^2 = 3x, 0 \le x \le 1$;

(d) $\varepsilon y'' + (x - x^2)y' + y^2 = 0, -1 \le x \le 1$;

(e) $\varepsilon y'' + (x + y)y' + y = 0, 1 \le x \le 2$.

Q2.22 *Transition layer near a fixed point.* In these problems, a transition layer exists at a fixed point, independent of the boundary values. Find, for (a), the first two terms, and for (b) the first term only, in an asymptotic expansion (as $\varepsilon \to 0^+$) valid away from the transition layer and the first term only of an expansion valid in this layer; match your expansions as necessary.

(a) $\varepsilon y'' + 2(x-1)y' = 8(x-1)^3, 0 \le x \le 2, y(0;\varepsilon) = 1, y(2;\varepsilon) = 2$;

(b) $\varepsilon y'' + x^{1/5}y' + \frac{5}{6}y^2 = 0, -1 \le x \le 1, y(-1;\varepsilon) = \frac{1}{2}, y(1;\varepsilon) = \frac{1}{3}$.

Q2.23 *Boundary layer or transition layer?* (This example is based on the one which is discussed carefully and extensively in Kevorkian & Cole, 1981 & 1996.) The equation is

$$\tfrac{1}{2}\varepsilon y'' + yy' + y = 0, \quad 0 \le x \le 1,$$

for $\varepsilon \to 0^+$ and given $y(0;\varepsilon) = \alpha \ (\neq 0)$, $y(1;\varepsilon) = \beta \ (\neq 0)$. Suppose that a transition layer exists near $x = x_0$, $0 < x_0 < 1$; find the leading terms in the asymptotic expansions valid outside the transition layer, and in the transition layer. Hence deduce that a transition layer is required if α and β are of opposite sign and $0 < \alpha + \beta + 1 < 2$, and then find the leading terms in all the relevant regions for:

(a) $\alpha = 1, \beta = -1$; (b) $\alpha = 1, \beta = 2$; (c) $\alpha = -1, \beta = 3$.

Q2.24 *Transition layers and turning points.* Consider the equation

$$\varepsilon y'' + a(x)y' + f(x, y; \varepsilon) = 0;$$

introduce $y(x;\varepsilon) = u(x;\varepsilon)v(x;\varepsilon)$ and find a choice of $u(x;\varepsilon)$ which produces an equation for $v(x;\varepsilon)$ in the form

$$\varepsilon^2 v'' + F(x, v; \varepsilon) = 0$$

and identify F. If F changes sign on the domain of the solution, then the point where this occurs is called a *turning point*; find the equation that defines the turning points in the case $f = \varepsilon^{-1} b(x;\varepsilon) y$.

Q2.25 *Transition layer at a turning point.* Consider the equation

$$\varepsilon^3 y'' + (x - 2x^2)y = 0, \quad 0 \le x \le 1;$$

find the position of the turning point and scale in the neighbourhood of the transition layer. Write down the general solution, to leading order, valid in the transition layer, as $\varepsilon \to 0^+$. (This solution is best written in terms of *Airy* functions. A uniformly valid solution is usually expressed using the *WKB* method; see Chapter 4.)

Q2.26 *Higher-order equations.* For these problems, find the first terms only in asymptotic expansions valid in each region of the solution, for $\varepsilon \to 0^+$.

(a) $\varepsilon^2 y'''' - (1 + x)^2 y'' = 1$, $0 \le x \le 1$, $y(0;\varepsilon) = y'(0;\varepsilon) = y(1;\varepsilon) = y'(1;\varepsilon) = 0$;

(b) $\varepsilon^2 y''' - xy' + y = 2x^3$, $1 \le x \le 4$, $y(1;\varepsilon) = 1$, $y'(1;\varepsilon) = 0$, $y(4;\varepsilon) = 0$;

(c) $\varepsilon^2 y''' - (1 + x)^2 y' + y^2 = 0$, $0 \le x \le 1$, $y(0;\varepsilon) = 1$, $y(1;\varepsilon) = \frac{2}{3}$, $y'(1;\varepsilon) = 0$.

Q2.27 *Vertical motion under gravity.* Consider an object that is projected vertically upwards from the surface of a planetary body (or, rather, for example, from our moon, because we will assume no atmosphere in this model). The height above the surface is $z(t)$, where t is time, and this function is a solution of

$$\ddot{z} = -\frac{gR^2}{(R + z)^2}, \quad t \ge 0,$$

where R is the distance from the centre of mass of the body to the point of projection, and g is the appropriate (constant) acceleration of gravity. (For our moon, $R \approx 1735$ km, $g \approx 1{\cdot}63$ m/s^2.) The initial conditions are $z(0) = 0$, $\dot{z}(0) = V\,(> 0)$; find the relevant solution in the form $t = t(z)$ (and you may assume that $V^2 < 2gR$).

(a) Write your solution, and the differential equation, in terms of the non-dimensional variables (Z, τ): $t = (V/g)\tau$, $z = (V^2/g)Z$, and introduce the parameter $\varepsilon = V^2/(Rg)$. Suppose that the limit of interest is $\varepsilon \to 0^+$ (which you may care to interpret); find the first two terms of an asymptotic expansion, valid for $\tau = O(1)$ (in the form $Z = Z(\tau; \varepsilon)$), directly from the governing equation. (You should compare this with the expansion of the exact solution.) From your results, find approximations to the time to reach the maximum height, the value of this height and the time to return to the point of projection.

(b) A better model, for motion through an atmosphere, is represented by the equation

$$\ddot{z} = -\frac{gR^2}{(R+z)^2} - \frac{k\dot{z}}{R+z},$$

where k (> 0) is a constant. (This is only a rather crude model for air resistance, but it has the considerable advantage that it is valid for both $\dot{z} > 0$ and $\dot{z} < 0$.) Non-dimensionalise this equation as in (a), and then write $(kV)/(Rg) = (V^2/(Rg))(k/V) = \varepsilon\delta$ where $\delta = O(1)$ as $\varepsilon \to 0^+$. Repeat all the calculations in (a).

(c) Finally, in the special case $\varepsilon = 2$ (i.e. $V = \sqrt{2Rg}$, the *escape speed*), find the first term in an asymptotic expansion valid as $\delta \to 0^+$. Now find the equation for the second term and a particular integral of it. On the assumption that the rest of the solution contributes only an exponentially decaying solution, show that your expansion breaks down at large distances; rescale and write down—do not solve—the equation valid in this new region.

Q2.28 *Earth-moon-spaceship (1D).* In this simple model for the passage of a spaceship moving from the Earth to our moon, we assume that both these objects are fixed in our chosen coordinate system, and that the trajectory is along the straight line joining the two centres of mass. (More complete and accurate models will be discussed in later exercises.) We take $x(t)$ to be the distance measured along this line from the Earth, and then Newton's Law of Gravitation gives the equation of motion as

$$m\ddot{x} = \frac{GmM_m}{(d-x)^2} - \frac{GmM_e}{x^2},$$

where m is the mass of the spaceship, M_e and M_m the masses of the Earth and Moon, respectively, G is the universal gravitational constant and d the distance between the mass centres. Non-dimensionalise this equation, using d as the distance scale and $\sqrt{d^3/(G(M_e + M_m))}$ as the time scale, to give the non-dimensional version of the equation (x and t now non-dimensional) as

$$\ddot{x} = \frac{\varepsilon}{(1-x)^2} - \frac{1-\varepsilon}{x^2}$$

where $\varepsilon = M_m/(M_e + M_m)$. We will construct an asymptotic solution, for $x \in (0, 1)$, as $\varepsilon \to 0^+$. (The actual values, for the Earth and Moon, give $\varepsilon \approx 0.012$, and a trajectory from surface to surface requires $x \in [0.017, 0.996]$, approximately.) Write down a first integral of the equation.

(a) Find the first two terms in an asymptotic expansion valid for $x = \mathrm{O}(1)$, by seeking $t = t(x; \varepsilon)$ (cf. Q2.27), and use the data $\dot{x}(\Delta; \varepsilon) = \alpha$, $t \to 0$ as $x \to 0$, and write

$$\frac{1}{2}\alpha^2 = -k + \frac{1 - \varepsilon - \Delta + 2\varepsilon\Delta}{\Delta(1 - \Delta)} \quad (k = \mathrm{O}(1), 0 < k < 1).$$

(Here, α is the non-dimensional initial speed away from the Earth, Δ is small ($\Delta \approx 0{\cdot}017$), and the condition on k ensures that the spaceship reaches the Moon, but not at such a high speed that it can escape to infinity.) Show that this expansion breaks down as $x \to 1$.

(b) Seek a scaling of the governing equation in the neighbourhood of $x = 1$ by writing $x = 1 - \delta X$, $t = T_0 + \delta T(X; \varepsilon)$ (which is consistent with the solution obtained in (a), where the first term, $T_0 =$ constant, provides the dominant contribution at $x = 1$). Find the first term in an asymptotic expansion of $T(X; \varepsilon)$, match to your solution from (a) and hence determine $T_0(k)$. (Be warned that $\ln \varepsilon$ terms appear in this problem.)

Q2.29 *Eigenvalues for a vibrating beam.* The (linearised) problem of an elastic beam clamped at each end is

$$\varepsilon^2 y'''' - y'' = \lambda^2 y, \quad 0 \le x \le 1,$$

for $\varepsilon \to 0^+$, with $y(0; \varepsilon) = y'(0; \varepsilon) = y(1; \varepsilon) = y'(1; \varepsilon) = 0$, where λ is the eigenvalue (which arises from the time-dependence), and $\varepsilon^2 \propto E$, *Young's modulus*. Find the first term in an asymptotic expansion of the eigenvalues. (This problem can be solved exactly, and then the exponents expanded for $\varepsilon \to 0^+$; this is an alternative that could be explored.)

Q2.30 *Heat transfer in 1D.* An equation which describes heat transfer in the presence of a one-dimensional, steady flow (Hanks, 1971) is

$$\varepsilon T'' + xT' - xT = 0, \quad 0 \le x \le 1, \quad \varepsilon > 0,$$

with temperature conditions $T(0; \varepsilon) = T_0$, $T(1; \varepsilon) = T_1$. Find the first two terms in an asymptotic expansion, valid for $x = \mathrm{O}(1)$ as $\varepsilon \to 0^+$, and the leading term valid in the boundary layer, matching as necessary.

Q2.31 *Self-gravitating annulus.* A particular model for the study of planetary rings is represented by the equation

$$\frac{\mathrm{d}}{\mathrm{d}r}\left[r\frac{\mathrm{d}\rho}{\mathrm{d}r}\right] + \alpha r \rho = \frac{1}{r^2} - \frac{\beta}{r^3}, \quad 1 - \varepsilon \le r \le 1 + \varepsilon,$$

where $\alpha > 0$, $\beta > 1$ and $0 < \varepsilon < 1$ are constants, with the density, ρ, satisfying $\rho(1-\varepsilon;\varepsilon) = \rho(1+\varepsilon;\varepsilon) = 0$. (This example is based on the more general equation given in Christodoulou & Narayan, 1992.) For $\varepsilon \to 0^+$ (the narrow annulus approximation), introduce $r = 1 + \varepsilon R$ $(-1 \le R \le 1)$ and then write the density as $\rho = \varepsilon^2 P(R;\varepsilon)$; find the first three terms in an asymptotic expansion for P. On the basis of this information, deduce that your expansion would appear to be uniformly valid for $\forall R \in [-1, 1]$.

Q2.32 *An elastic displacement problem.* A simplified version of an equation which describes the displacement of a (weakly) nonlinear string, in the presence of forcing, which rests on an elastic bed, is

$$[y'(1-\varepsilon y')]' - y = x, \quad 0 \le x \le 1,$$

where ε is a constant, with $y(0;\varepsilon) = y(1;\varepsilon) = 0$. Find the first two terms in an asymptotic expansion, for $\varepsilon \to 0^+$, and use this evidence to deduce that this expansion would appear to be uniformly valid for $\forall x \in [0, 1]$.

Q2.33 *Laminar flow through a channel.* A model for laminar flow through a channel which has porous walls, through which suction occurs, can be reduced to

$$\varepsilon y''' - yy'' + (y')^2 = A(\varepsilon), \quad 0 \le x \le 1,$$

where $A(\varepsilon)$ is an arbitrary constant of integration, with $y(0;\varepsilon) = k$ $(0 < k < 1)$, $y'(0;\varepsilon) = 0$, $y(1;\varepsilon) = 1$, $y'(1;\varepsilon) = 0$. (This is taken from Proudman, 1960; see also Terrill & Shrestha, 1965, and McLeod in Segur, *et al.*, 1991; here, the stream function is proportional to the function $y(x;\varepsilon)$ and $\varepsilon \propto 1/(\text{Reynolds' Number})$.) Assume that $A(0)$ exists and is non-zero, and then find the first term in an asymptotic expansion for $y(x;\varepsilon)$, and for $A(\varepsilon)$, valid for $x = O(1)$, and then the first two terms valid in the boundary layer (the first being simply the boundary value there).

Q2.34 *Slider bearing.* The pressure, p, within the fluid film of a slider bearing, based on Reynolds' equation, can be reduced to the equation

$$\varepsilon(h^3 pp')' = (hp)', \quad 0 \le x \le 1,$$

written in non-dimensional form; here, $\varepsilon > 0$ is a constant and $h(x)$ (> 0) is the given (smooth) film thickness, with $p(0;\varepsilon) = p(1;\varepsilon) = 1$ (and $h(0) \ne h(1)$). Find the first two terms in an asymptotic expansion, for $\varepsilon \to 0^+$, valid for $x = O(1)$, and then the first term only in the boundary layer, matching as necessary. (The first term in the boundary layer can be written only in implicit form, but this is sufficient to allow matching.)

Q2.35 *An enzyme reaction.* The concentration, $c(r;\varepsilon)$, of oxygen in an enzyme reaction can be modelled by the equation

$$\varepsilon\left(c'' + \frac{2}{r}c'\right) = \frac{c}{c+k}, \quad 0 \le r \le 1,$$

with $\varepsilon \to 0^+$, where k (> 0) is a constant. The boundary conditions specify the concentration on $r = 1$ and that the flux of oxygen must be zero at $r = 0$; these are expressed as

$$c(1;\varepsilon) = 1 \quad \text{and} \quad c'(r;\varepsilon) \to 0 \quad \text{as } r \to 0.$$

(a) Find the size of the boundary layer near $r = 1$ (in the form $r = 1 - \delta R$, for suitable δ) and hence show that $c(1 - \delta R; \varepsilon) \sim C_0(R)$ satisfies

$$\frac{1}{\sqrt{2}} \int_{C_0(R)}^{1} [c - k\ln(1 + c/k)]^{-1/2} \mathrm{d}c = R,$$

where we have assumed that $C_0(R) \to 0$ and $C_0'(R) \to 0$ as $R \to \infty$ (which is consistent with the equation).

(b) From the result in (a), deduce that $C_0(R)$ is *exponentially* small as $R \to \infty$; for $1 - r = O(1)$, seek a solution (which is exponentially small) in terms of the scaled variable $\rho = r/\sqrt{\varepsilon}$, and show that $c(r;\varepsilon) \sim A(\varepsilon)s(\rho)$, where $s(\rho)$ satisfies

$$s'' + \frac{2}{\rho}s' = \frac{s}{k}.$$

Solve this equation, apply the relevant boundary condition, match and hence show that

$$c(r;\varepsilon) \sim \frac{A_0}{r} \mathrm{e}^{-1/\sqrt{\varepsilon k}} \sinh\left(r/\sqrt{\varepsilon k}\right)$$

where A_0 is a constant (independent of ε) which should be identified.

3. FURTHER APPLICATIONS

The ideas and techniques developed in Chapter 2 have taken us beyond elementary applications, and as far as some methods that enable us to construct (asymptotic) solutions of a few types of ordinary differential equation. The aim now is to extend these methods, in particular, to partial differential equations. The first reaction to this proposal might be that the move from ordinary to partial differential equations is a very big step—and it can certainly be argued thus if we compare the solutions, and methods of solution, for these two categories of equation. However, in the context of singular perturbation theory, this is a misleading position to adopt. Without doubt we must have some skills in the methods of solution of partial differential equations (albeit usually in a reduced, simplified form), but the fundamental ideas of singular perturbation theory are essentially the same as those developed for ordinary differential equations. The only adjustment, because the solution will now sit in a domain of two or more dimensions, is that an appropriate scaling may apply, for example, in only one direction and not in the others, or in time and not in space.

In this chapter we will examine some fairly straightforward problems that are represented by partial differential equations, starting with an example of a regular problem. The approach that we adopt will emphasise how the methods for ordinary differential equations carry over directly to partial differential equations. In addition, we will take the opportunity to write a little more about more advanced aspects of the solution of ordinary differential equations, in part as a preparation for the very powerful and general methods introduced in Chapter 4.

3.1 A REGULAR PROBLEM

A simple, classical problem in elementary fluid mechanics is that of uniform flow of an incompressible, inviscid fluid past a circle. (This is taken as a two-dimensional model for a circular cylinder placed in the uniform flow.) Represented as a *complex potential* (w), the solution of this problem can be written as

$$w(z) = \phi(x, y) + \mathrm{i}\psi(x, y) = U\left(z + \frac{a^2}{z}\right),$$

where ϕ is the velocity potential, ψ the stream function, U the constant speed of the uniform flow (moving parallel to the x-axis) and a is the radius of the circle, centred at the origin. In terms of complex potentials, this solution is constructed from the potential for the uniform flow (Uz) and that for a dipole at the origin (Ua^2/z); the complex variable is $z = x + \mathrm{i}y = r\mathrm{e}^{\mathrm{i}\theta}$. Both ϕ and ψ satisfy Laplace's equation in two dimensions:

$$\nabla^2 u = 0 \quad \text{i.e.} \quad \frac{\partial^2 u}{\partial x^2} + \frac{\partial^2 u}{\partial y^2} = 0 \quad \text{or} \quad \frac{\partial^2 u}{\partial r^2} + \frac{1}{r}\frac{\partial u}{\partial r} + \frac{1}{r^2}\frac{\partial^2 u}{\partial \theta^2} = 0$$

and, if we elect to work with the stream function (as is usual), then we have

$$\psi(r, \theta) = U\left(r - \frac{a^2}{r}\right)\sin\theta, \quad r \geq a, \tag{3.1}$$

expressed in plane polar coordinates. We now formulate a variant of this problem: uniform flow past a slightly distorted circle.

Let the distorted circle be represented by

$$r = r_0(\theta) = a\left[1 + \varepsilon f(\theta)\right], \tag{3.2}$$

where ε will be our small parameter; in terms of $\psi(r, \theta; \varepsilon)$, the problem is to solve

$$\psi_{rr} + \frac{1}{r}\psi_r + \frac{1}{r^2}\psi_{\theta\theta} = 0 \tag{3.3}$$

(where subscripts denote partial derivatives) with

$$\frac{\psi}{r} \to U\sin\theta \quad \text{as } r \to \infty \tag{3.4}$$

and

$$\psi\left[a(1 + \varepsilon f(\theta)), \theta; \varepsilon\right] = 0. \tag{3.5}$$

The condition (3.4) ensures that there is the prescribed uniform flow at infinity i.e. $\psi \sim Ur\sin\theta$ as $r \to \infty$ (see (3.1)), and (3.5) states that the surface of the distorted circle is a streamline (designated $\psi = 0$) of the flow. We see, immediately, that there is a small complication here: ε is embedded in the second boundary condition, (3.5). In order to use our familiar methods, we must first reformulate this condition.

We assume (and this must be checked at the conclusion of the calculation) that ψ on $r = r_0(\theta)$ can be expanded as a Taylor series about $r = a$ i.e.

$$\psi(a, \theta; \varepsilon) + \varepsilon a f(\theta)\psi_r(a, \theta; \varepsilon) + \tfrac{1}{2}\left[\varepsilon a f(\theta)\right]^2 \psi_{rr}(a, \theta; \varepsilon) + \cdots = 0. \tag{3.6}$$

Now the problem—albeit *via* a boundary condition—contains the parameter ε in a form which suggests that we may seek a solution for ψ based on the asymptotic sequence $\{\varepsilon^n\}$. Thus we write

$$\psi(r, \theta; \varepsilon) \sim \sum_{n=0}^{\infty} \varepsilon^n \psi_n(r, \theta)$$

and then
$$\psi_{nrr} + \frac{1}{r}\psi_{nr} + \frac{1}{r^2}\psi_{n\theta\theta} = 0, \quad n = 0,\ 1,\ 2\ldots,$$

with
$$\psi_0/r \to U \sin\theta \quad \text{and} \quad \psi_n \to 0\ (n = 1,\ 2, \ldots) \quad \text{as } r \to \infty$$

and from (3.6):

$$\psi_0(a, \theta) = 0; \quad \psi_1(a, \theta) + a f(\theta)\psi_{0r}(a, \theta) = 0,$$

and so on. The problem for $\psi_0(r, \theta)$ is precisely that for the undistorted circle, so

$$\psi_0(r, \theta) = U\left(r - \frac{a^2}{r}\right)\sin\theta,$$

as given in (3.1).

The problem for $\psi_1(r, \theta)$ now becomes that of finding a solution of Laplace's equation which satisfies

$$\psi_1(a, \theta) = -a f(\theta)\psi_{0r}(a, \theta) = -2aUf(\theta)\sin\theta$$

and $\psi_1 \to 0$ as $r \to \infty$. The most natural way to proceed is to represent $f(\theta)$ as a Fourier Series, and then a solution for ψ_1 follows directly by employing the method of separation of variables. As a particularly simple example of this, let us suppose that $f(\theta) = \cos\theta$, and so

$$\psi_1(a, \theta) = -aU\sin 2\theta;$$

the relevant solution is then

$$\psi_1(r, \theta) = -\frac{a^3 U}{r^2}\sin 2\theta$$

and so we have

$$\psi(r,\theta;\varepsilon) \sim U\left(r - \frac{a^2}{r}\right)\sin\theta - \varepsilon\frac{a^3 U}{r^2}\sin 2\theta. \tag{3.7}$$

This two-term asymptotic expansion is clearly uniformly valid for $\forall\theta$ and for $r \geq a$, and it is also analytic in this domain, so the use of the Taylor expansion to generate (3.6) is justified. It is left as an exercise (Q3.1) to find the next term in this asymptotic expansion and then to discuss further the validity of the expansion; it is indeed uniformly valid. A few other examples of regular expansions obtained from problems posed using partial differential equations can be found in Q3.2 and Q3.3.

This exercise has demonstrated that, as with analogous problems based on ordinary differential equations, we may encounter asymptotic expansions that are essentially uniformly valid i.e. the problem is *regular*. However, this is very much a rarity: most problems that we meet, and that are of interest, turn out to be *singular* perturbation problems. We now discuss this aspect, in the context of partial differential equations, and highlight the two main types of non-uniformity that can arise.

3.2 SINGULAR PROBLEMS I

The most straightforward type of non-uniformity, as we have seen for ordinary differential equations, arises when the asymptotic expansion that has been obtained breaks down and thereby leads to the introduction of a new, scaled variable. This situation is typical of some wave propagation problems, for which an asymptotic expansion valid near the initial data becomes non-uniform for later times/large distances. Indeed, the general structure of such problems is readily characterised by an expansion of the form

$$u(x,t;\varepsilon) \sim f(x-ct) + \varepsilon\{tg(x-ct) + h(x-ct)\} \tag{3.8}$$

where c is the speed of the wave and the functions f, g and h are bounded (and typically well-behaved, often decaying for $|x-ct| \to \infty$). However, for a solution defined in $t \geq 0$ (which is expected in wave problems), we clearly have a breakdown when $\varepsilon t = \mathrm{O}(1)$, irrespective of the value (size) of $(x-ct)$. Thus we would need to examine the problem in the *far-field*, defined by the new variables $\tau = \varepsilon t$, $\xi = x - ct$. (In this example, we have used $x - ct$ $(c > 0)$ for right-running waves; correspondingly, for left-running waves, we would work with $x + ct$.) We present an example of this type of singular perturbation problem.

E3.1 Nonlinear, dispersive wave propagation

A model equation which describes small-amplitude, weakly dispersive water waves can be written as

$$u_{tt} - u_{xx} = \varepsilon\left(u_{xx} - 3u^2\right)_{xx}, \quad -\infty < x < \infty, \quad t \geq 0, \tag{3.9}$$

where we have again used subscripts to denote partial derivatives, and here the speed (associated with the left side of the equation) is one. (This equation is usually called the *Boussinesq equation* and it happens to be one of the equations that is *completely integrable*, in the sense of *soliton theory*, for all $\varepsilon > 0$; see Johnson, 1997, and Drazin & Johnson, 1992.) Our intention is to find an asymptotic solution of (3.9), valid as $\varepsilon \to 0^+$, subject to the initial data

$$u(x, 0; \varepsilon) = f(x), \quad u_t(x, 0; \varepsilon) = -f(x) \tag{3.10}$$

where $f(x) \to 0$ as $|x| \to \infty$ (and, further, all relevant derivatives of $f(x)$ satisfy this same requirement).

The equation for $u(x, t; \varepsilon)$ (where u is the amplitude of the wave), on using our familiar 'iteration' argument, suggests that we may seek a solution in the form

$$u(x, t; \varepsilon) \sim \sum_{n=0}^{\infty} \varepsilon^n u_n(x, t), \tag{3.11}$$

for some xs (distance) and some ts (time). Thus we obtain from (3.9)

$$u_{0tt} - u_{0xx} + \varepsilon \left\{u_{1tt} - u_{1xx} - \left[u_{0xx} - 3u_0^2\right]_{xx}\right\} + \cdots = 0$$

(where '$= 0$' means zero to all orders in ε), which gives

$$u_{0tt} - u_{0xx} = 0; \quad u_{1tt} - u_{1xx} = \left(u_{0xx} - 3u_0^2\right)_{xx}, \tag{3.12a,b}$$

and so on. The initial data, (3.10), then requires

$$u_0(x, 0) = f(x), \quad u_{0t}(x, 0) = -f(x) \tag{3.13}$$

and

$$u_n(x, 0) = u_{nt}(x, 0) = 0, \; n = 1, 2 \ldots. \tag{3.14}$$

Equation (3.12a) is the classical wave equation with the general solution (*d'Alembert's solution*)

$$u(x, t) = F(x - t) + G(x + t),$$

for arbitrary functions F and G; application of the initial data for u_0 (given in (3.13)) then produces

$$u_0(x, t) = f(x - t). \tag{3.15}$$

(We note, in this example, that the particular initial data which we have been given produces a wave moving only to the right (with speed 1), at this order.)

The equation for u_1, (3.12b), is most conveniently written in terms of the *characteristic variables* $\xi = x - t$, $\eta = x + t$, so that we have the operator identities

$$\frac{\partial}{\partial x} \equiv \frac{\partial}{\partial \xi} + \frac{\partial}{\partial \eta} \quad \text{and} \quad \frac{\partial}{\partial t} \equiv \frac{\partial}{\partial \eta} - \frac{\partial}{\partial \xi}.$$

Thus (3.12b) becomes

$$-4u_{1\xi\eta} = \left(\frac{\partial}{\partial \xi} + \frac{\partial}{\partial \eta}\right)^4 u_0 - 3\left(\frac{\partial}{\partial \xi} + \frac{\partial}{\partial \eta}\right)^2 \left(u_0^2\right) = f^{iv}(\xi) - 6\{f(\xi)f'(\xi)\}'$$

which can be integrated directly to give

$$u_1 = \eta\left\{\tfrac{3}{2}f(\xi)f'(\xi) - \tfrac{1}{4}f'''(\xi)\right\} + J(\xi) + K(\eta),$$

where J and K are arbitrary functions. The initial data, (3.14), requires

$$J(x) + K(x) + x\left\{\tfrac{3}{2}f(x)f'(x) - \tfrac{1}{4}f'''(x)\right\} = 0$$

and

$$K'(x) - J'(x) + \tfrac{3}{2}f(x)f'(x) - \tfrac{1}{4}f'''(x) - x\left\{\tfrac{3}{2}f(x)f'(x) - \tfrac{1}{4}f'''(x)\right\}' = 0$$

or

$$K(x) - J(x) + \tfrac{3}{2}f^2 - \tfrac{1}{2}f'' - x\left\{\tfrac{3}{2}ff' - \tfrac{1}{4}f'''\right\} = 2A,$$

where A is an arbitrary constant. These two equations enable us to find J and K as

$$K(x) = \tfrac{1}{4}f'' - \tfrac{3}{4}f^2 + A, \quad J(x) = \tfrac{3}{4}f^2 - \tfrac{1}{4}f'' - x\left\{\tfrac{3}{2}ff' - \tfrac{1}{4}f'''\right\} - A;$$

hence the solution for u_1 becomes

$$u_1 = 2t\left\{\tfrac{3}{2}f(\xi)f'(\xi) - \tfrac{1}{4}f'''(\xi)\right\} + \tfrac{3}{4}\left\{f^2(\xi) - f^2(\eta)\right\} + \tfrac{1}{4}\{f''(\eta) - f''(\xi)\}, \qquad (3.16)$$

where we have written $\eta - \xi = x + t - (x - t) = 2t$. For convenience, let us set

$$H(x) = \tfrac{3}{4}f^2(x) - \tfrac{1}{4}f''(x),$$

then we have the asymptotic solution (to this order):

$$u(x, t; \varepsilon) \sim f(x - t) + \varepsilon\left\{2tH'(x - t) + H(x - t) - H(x + t)\right\}. \qquad (3.17)$$

This two-term asymptotic expansion, (3.17), contains terms f, H and H', all of which are bounded and decay as their arguments approach infinity (because of our given $f(x)$), and so there is no non-uniformity associated with these. However, if we follow the right-going wave (by selecting any $x - t = $ constant) then, as t increases indefinitely, we will encounter a breakdown when $\varepsilon t = \mathrm{O}(1)$. This leads us to introduce a new variable $\tau = \varepsilon t$, and otherwise we may use $\xi = x - t$ (because we are following the right-running wave) and we observe that no scaling is associated with this variable. Thus we transform from (x, t) variables (the *near-field*) to (ξ, τ) variables (the *far-field*), i.e.

$$\frac{\partial}{\partial x} \equiv \frac{\partial}{\partial \xi} \quad \text{and} \quad \frac{\partial}{\partial t} \equiv \varepsilon \frac{\partial}{\partial \tau} - \frac{\partial}{\partial \xi}, \tag{3.18}$$

and so our original equation, (3.9), becomes

$$\varepsilon U_{\tau\tau} - 2U_{\tau\xi} = \left(U_{\xi\xi} - 3U^2\right)_{\xi\xi}, \tag{3.19}$$

where $$u(\xi + \tau/\varepsilon, \tau/\varepsilon; \varepsilon) \equiv U(\xi, \tau; \varepsilon) = \mathrm{O}(1).$$

The nature of the appearance of ε in this equation is identical to the original equation, suggesting that again we may seek a solution in the form

$$U(\xi, \tau; \varepsilon) \sim \sum_{n=0}^{\infty} \varepsilon^n U_n(\xi, \tau)$$

which gives $$2U_{0\tau\xi} + \left(U_{0\xi\xi} - 3U_0^2\right)_{\xi\xi} = 0, \tag{3.20}$$

and so on. This equation can be integrated once in ξ and, when we invoke decay conditions at infinity (i.e. U_0, $U_{0\tau}$, $U_{0\xi}$, $U_{0\xi\xi}$ all $\to 0$ as $|\xi| \to \infty$, which mirrors our given conditions on $f(x)$), we obtain

$$2U_{0\tau} - 6U_0 U_{0\xi} + U_{0\xi\xi\xi} = 0. \tag{3.21}$$

This equation is very different from our previous leading-order equation (in the near-field): that was simply the classical wave equation. Our dominant equation in the far-field, (3.21), describes the time evolution of the wave in terms of the wave's nonlinearity ($U_0 U_{0\xi}$) and its dispersive character ($U_{0\xi\xi\xi}$). Equation (3.21) is a variant of the famous Korteweg-de Vries (KdV) equation; its solutions, and method of solution, initiated (from the late 1960s) the important studies in soliton theory. For solutions that decay as $|\xi| \to \infty$, it can be demonstrated that the later terms (U_n) in the asymptotic expansion contribute uniformly small corrections to U_0, as $\tau \to \infty$. (This is a far-from-trivial exercise and is not undertaken in this text; the interested reader who wishes to explore this further should consult any good text on wave propagation e.g. Whitham, 1974.)

In summary, we have seen that this example (which we have worked through rather carefully) describes a predominantly *linear* wave in the near-field (where $t = \mathrm{O}(1)$ or smaller) but, for $t = \mathrm{O}(\varepsilon^{-1})$ (the far-field), the wave is, to leading order, described by a *nonlinear* equation. Equation (3.21) can be solved exactly and, further, we may impose initial data at $\tau = 0$; in terms of matching (first term to first term), this is equivalent to solving (3.21) with $U_0 \to F(\xi)$ as $\tau \to 0$ (where F is a suitable arbitrary function) and hence, for the right-going wave, we select $F(\xi) = f(\xi)$ and the solution $U \sim U_0(\xi, \tau)$ is uniformly valid for $\tau \geq 0$ i.e. $t \geq 0$. This uniform validity generated by the far-field solution is an unlooked-for bonus, but it should be remembered that the basic expansion (3.11) is not uniformly valid, and so we certainly have a singular perturbation problem.

Other examples of wave propagation problems, which exhibit a breakdown and consequent rescaling, are set as exercises in Q3.4–3.8. In addition, we present one further example which embodies this same mathematical structure, but which is a little more involved. However, this is a classical problem which should appear in any standard text and, further, it has various different limits that are of practical interest (and some of these will be discussed later; see also Q3.10 and Q3.11).

E3.2 Supersonic, thin-aerofoil theory

We consider irrotational, steady, supersonic flow of a compressible fluid past a (two-dimensional) thin aerofoil. The equations of mass and momentum conservation can be reduced to the single equation

$$(u^2 - a^2)\hat{\phi}_{xx} + 2uv\hat{\phi}_{xy} + (v^2 - a^2)\hat{\phi}_{yy} = 0$$

where a is the local sound speed in the gas, and the velocity is $\mathbf{u} \equiv (u, v) = (\hat{\phi}_x, \hat{\phi}_y)$. The corresponding energy equation (*Bernoulli's equation*) is

$$\frac{1}{2}(u^2 + v^2) + \frac{a^2}{\gamma - 1} = \frac{1}{2}U^2 + \frac{a_0^2}{\gamma - 1}$$

where $\mathbf{u} \to (U, 0)$ and $a \to a_0$ as $x \to -\infty$, and γ is a constant describing the nature of the gas: pressure $\propto$ (density)$^\gamma$ (an *isentropic* gas). The aerofoil is described by $y = F_\pm(x)$, $0 \leq x \leq \ell$, with $F_\pm(0) = F_\pm(\ell) = 0$, where the upper/lower surfaces are denoted by $+/-$. Using U and ℓ to non-dimensionalise the variables, eliminating a^2 and then writing the resulting non-dimensional velocity potential as $(x + \varepsilon\phi)$, we obtain

$$\begin{aligned}\phi_{yy} - (M_0^2 - 1)\phi_{xx} = \varepsilon M_0^2 \big[&\tfrac{1}{2}(\gamma - 1)\left(2\phi_x + \varepsilon\phi_x^2 + \varepsilon\phi_y^2\right)(\phi_{xx} + \phi_{yy}) \\ &+ 2(1 + \varepsilon\phi_x)\,\phi_y\phi_{xy} + \varepsilon\phi_y^2\phi_{yy} + (2 + \varepsilon\phi_x)\phi_x\phi_{xx}\big] \qquad (3.22)\end{aligned}$$

where $M_0 = U/a$ (> 1) is the *Mach number* of the oncoming flow from infinity. The aerofoil is now written as $y = \varepsilon f_\pm(x)$, $0 \leq x \leq 1$, (which defines ε for this problem) and so $\varepsilon \to 0$ ensures that we have a *thin* aerofoil. (Thus $\varepsilon = 0$ implies that there is no aerofoil present, and then the non-dimensional velocity potential is simply x i.e.

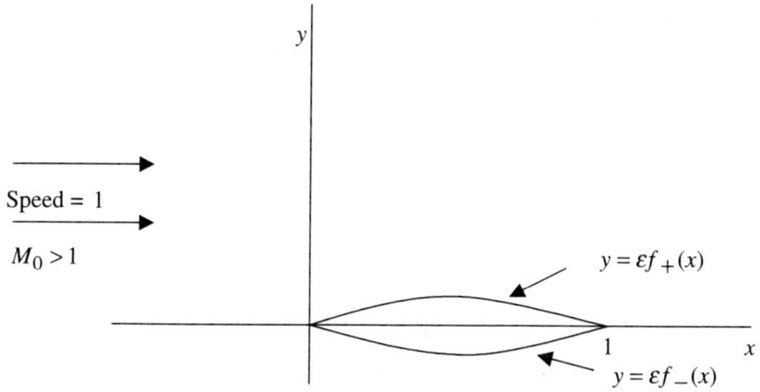

Figure 4. Sketch of the uniform flow (speed = 1, Mach number >1) past the aerofoil $y = \varepsilon f_\pm(x)$, $0 \le x \le 1$.

$\mathbf{u} \equiv (1, 0)$ everywhere.) The non-dimensional version of the physical configuration is shown in figure 4. Finally, the boundary conditions are

$$\phi \to 0 \quad \text{as } x \to -\infty \tag{3.23}$$

and

$$\frac{v}{u} = \frac{\varepsilon\phi_y}{1 + \varepsilon\phi_x} = \varepsilon f'_\pm(x) \quad \text{on} \quad y = \varepsilon f_\pm(x) \quad \text{for } 0 \le x \le 1, \tag{3.24}$$

where the first, (3.23), ensures that, upstream, we have only the given uniform flow ($u \to 1$ with $M_0 > 1$), and the second, (3.24), states that the surface of the aerofoil is a streamline of the flow.

Equation (3.22), with $\varepsilon \to 0$ and $M_0 > 1$, is predominantly a wave equation (but here expressed in spatial variables, x and y); the form of this equation suggests that we seek an asymptotic solution

$$\phi(x, y; \varepsilon) \sim \sum_{n=0}^{\infty} \varepsilon^n \phi_n(x, y) \tag{3.25}$$

and then we obtain

$$\phi_{0yy} - \left(M_0^2 - 1\right)\phi_{0xx} = 0; \tag{3.26}$$

$$\phi_{1yy} - \left(M_0^2 - 1\right)\phi_{1xx} = M_0^2[(\gamma - 1)\phi_{0x}(\phi_{0xx} + \phi_{0yy}) + 2\phi_{0y}\phi_{0xy} + 2\phi_{0x}\phi_{0xx}], \tag{3.27}$$

and so on. Correspondingly, the boundary conditions give

$$\phi_n(x, y) \to 0 \quad \text{as } x \to -\infty, \quad n = 0, 1, 2 \ldots \tag{3.28}$$

and

$$\phi_{0y} + \varepsilon\phi_{1y} + \cdots = (1 + \varepsilon\phi_{0x} + \cdots) f'_\pm(x) \quad \text{on} \quad y = \varepsilon f_\pm(x), \quad 0 \le x \le 1.$$

This second condition is rewritten as an evaluation on $y = 0$, by assuming that ϕ possesses a Taylor expansion (cf. §3.1), so we now obtain

$$\phi_{0y} + \varepsilon f_{\pm}(x)\phi_{0yy} + \varepsilon\phi_{1y} + \cdots = (1 + \varepsilon\phi_{0x} + \cdots) f'_{\pm}(x) \quad \text{on} \quad y = 0, \quad 0 \le x \le 1$$

or

$$\left.\begin{aligned} \phi_{0y} &= f'_{\pm}(x) \\ \phi_{1y} + f_{\pm}(x)\phi_{0yy} &= f'_{\pm}(x)\phi_{0x} \end{aligned}\right\} \quad \text{on} \quad y = 0, \quad 0 \le x \le 1, \tag{3.29a,b}$$

and so on. It is not possible to solve, in any simple compact way, for both the upper and lower surfaces together (as will become clear), so we will consider only the upper surface, $y = \varepsilon f_+(x)$. (The problem for the lower surface follows a similar, but different, construction.)

The general solution of equation (3.26), for the case of supersonic flow ($M_0 > 1$), is given by d'Alembert's solution:

$$\phi_0(x, y) = F(x - \beta y) + G(x + \beta y),$$

where $\beta = \sqrt{M_0^2 - 1}$ and F and G are arbitrary functions. The contribution to the solution from G, in the upper half-plane, extends into $x < 0$ and so, in order to satisfy (3.23), we must have $G \equiv 0$. (The contribution from F extends into $x > 0$ for $y > 0$.) Condition (3.29a) (upper surface) now requires that

$$-\beta F'(x) = \begin{cases} f'_+(x), & 0 \le x \le 1 \\ 0 & \text{otherwise} \end{cases}$$

and so $F(x) = -f_+(x)/\beta$ $(0 \le x \le 1)$ to within an arbitrary constant, which may be ignored in the determination of a velocity potential (because such a constant cannot contribute to the velocity field). Thus we have

$$\phi_0(x, y) = \begin{cases} -\frac{1}{\beta} f_+(x - \beta y), & 0 \le x - \beta y \le 1 \\ 0, & \text{otherwise} \end{cases} \tag{3.30}$$

in $y > 0$. (It will be noted that the corresponding problem in the lower half-plane requires the retention of $G(x + \beta y)$ with $F \equiv 0$.)

The solution of equation (3.27), for ϕ_1, is approached in the same way that we adopted for the similar exercise in E3.1: we introduce characteristic variables, which here are $\xi = x - \beta y$, $\eta = x + \beta y$. With these variables, equation (3.27) becomes

$$\begin{aligned} -4\beta^2\phi_{1\xi\eta} &= \frac{M_0^2}{\beta^2}\left[(\gamma - 1)(1 + \beta^2) f'_+(\xi) f''_+(\xi) + 2\beta^2 f'_+(\xi) f''_+(\xi) + 2 f'_+(\xi) f''_+(\xi)\right] \\ &= \frac{(1 + \gamma)M_0^4}{\beta^2} f'_+(\xi) f''_+(\xi) \end{aligned} \tag{3.31}$$

for $0 \le \xi \le 1$ (where we have used our solution for ϕ_0), and otherwise $\phi_{1\xi\eta} = 0$, which generates only the zero solution when the boundary conditions are applied. The boundary conditions relevant to the solution of (3.31) are (from (3.28) and (3.29b))

$$\phi_1 \to 0 \quad \text{as } x \to -\infty \tag{3.32}$$

and
$$\phi_{1y} = \beta f_+(x) f_+''(x) - \frac{1}{\beta}[f_+'(x)]^2 \quad \text{on} \quad y = 0, \quad 0 \le x \le 1; \tag{3.33}$$

because of the form of this latter boundary condition, it is a little more convenient to find the asymptotic solution for ϕ_y, as $\varepsilon \to 0$. (Of course, from this it is then possible to deduce both ϕ and ϕ_x, if these are required; see later.) From (3.31) we obtain directly that

$$\phi_1 = -\frac{(1+\gamma)M_0^4}{8\beta^4}\eta[f_+'(\xi)]^2 + J(\xi) + K(\eta)$$

where J and K are arbitrary functions, and condition (3.32) then requires that $K \equiv 0$; cf. the solution for ϕ_0. The boundary condition (3.33) is satisfied if

$$-\frac{(1+\gamma)M_0^4}{8\beta^4}\left\{\beta[f_+'(x)]^2 - 2\beta x f_+'(x) f_+''(x)\right\} - \beta J'(x) = \beta f_+(x) f_+''(x) - \frac{1}{\beta}[f_+'(x)]^2$$

and then we obtain directly (because we may use J')

$$\phi_{1y} = \frac{(1+\gamma)M_0^4}{2\beta^2} y f_+'(\xi) f_+''(\xi) + \beta f_+(\xi) f_+''(\xi) - \frac{1}{\beta}[f_+'(\xi)]^2 \tag{3.34}$$

(for $0 \le \xi \le 1$), where we have written $\eta - \xi = 2\beta y$. The two-term asymptotic expansion for ϕ_y is therefore

$$\phi_y \sim f_+'(\xi) + \varepsilon\left\{\frac{(1+\gamma)M_0^4}{2\beta^2} y f_+'(\xi) f_+''(\xi) + \beta f_+(\xi) f_+''(\xi) - \frac{1}{\beta}[f_+'(\xi)]^2\right\} \tag{3.35}$$

as $\varepsilon \to 0$, for $0 \le \xi \le 1$ and $y = O(1)$.

It is clear, for f_+, f_+' and f_+'' bounded, that the expansion (3.35) breaks down where $y = O(\varepsilon^{-1})$—the far-field. (This same property is exhibited by the expansions for ϕ and ϕ_x.) The assumption that we have f_+' and f_+'' bounded (and correspondingly, f_-' and f_-'' for the lower surface), for $0 \le \xi \le 1$, implies that these aerofoils are *sharp* at both the leading and trailing edges—which is certainly what is aimed for in their design and construction. However, if these edges are suitably magnified than it will become evident that a real aerofoil must have rounded edges on some scale. This in turn implies that stagnation points exist, where $\mathbf{u} \equiv (\phi_x, \phi_y) = \mathbf{0}$, and then the asymptotic expansion, (3.35), certainly cannot be uniformly valid near to $\xi = 0$ and $\xi = 1$, even for $y = O(1)$: a boundary-layer-type structure is required near $\xi = 0$ and near $\xi = 1$.

This, and other aspects of compressible flow past aerofoils, will be put to one side in the current discussion (but some additional ideas are addressed in Q3.10 and 3.11). Here, we will work with the mathematical model for the flow in which f_+, f'_+ and f''_+ *are* bounded, so that the only non-uniformity arises as $y \to \infty$.

The solution in the far-field is written in terms of the variables $Y = \varepsilon y$ and $\xi = x - \beta y$, with $\phi(x, y; \varepsilon) = \phi(\xi + \beta Y/\varepsilon, Y/\varepsilon; \varepsilon) \equiv \Phi(\xi, Y; \varepsilon) = \mathrm{O}(1)$. Equation (3.22) now becomes

$$\varepsilon \Phi_{YY} - 2\beta \Phi_{\xi Y} = M_0^2 \left[(\gamma - 1)(1 + \beta^2)\Phi_\xi \Phi_{\xi\xi} + (1 + 2\beta^2)\Phi_\xi \Phi_{\xi\xi} + \mathrm{O}(\varepsilon)\right] \qquad (3.36)$$

where, for simplicity, we have written down only the leading terms on the right-hand side of the equation. We seek a solution

$$\Phi(\xi, Y; \varepsilon) \sim \sum_{n=0}^{\infty} \varepsilon^n \Phi_n(\xi, Y) \qquad (3.37)$$

and then Φ_0 satisfies the nonlinear equation

$$2\beta \Phi_{0\xi Y} + (1 + \gamma) M_0^4 \Phi_{0\xi} \Phi_{0\xi\xi} = 0 \qquad (3.38)$$

or
$$U_{0Y} + \frac{(1 + \gamma) M_0^4}{2\beta} U_0 U_{0\xi} = 0 \quad \text{where} \quad U_0 = \Phi_{0\xi},$$

and note that $u = \phi_x \equiv \Phi_\xi \sim U_0$. The general solution for U_0 is

$$U_0 = H\left(\xi - \frac{(1 + \gamma) M_0^4}{2\beta} Y U_0\right), \qquad (3.39)$$

where H is an arbitrary function; solution (3.39) provides, for general H, an *implicit* representation only for U_0 (which has far-reaching consequences, as we shall see shortly).

We determine H by matching, and this is most easily done by matching ϕ_x and Φ_ξ—but for this we require the expansion of ϕ_x. It is left as an exercise to show that, from (3.35), we obtain

$$\phi_x \sim -\frac{1}{\beta} f'_+(\xi) + \varepsilon \left\{ \left[\frac{1}{\beta} f'_+(\xi)\right]^2 - f_+(\xi) f''_+(\xi) - \frac{(1 + \gamma) M_0^4}{4\beta^4} [f'_+(\xi)]^2 - \frac{(1 + \gamma) M_0^4}{2\beta^3} y f'_+(\xi) f''_+(\xi) \right\} \qquad (3.40)$$

which gives
$$\Phi_\xi \sim -\frac{1}{\beta} f'_+(\xi) - \frac{(1 + \gamma) M_0^4}{2\beta^3} Y f'_+(\xi) f''_+(\xi) \qquad (3.41)$$

when we retain the O(1) term only. From (3.39), we find that

$$\phi_x \sim H(\xi) - \frac{(1+\gamma)M_0^4}{2\beta}\varepsilon\gamma H(\xi)H'(\xi) \tag{3.42}$$

where we have retained the O(1) and O(ε) terms (and we have assumed that H possesses a suitable Taylor expansion). These two expansions, (3.41) and (3.42), match precisely when we select $H(\xi) = -f_+'(\xi)/\beta$, and then

$$\Phi_\xi = U \sim U_0 \quad \text{where} \quad U_0 = -\frac{1}{\beta}f_+'\left(\xi - \frac{(1+\gamma)M_0^4}{2\beta}YU_0\right). \tag{3.43}$$

We conclude this important example by making a few observations.

First, the behaviour of the far-field solution, (3.43), as $Y \to 0$, recovers the near-field solution, and so (to this order, at least) the far-field solution is uniformly valid in $Y \geq 0$ (although our original expansion, (3.25), exhibits a singular behaviour). The solution (3.43), as Y increases, can be completely described by the characteristic lines, in the form

$$\xi - \frac{(1+\gamma)M_0^4}{2\beta}YU_0 = \text{constant},$$

along which $U_0 = -f_+'/\beta = \text{constant}$. These lines are therefore straight, but not parallel; they first intersect for some $Y = Y_c$ (which depends on the details of the function f_+') and then the solution becomes multi-valued in $Y \geq Y_c$—which is unacceptable, unless we revert to an integral form of equation (3.38) and then admit a discontinuous solution. This discontinuity, at a distance O(ε^{-1}) from the surface of the aerofoil, manifests itself in the physical world as heralding the formation of a *shock wave*. It should be no surprise that the characteristic variable plays a significant rôle in the solution of this wave-type (hyperbolic) equation; indeed, the results described here can be obtained by seeking asymptotic expansions for these, rather than for the functions themselves (see Q3.9). A final point, which embodies an important idea, is to note that the near-field solution takes essentially the correct form (to leading order) even for the far-field, in the sense that the solution $\{-f_+'(\xi)/\beta\}$ is replaced by $\{-f_+'(\hat{\xi})/\beta\}$, where $\hat{\xi}$ is the appropriate approximation to the characteristic variable. One way to interpret this is to regard $\{-f_+'(\xi)/\beta\}$ as the correct solution, but that it is in the 'wrong place' i.e. it is not constant along lines $\xi = \text{constant}$, but rather, along lines $\hat{\xi} = \text{constant}$.

We have seen, in these two somewhat routine examples, and the similar problems in the exercises, how the simplest type of singular perturbation problem can arise. The other fundamentally different problem that we may encounter, just as for ordinary differential equations, is where the small parameter multiplies the highest derivative:

the boundary- (or transition-) layer problem. We now turn to an examination of this classical singular perturbation problem in the context of partial differential equations.

3.3 SINGULAR PROBLEMS II

There are two partial differential equation-types that are often encountered with small parameters multiplying the highest derivative(s): the *elliptic* equation (e.g. Laplace's equation) and the *parabolic* equation (e.g. the heat conduction, or diffusion, equation). These two, together with the wave equation (i.e. of *hyperbolic* type) discussed in §3.2, complete the set of the three that constitutes the classification of second-order partial differential equations. The two new equations are exemplified by

$$u_{xx} + \varepsilon u_{yy} = 0 \quad \text{and} \quad u_t = \varepsilon u_{xx},$$

respectively. Of course, the use of singular perturbation methods to solve these particular examples is somewhat redundant, because we are able to solve them exactly (for $\varepsilon > 0$) using standard techniques. Thus we will discuss two simple—but not completely trivial—extensions of these basic equations.

E3.3 Laplace's equation with nonlinearity

We are going to find an asymptotic solution, for $\varepsilon \to 0^+$, of the equation

$$u_{xx} + \varepsilon u_{yy} = \sqrt{\varepsilon} u u_x, \quad 0 \le x \le a, \quad 0 \le y \le b, \tag{3.44}$$

where u is prescribed on the boundary of the region. First, we will make a few general observations about this problem and then obtain some of the details in a particular case.

The appearance of ε in equation (3.44) suggests that we may seek a solution

$$u(x, y; \varepsilon) \sim \sum_{n=0}^{\infty} \varepsilon^{n/2} u_n(x, y), \tag{3.45}$$

which turns out to be consistent with the matching requirements to the boundary layers; thus we obtain

$$u_{0xx} = 0; \quad u_{1xx} = u_0 u_{0x}, \tag{3.46a,b}$$

and so on. Immediately we see that we can find $u_0(x, y)$ which satisfies the given data on $x = 0$ and $x = a$, but only in very special circumstances will this also satisfy the data on $y = 0$ and $y = b$: the solution will (in general) require boundary layers near $y = 0$ and near $y = b$. No such layers exist near $x = 0$ and $x = a$. All this follows from the term εu_{yy}; the term $\varepsilon u u_x$ simply contributes a (small) nonlinear adjustment to the solution. To proceed, let us be given, as an example, the specific data:

$$u(0, y; \varepsilon) = y(1-y) \quad \text{and} \quad u(1, y; \varepsilon) = y \quad \text{both for } 0 \le y \le 1; \tag{3.47}$$

$$u(x, 0; \varepsilon) = \sin(\pi x) \quad \text{and} \quad u(x, 1; \varepsilon) = x^2 \quad \text{both for } 0 \le x \le 1, \tag{3.48}$$

and we will find the first two terms in the solution valid away from the boundary layer, and then the first terms in each boundary-layer solution. From (3.46a), we obtain

$$u_0(x, y) = A(y) + B(y)x$$

where A and B are arbitrary functions; the available data, (3.47), requires that

$$y(1-y) = A(y) \quad \text{and} \quad y = A(y) + B(y)$$

and so we have

$$u_0(x, y) = y(1-y) + y^2 x, \quad \text{for } 0 \le x \le 1.$$

The next term, $u_1(x, y)$, satisfies (3.46b) with

$$u_1(0, y) = u_1(1, y) = 0, \quad 0 \le y \le 1.$$

Equation (3.46b) can be written

$$u_{1xx} = u_0 u_{0x} = \left\{ y(1-y) + y^2 x \right\} y^2$$

and so the solution for u_1 is

$$u_1(x, y) = \frac{1}{2} y^3 (1-y)(x^2 - x) + \frac{1}{6} y^4 (x^3 - x),$$

and we have the asymptotic solution

$$u(x, y; \varepsilon) \sim y(1-y) + y^2 x + \frac{\sqrt{\varepsilon}}{6} \left\{ 3y^3(1-y)(x^2 - x) + y^4(x^3 - x) \right\}. \tag{3.49}$$

This two-term asymptotic expansion does not satisfy the given data on $y = 0$ or on $y = 1$; thus we require (thin) layers near these two boundaries of the domain.

The first stage involves finding the size of the boundary layers; let us introduce $y = \delta(\varepsilon) Y$ with $\delta \to 0$ as $\varepsilon \to 0$, for the boundary layer near $y = 0$. Further, we note that $u = \mathrm{O}(1)$ here and, of course, there is no scaling in x. Thus, with $u(x, \delta Y; \varepsilon) \equiv U(x, Y; \varepsilon) = \mathrm{O}(1)$, equation (3.44) becomes

$$U_{xx} + \frac{\varepsilon}{\delta^2} U_{YY} = \sqrt{\varepsilon} U U_x \tag{3.50}$$

and we must choose $\delta = \mathrm{O}(\sqrt{\varepsilon})$, or simply $\delta = \sqrt{\varepsilon}$, and then seek a solution

$$U(x, Y; \varepsilon) \sim U_0(x, Y)$$

with
$$U_{0xx} + U_{0YY} = 0. \tag{3.51}$$

The matching condition, from (3.98), gives $U \sim \sqrt{\varepsilon}Y$, so

$$U_0 \to 0 \quad \text{as } Y \to \infty. \tag{3.52}$$

The simplest solution available for U_0 involves using the *method of separation of variables* and so, noting the boundary value on $y = 0$, i.e. $Y = 0$, given in (3.48), we write

$$U_0(x, Y) = V(Y)\sin(\pi x).$$

Thus we obtain $\quad V'' - \pi^2 V = 0 \quad \text{or} \quad V(Y) = Ce^{\pi Y} + De^{-\pi Y}$

where C and D are arbitrary constants; a bounded solution valid away from the boundary layer ($Y \to \infty$) requires $C = 0$ and then $U_0 \to 0$ as $Y \to \infty$. The condition on $Y = 0$ is then satisfied if we select $D = 1$, and therefore the solution in the boundary layer near $y = 0$ is

$$U(x, Y; \varepsilon) \sim e^{-\pi Y}\sin(\pi x).$$

Near $y = 1$, the boundary layer is clearly the same thickness, so here we write $y = 1 - \sqrt{\varepsilon}\hat{Y}$ and set $u(x, 1 - \sqrt{\varepsilon}\hat{Y}; \varepsilon) \equiv \hat{U}(x, \hat{Y}; \varepsilon) = \mathrm{O}(1)$, to obtain the equation

$$\hat{U}_{xx} + \hat{U}_{\hat{Y}\hat{Y}} = \sqrt{\varepsilon}\hat{U}\hat{U}_x;$$

cf. equation (3.50) with $\delta = \sqrt{\varepsilon}$. The matching condition this time, again from (3.98), is

$$\hat{U} \sim x;$$

we seek a solution $\hat{U}(x, \hat{Y}; \varepsilon) \sim \hat{U}_0(x, \hat{Y})$ where

$$\hat{U}_{0xx} + \hat{U}_{0\hat{Y}\hat{Y}} = 0$$

with
$$U_0(x, 0) = x^2 \quad \text{and} \quad \hat{U}_0 \to x \quad \text{as } \hat{Y} \to \infty.$$

The solution is obtained altogether routinely by writing $\hat{U}_0 = x + W(x, \hat{Y})$, so that

$$W_{xx} + W_{\hat{Y}\hat{Y}} = 0$$

with
$$W \to 0 \quad \text{as } \hat{Y} \to \infty \quad \text{and} \quad W(x, 0) = x^2 - x,$$

which produces

$$W = \sum_{n=1}^{\infty} A_n \mathrm{e}^{-n\hat{Y}} \sin(nx) \quad \text{with} \quad x^2 - x = \sum_{n=1}^{\infty} A_n \sin(nx)$$

where the constants A_n are determined as the coefficients of the Fourier-series representation of $(x^2 - x)$; this is left as an exercise. Thus the solution in the boundary layer near $y = 1$ is

$$\hat{U}(x, \hat{Y}; \varepsilon) \sim x + \sum_{n=1}^{\infty} A_n e^{-n\hat{Y}} \sin(nx),$$

where the A_ns are known. This completes, for our purposes, the analysis of this boundary-layer problem.

This example has admirably demonstrated, we submit, how the ideas of singular perturbation theory (here exhibited by the existence of boundary layers) developed for ordinary differential equations, carry over directly to partial differential equations. The boundary layers have been required in the y-direction, by virtue of the presence of the parameter ε, but not in the x-direction. Of course, the method of solution has required some knowledge of the methods for solving partial differential equations, but that was to be expected; otherwise no other complications have arisen in the calculations.

For our second example, we consider a physically-based problem: heat conduction, and so the governing equation will now be parabolic.

E3.4 Heat transfer to fluid flowing through a pipe

We consider a circular pipe of radius $r = 1$ (we will describe this problem, from the outset, in terms of non-dimensional variables), extending in a straight line in $x \geq 0$; the rôle played by the length of the pipe will be discussed later. Through the pipe flows a fluid, with a known velocity profile represented by $u = u(r)$, $0 \leq r \leq 1$; the equation for the temperature, $t(r, x; \varepsilon)$, of the fluid is

$$u(r)t_x = \varepsilon \left(t_{rr} + \frac{1}{r} t_r + t_{xx} \right), \tag{3.53}$$

where we have assumed no variation in the angular variable around the pipe. The non-dimensional parameter, $\varepsilon > 0$, is proportional to the thermal conductivity of the fluid. We seek a solution of equation (3.53), for $\varepsilon \to 0$, subject to the boundary conditions

$$t(r, 0; \varepsilon) = t_i(r), \quad 0 \leq r \leq 1, \quad \text{and} \quad t(1, x; \varepsilon) = t_p(x), \quad x \geq 0. \tag{3.54a,b}$$

The fluid enters the pipe (at $x = 0$) with an initial temperature distribution ($t_i(r)$) and the temperature of the pipe wall ($r = 1$) is prescribed along the pipe ($t_p(x)$) i.e. heat is transmitted to (or possibly lost by) the fluid as it flows along the pipe. (Note that, in order to avoid a discontinuity in temperature at the start of the pipe—which is not an essential requirement in the formulation of the problem—then we must have $t_i(1) = t_p(0)$.) Finally, observe that ε multiplies the highest derivative terms, so we must expect a boundary-layer structure.

We will choose the velocity profile to be that associated with a laminar, viscous flow i.e. $u(r) = 1 - r^2$, and then we seek a solution

$$t(r, x; \varepsilon) = t_0(r, x) + \mathrm{o}(1) \quad \text{as } \varepsilon \to 0^+$$

where we have been careful not to commit ourselves to the second term in this asymptotic expansion. Thus we have, from (3.53),

$$(1 - r^2)t_{0x} = 0 \quad \text{and so} \quad t_0(r, x) = t_i(r) \tag{3.54a,b}$$

when we invoke the boundary condition (3.54a); this solution is, apparently, valid for all $x \geq 0$, but (in general) it cannot possibly accommodate the boundary condition on $r = 1$ in $x > 0$, (3.54b). This observation, together with the form of the governing equation, (3.53), suggests that we need a boundary layer near $r = 1$; let us set $r = 1 - \delta(\varepsilon)R$, with $\delta \to 0^+$ as $\varepsilon \to 0^+$, and write $t(1 - \delta R, x; \varepsilon) \equiv T(R, x; \varepsilon) = \mathrm{O}(1)$. Thus equation (3.53) becomes

$$\left(2\delta R - \delta^2 R^2\right) T_x = \varepsilon \left\{ \frac{1}{\delta^2} T_{RR} - \frac{1}{1 - \delta R} \frac{1}{\delta} T_R + T_{xx} \right\}$$

and, as $\delta \to 0$, we must use the balance (using the 'old' term/'new' term concept): $2\delta = \mathrm{O}(\varepsilon/\delta^2)$, which is satisfied by the choice $\delta = \varepsilon^{1/3}$, and so we have

$$\left(2R - \varepsilon^{1/3} R^2\right) T_x = T_{RR} - \frac{\varepsilon^{1/3}}{1 - \varepsilon^{1/3} R} T_R + \varepsilon^{2/3} T_{xx}. \tag{3.56}$$

We seek a solution of this equation in the form

$$T(R, x; \varepsilon) \sim \sum_{n=0}^{\infty} \varepsilon^{n/3} T_n(R, x)$$

so that
$$2RT_{0x} = T_{0RR} \tag{3.57}$$

with $T_0(0, x) = t_p(x)$, $T_0 \to t_i(1)$ as $x \to 0^+$, and a matching condition for $R \to \infty$. Although it is possible to find the appropriate solution of (3.57), satisfying the given boundary conditions, it is somewhat involved and we are likely to lose much of the transparency of the results. Thus we will complete the solution in the special case: constant wall temperature $t_p(x) = 1$, $x > 0$, $t_i(1) \neq 1$, and we will seek a solution in $x > 0$, thereby ignoring the discontinuity that is evident as $x \to 0^+$, $r \to 1^-$.

(This difficulty in the solution can be discussed separately—but not here—or the discontinuity can be replaced by a smooth function that takes $t_p(x)$ from $t_p = 1$, at $x = x_0 > 0$, to $t_p(0) = t_i(1)$.) Now, since

$$t_i(1 - \varepsilon^{1/3} R) \to t_i(1) \quad (\neq 1) \quad \text{as } \varepsilon \to 0,$$

we must find a solution of (3.57), subject to

$$T_0(0, x) = 1, \quad T_0 \to t_i(1) \quad \text{as } R \to \infty \quad \text{and} \quad \text{as } x \to 0^+.$$

To proceed with the solution of this problem, we write $T_0 = t_i(1) + V(R, x)$, and so

$$2RV_x = V_{RR} \quad \text{with} \quad V(0, x) = 1 - t_i(1) = k \quad \text{and} \quad V \to 0 \quad \text{as } R \to \infty \quad \text{and} \quad \text{as } x \to 0^+;$$

this is solved very simply by introducing the *similarity solution*. Set $V(R, x) = F(x^m R)$, for some m, then we find directly that $m = -1/3$ and

$$F(\eta) = A \int_a^{\eta} \exp\left(-\tfrac{2}{9} y^3\right) \mathrm{d}y, \quad \eta = \frac{R}{x^{1/3}},$$

where A and a are arbitrary constants, and then all the boundary conditions are satisfied by

$$F(\eta) = k \int_{\eta}^{\infty} \exp\left(-\tfrac{2}{9} y^3\right) \mathrm{d}y \bigg/ \int_{\eta}^{\infty} \exp\left(-\tfrac{2}{9} y^3\right) \mathrm{d}y.$$

Thus the solution in the boundary layer, $R = \mathrm{O}(1)$, is

$$T_0(R, x) = t_i(1) + \{1 - t_i(1)\} \frac{\int_{R/x^{1/3}}^{\infty} \exp\left(-\tfrac{2}{9} y^3\right) \mathrm{d}y}{\int_0^{\infty} \exp\left(-\tfrac{2}{9} y^3\right) \mathrm{d}y}, \tag{3.58}$$

in this special case. We have the first terms in each of the *outer* region (away from the pipe wall) and the region close to the pipe wall (the boundary layer): (3.55b) and (3.58), respectively.

We have completed all the detailed calculations that we will present, although we make a few concluding comments that highlight some intriguing issues. First, knowing the temperature in the fluid near the pipe wall enables us to find the heat transferred to ($t_i(1) < 1$) or from ($t_i(1) > 1$) the fluid, if that is a property of particular interest. Second, a more technical matter: what is the form of the solution in the outer region, i.e. what terms in the asymptotic expansion must be present in order to match to the boundary-layer solution? To answer this, we need to know the behaviour of the boundary-layer solution as $R \to \infty$ (because we have $R = (1 - r)/\varepsilon^{1/3}$ for $1 - r = \mathrm{O}(1)$ as $\varepsilon \to 0$). For our similarity solution, (3.58), this can be obtained by using a

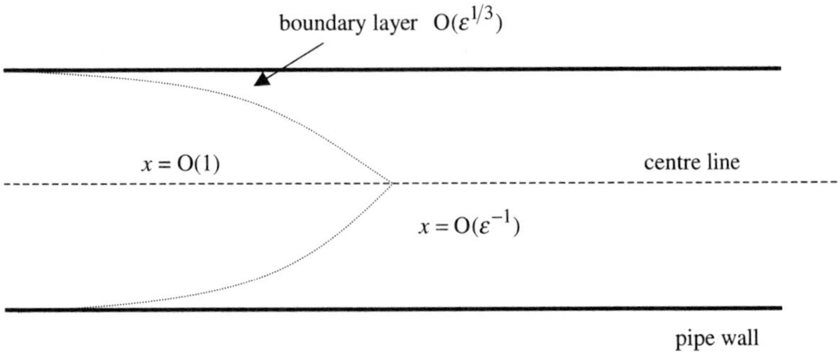

Figure 5. Schematic of the boundary-layer region, which grows from the pipe wall, and the 'fully developed' region for $x = O(\varepsilon^{-1})$ and larger.

suitable integration by parts:

$$\int_\eta^\infty y^2 \frac{1}{y^2} \exp\left(-\tfrac{2}{9}y^3\right) dy = \left[-\frac{3}{2}\frac{1}{y^2}\exp\left(-\tfrac{2}{9}y^3\right)\right]_\eta^\infty - 3\int_\eta^\infty \frac{1}{y^3}\exp\left(-\tfrac{2}{9}y^3\right) dy$$

$$\sim \frac{3}{2}\frac{1}{\eta^2}\exp\left(-\tfrac{2}{9}\eta^3\right) \quad \text{as } \eta = R/x^{1/3} \to \infty.$$

Thus we will need to match to

$$T \sim t \sim t_i(1) + \frac{\{1 - t_i(1)\}}{\int_0^\infty \exp\left(-\frac{2}{9}y^3\right) dy} \frac{3}{2} \frac{(\varepsilon x)^{2/3}}{(1-r)^2} \exp\left\{-\frac{2}{9}\frac{(1-r)^3}{\varepsilon x}\right\} \tag{3.59}$$

as $\varepsilon \to 0^+$, for $x = O(1)$ and $1 - r = O(1)$. That is, the asymptotic expansion valid away from the boundary layer must include the exponentially small term of the form

$$\varepsilon^{2/3} f_0(r, x) \exp\left\{-\frac{g(r, x)}{\varepsilon}\right\},$$

for suitable functions $f_0(r, x)$ and $g(r, x)$. (It turns out that g satisfies a *nonlinear*, first order partial differential equation, and that f_0 is governed by a *linear*, first order partial differential equation, with coefficients dependent on g.) This result has particularly dramatic consequences in the case $t_i(r) = 1, 0 \le r \le 1$, for then the solution away from the pipe wall is apparently exactly $t(r, x; \varepsilon) = 1$. However, the requirement to match to the boundary-layer solution introduces an exponentially small correction to the outer solution—and this is the sole effect of the presence of the different temperature at the pipe wall, at least for $x = O(1)$.

Finally, we address the problem-and there is one-associated with the length of the pipe. As we have just commented (and we will relate all this only to the simple case of the similarity solution with $t_i(r) \equiv 1$), if the total length of the pipe is $x = \ell = O(1)$, as measured in terms of $\varepsilon \to 0$, then the flow away from the wall is $t = 1$, with only exponentially small corrections. The boundary layer remains thin ($O(\varepsilon^{1/3})$) along the full length of the pipe, but note that lines of constant temperature, emanating from the neighbourhood of the wall, are the lines $R/x^{1/3} =$ constant i.e. $R \propto x^{1/3}$. Thus, defining the boundary-layer thickness in terms of a particular temperature, this thickness increases as x increases, although it remains $O(\varepsilon^{1/3})$ for $x = O(1)$. However, if the pipe is so long that $\ell = O(\varepsilon^{-1})$, then the exponential term in (3.59) becomes $O(1)$, and the temperature now has an $O(1)$ correction. In other words, the $O(1)$ temperature at the pipe wall has caused heat to be conducted through the fluid to affect the whole pipe flow, $0 \le r \le 1$. Indeed, we see that the scaling $x = X/\varepsilon$ in our original equation, (3.53), balances dominant terms from *both* sides of the equation for $0 \le r \le 1$: there is no longer a boundary layer at the wall; this is shown schematically in figure 5.

The two examples presented above have shown how the notion of a boundary layer, as developed for ordinary differential equations, is relevant to partial differential equations—and essentially without any adjustment to the method; some similar examples can be found in exercises Q3.12–3.14. We have now seen the two basic types of problem (breakdown and rescale; select a scaling relevant to a layer), although the equations that we have introduced as the vehicles for these demonstrations—quite deliberately—have been relatively uncomplicated. We conclude this section with an example that, ultimately, possesses a simple perturbation structure (as in §3.2), but which involves a set of four, coupled, nonlinear partial differential equations. As before, the purpose of the example is to exhibit the power (and inherent simplicity) of the singular perturbation approach.

E3.5 Unsteady, one-dimensional flow of a viscous, compressible gas

The flow of a compressible gas, with temperature variations and viscosity, is described by the equations

$$u'_{t'} + u'u'_{x'} + \frac{1}{\rho'}p'_{x'} = \frac{4}{3}\frac{\mu}{\rho'}u'_{x'x'}; \tag{3.60}$$

$$\rho'_{t'} + u'\rho'_{x'} + \rho'u'_{x'} = 0; \tag{3.61}$$

$$\frac{1}{\gamma - 1}\left\{p'_{t'} + u'p'_{x'} - \gamma\frac{p'}{\rho'}(\rho'_{t'} + u'\rho'_{x'})\right\} = \frac{4}{3}\mu(u'_{x'})^2 + \kappa T'_{x'x'}; \tag{3.62}$$

$$p' = \rho' R' T', \tag{3.63}$$

which are the equations of momentum, mass conservation, energy and state, respectively. Here, μ is the coefficient of (Newtonian) viscosity, κ the thermal conductivity, R' the gas constant and γ is associated with the isentropic-gas model (see E3.2); we shall take all these parameters to be constant. Any movement of the gas is in the x' direction—we use primes here to denote physical variables—with no variation in other directions, so any disturbance generated in the gas is assumed to propagate (and possibly change) only in x'; t' is time. The speed of the gas is u', and its pressure, density and temperature are p', ρ' and T', respectively.

First, we suppose that the gas in its stationary ($u' = 0$), undisturbed state is described by

$$p' = \hat{p}_0, \quad \rho' = \hat{\rho}_0, \quad T' = \hat{T}_0,$$

all constant. The gas is now disturbed, thereby producing a weak *pressure wave* (often called an *acoustic wave*, although this is usually treated with temperature fixed); the size of the initiating disturbance will be measured by the parameter ε ($\to 0$). We introduce the sound speed, a_0, of the gas in its undisturbed state, defined by

$$a_0^2 = \frac{\gamma \hat{p}_0}{\hat{\rho}_0} = \gamma R' \hat{T}_0 \tag{3.64}$$

and then we move to non-dimensional variables (without the primes) by writing

$$p' = \hat{p}_0 (1 + \varepsilon p), \quad \rho' = \hat{\rho}_0 (1 + \varepsilon \rho), \quad T' = \hat{T}_0 (1 + \varepsilon T), \quad u' = \varepsilon a_0 u;$$

we let a typical or appropriate length scale (e.g. an average wave length) be ℓ, and also define

$$x = x'/\ell, \quad t = a_0 t'/\ell.$$

The governing equations, (3.60)–(3.63), therefore become

$$u_t + \varepsilon u u_x + \frac{1}{\gamma}(1 + \varepsilon \rho)^{-1} p_x = \frac{4}{3}\frac{\varepsilon}{R_e}(1 + \varepsilon \rho)^{-1} u_{xx}; \tag{3.65}$$

$$\rho_t + \varepsilon u \rho_x + (1 + \varepsilon \rho) u_x = 0; \tag{3.66}$$

$$\frac{1}{\gamma(\gamma - 1)}\left\{p_t + \varepsilon u p_x - \gamma \left(\frac{1 + \varepsilon p}{1 + \varepsilon \rho}\right)(\rho_t + \varepsilon u \rho_x)\right\} = \frac{4}{3}\frac{1}{R_e}\varepsilon^2 u_x^2 + \frac{1}{\gamma - 1}\frac{\varepsilon}{R_e}\frac{1}{P_r} T_{xx}; \tag{3.67}$$

$$p = \rho + T + \varepsilon \rho T, \tag{3.68}$$

where the *Reynolds Number* is $\mathrm{R_e} = \varepsilon a_0 \hat{\rho}_0 \ell / \mu$ and the *Prandtl Number* is $\mathrm{P_r} = \mu c_p \kappa$, with $R' = c_p(1 - \gamma^{-1})$. (Note that we have elected to define the speed in the definition of $\mathrm{R_e}$ as εa_0, which is proportional to the scale of the speed generated by the disturbance; a suitable choice of $\mathrm{R_e}$ is a crucial step in ensuring that we obtain the limit of interest.)

We now seek a solution of the set of equations (3.65)–(3.68), for γ, $\mathrm{R_e}$ and $\mathrm{P_r}$ fixed as $\varepsilon \to 0$, by writing

$$q(x, t; \varepsilon) \sim \sum_{n=0}^{\infty} \varepsilon^n q_n(x, t), \tag{3.69}$$

where q (and correspondingly q_n) represents each of p, ρ, T and u. The first terms in each of these asymptotic expansions satisfy the equations

$$u_{0t} + \frac{1}{\gamma} p_{0x} = 0; \quad \rho_{0t} + u_{0x} = 0; \quad p_{0t} - \gamma \rho_{0t} = 0; \quad p_0 = \rho_0 + T_0,$$

which follow from (3.65)–(3.68), respectively. These equations then give

$$u_{0tt} = -\frac{1}{\gamma} p_{0xt} = -\rho_{0xt} = u_{0xx} \quad \text{or} \quad u_{0tt} - u_{0xx} = 0$$

and then, selecting the right-going wave (for simplicity), we have

$$u_0(x, t) = f(x - t), \tag{3.70}$$

for some $f(x)$ at $t = 0$. For a solution-set which recovers the undisturbed state for $f(x) \equiv 0$, we also have

$$p_0 = \gamma u_0 = \gamma f(x - t); \quad \rho_0 = u_0 = f(x - t); \quad T_0 = (\gamma - 1)\rho_0 = (\gamma - 1) f(x - t). \tag{3.71}$$

However, our experience with hyperbolic problems (see §3.2) is that asymptotic expansions like (3.69) are not uniformly valid as t (or x) $\to \infty$ for $x - t = \mathrm{O}(1)$ i.e. in the far-field. Let us investigate the result of the non-uniformity directly, without examining the details of the breakdown (which is left as an exercise).

The variables that we choose to use in the far-field are

$$\xi = x - t \quad \text{and} \quad \tau = \varepsilon t \quad \text{with} \quad q(\xi + \tau/\varepsilon, \tau/\varepsilon; \varepsilon) \equiv Q(\xi, \tau; \varepsilon)$$

for each q, so we have the identities

$$\frac{\partial}{\partial x} \equiv \frac{\partial}{\partial \xi} \quad \text{and} \quad \frac{\partial}{\partial t} \equiv \varepsilon \frac{\partial}{\partial \tau} - \frac{\partial}{\partial \xi};$$

equations (3.65)–(3.68) become (when we retain only those terms relevant to the determination of the dominant contributions to each Q):

$$\varepsilon U_\tau - U_\xi + \varepsilon U U_\xi + \frac{1}{\gamma}(1 + \varepsilon R)^{-1} P_\xi = \frac{4}{3} \frac{\varepsilon}{\mathrm{R_e}} U_{\xi\xi} + \mathrm{O}(\varepsilon^2); \tag{3.72}$$

$$\varepsilon R_\tau - R_\xi + \varepsilon U R_\xi + (1 + \varepsilon R) U_\xi = 0; \tag{3.73}$$

$$\frac{1}{\gamma(\gamma-1)}\left\{\varepsilon P_\tau - P_\xi + \varepsilon U P_\xi + \gamma(1+\varepsilon P - \varepsilon R)R_\xi - \varepsilon\gamma(R_\tau + U R_\xi)\right\}$$
$$= \frac{\varepsilon}{(\gamma-1)\mathrm{R_e P_r}}\mathfrak{T}_{\xi\xi} + \mathrm{O}(\varepsilon^2); \qquad (3.74)$$

$$P = R + \mathfrak{T} + \varepsilon R\mathfrak{T}, \qquad (3.75)$$

where $(p, \rho, T, u) \equiv (P, R, \mathfrak{T}, U)$. Again we expand

$$Q(\xi, \tau; \varepsilon) \sim \sum_{n=0}^{\infty} \varepsilon^n Q_n(\xi, \tau) \qquad (3.76)$$

for each Q; immediately we see that the leading-order terms satisfy

$$U_0 = R_0 = \frac{1}{\gamma}P_0 = \frac{1}{\gamma-1}\mathfrak{T}_0 \qquad (3.77)$$

(cf. (3.71)), but these functions are otherwise unknown.

The terms Q_1 are defined by the equations

$$U_{0\tau} - U_{1\xi} + U_0 U_{0\xi} + \frac{1}{\gamma}(P_{1\xi} - R_0 P_{0\xi}) = \frac{4}{3}\frac{1}{\mathrm{R_e}}U_{0\xi\xi};$$

$$R_{0\tau} - R_{1\xi} + U_0 R_{0\xi} + U_{1\xi} + R_0 U_{0\xi} = 0;$$

$$\frac{1}{\gamma(\gamma-1)}\left\{P_{0\tau} - P_{1\xi} + U_0 P_{0\xi} + \gamma(R_{1\xi} + P_0 R_{0\xi} - R_0 R_{0\xi} - R_{0\tau} - U_0 R_{0\xi})\right\}$$
$$= \frac{1}{(\gamma-1)\mathrm{R_e P_r}}\mathfrak{T}_{0\xi\xi};$$

$$P_1 = R_1 + \mathfrak{T}_1 + R_0\mathfrak{T}_0,$$

from (3.72)–(3.75), respectively, and this last equation is used only to define $\mathfrak{T}_1$. The first three equations involve the combinations of terms:

$$U_{1\xi} - \frac{1}{\gamma}P_{1\xi}; \quad U_{1\xi} - R_{1\xi}; \quad P_{1\xi} - \gamma R_{1\xi},$$

respectively, and so we may eliminate all of U_1, P_1 and R_1 between them, which we do. The resulting single equation involving U_0, P_0, R_0 and $\mathfrak{T}_0$ is written in terms of one function—U_0 say—by using (3.77); this gives the leading term (for U) in the far-field as the solution of

$$U_{0\tau} + \frac{1}{2}(\gamma+1)U_0 U_{0\xi} = \frac{1}{2\mathrm{R_e}}\left(\frac{4}{3} + \frac{\gamma-1}{\mathrm{P_r}}\right)U_{0\xi\xi}. \qquad (3.78)$$

Thus $U_0(\xi, \tau)$ is described by an equation in which the time (τ)-evolution of the solution is controlled by both nonlinearity ($U_0 U_{0\xi}$) and dissipation ($U_{0\xi\xi}$); cf. the

KdV equation, (3.21). Equation (3.78), like the KdV equation, is also an important equation; it is the *Burgers equation* (Burgers, 1948) which can be solved exactly by applying the *Hopf-Cole transformation* (Hopf, 1950; Cole, 1951) which transforms the equation into the heat conduction (diffusion) equation. As with the corresponding KdV problem (in E3.1), the matching condition is $U_0 \to f(\xi)$ as $\tau \to 0$ and, for suitable $f(\xi)$, the asymptotic expansions (3.76) are uniformly valid as $\tau \to \infty$. The solution that we have obtained describes a weak pressure wave moving through the near-field ($t = \mathrm{O}(1)$) and, as t increases into the far-field, the wave-front steepens, but this effect is eventually balanced by the diffusion (when $t = \mathrm{O}(\varepsilon^{-1})$). Finally the wave will settle to some steady-state profile—a profile which is regarded as a model for a shock wave in which the discontinuity is *smoothed*; see Q3.5.

This concludes all that we shall present here, as examples of fairly routine singular perturbation problems in the context of partial differential equations. Further ideas—very powerful ideas—which are applicable to both ordinary and partial differential equations will be developed in the next chapter. We complete this chapter on some further applications by examining two rather more sophisticated problems that involve *ordinary* differential equations. The first employs the asymptotic expansion in a parameter in order to study an important equation in the theory of ordinary differential equations: *Mathieu's equation*. The second develops a technique, which is an extension of one of our earlier problems associated with wave propagation, that enables the asymptotic solution of certain ordinary differential equations to be written in a particularly compact and useful form—indeed, one that exhibits uniform validity when none appears to exist.

3.4 FURTHER APPLICATIONS TO ORDINARY DIFFERENTIAL EQUATIONS

The Mathieu equation, for $x(t)$,

$$\frac{\mathrm{d}^2 x}{\mathrm{d}t^2} + (\delta + \varepsilon \cos t)x = 0, \quad t \geq 0,$$

where δ and ε are constants, has a long and exalted history; it arose first in the work of E-L. Mathieu (1835–1900) on the problem of vibrations of an elliptical membrane. It also applies to the problem of the classical pendulum in which the pivot point is oscillated along a vertical line, one result being that, for certain amplitudes and frequencies of this oscillation (which corresponds to certain δ and ε), the pendulum becomes stable in the *up* position! It is also relevant to some problems in electromagnetic-wave propagation (in a medium with a periodic structure), some electrical circuits with special oscillatory properties and in celestial mechanics. The equation is conventionally analysed using *Floquet theory* (see e.g. Ince, 1956) in which the solution $x = \mathrm{e}^{\alpha t} y(t)$, for α, in general, a complex constant, has $y(t)$ periodic (with period 2π or 4π) for certain $\delta(\varepsilon)$. We will show how some of the properties of this equation are readily accessible, at least in the case $\varepsilon \to 0$.

E3.6 Mathieu's equation for $\varepsilon \to 0$

We consider the equation

$$\ddot{x} + (\delta + \varepsilon \cos t)x = 0, \quad t \geq 0, \tag{3.79}$$

where we use the over-dot to denote the derivative with respect to t; special curves in the (δ, ε)-plane separate stable (oscillatory) from unstable (exponentially growing) solutions: on these curves there exist both oscillatory and linearly growing solutions. We will seek these curves in the case $\varepsilon \to 0$.

First, with $\varepsilon = 0$, we obtain

$$\ddot{x} + \delta x = 0$$

which has periodic solutions, with period 2π or 4π, only if $\delta = n^2/4$ $(n = 0, 1, 2\ldots)$ although $n = 0$ might be thought an unimportant exceptional case. The form of the equation suggests that we should seek a solution in the form

$$x(t;\varepsilon) \sim \sum_{m=0}^{\infty} \varepsilon^m x_m(t) \quad \text{and} \quad \delta(\varepsilon) \sim \frac{n^2}{4} + \sum_{m=1}^{\infty} \varepsilon^m \delta_m \tag{3.80a,b}$$

and invoke the requirement that each $x_m(t)$ be periodic; each δ_m is a constant independent of ε. We will explore the cases $n = 0$ and $n = 1$ (and the case $n = 2$ is left as an exercise).

(a) Case n = 0

Equation (3.79), with (3.80a,b), becomes

$$\ddot{x}_0 + \varepsilon \ddot{x}_1 + \varepsilon^2 \ddot{x}_2 + \cdots + (\varepsilon\delta_1 + \varepsilon^2\delta_2 + \cdots + \varepsilon \cos t)(x_0 + \varepsilon x_1 + \cdots) = 0$$

where, as is our convention, '= 0' means zero to all orders in ε. Thus we have the sequence of equations

$$\ddot{x}_0 = 0; \quad \ddot{x}_1 + (\delta_1 + \cos t)x_1 = 0; \quad \ddot{x}_2 + \delta_2 x_0 + (\delta_1 + \cos t)x_1 = 0, \tag{3.81a,b,c}$$

and so on. The only periodic solution for x_0—and trivially so—is $x_0 =$ constant, which we will normalise to $x_0 = 1$. Then equation (3.81b) becomes

$$\ddot{x}_1 = -(\delta_1 + \cos t)$$

and the solution for $x_1(t)$ which is periodic requires $\delta_1 = 0$; thus

$$x_1(t) = \cos t + A,$$

where A is an arbitrary constant. Finally, from equation (3.81c), we obtain

$$\ddot{x}_2 = -\delta_2 - (A + \cos t)\cos t = -\left(\delta_2 + \tfrac{1}{2}\right) - A\cos t - \tfrac{1}{2}\cos 2t$$

which has a periodic solution for $x_2(t)$ only if $\delta_2 = -1/2$, and so the method proceeds. Thus the *transitional curve* in the (δ, ε)-plane, along which a 2π- or 4π-periodic solution exists, is

$$\delta(\varepsilon) \sim -\frac{1}{2}\varepsilon^2 \quad (n = 0). \tag{3.82}$$

(b) Case n = 1

This time, although the general procedure is essentially the same, the appearance of non-trivial periodic solutions at leading order complicates things somewhat. Equation (3.79) now becomes

$$\ddot{x}_0 + \varepsilon\ddot{x}_1 + \varepsilon^2\ddot{x}_2 + \cdots + \left(\tfrac{1}{4} + \varepsilon\delta_1 + \varepsilon^2\delta_2 + \cdots + \cos t\right)(x_0 + \varepsilon x_1 + \varepsilon^2 x_2 + \cdots) = 0,$$

and thus we obtain the equations

$$\ddot{x}_0 + \tfrac{1}{4}x_0 = 0; \quad \ddot{x}_1 + \tfrac{1}{4}x_1 + (\delta_1 + \cos t)x_0 = 0; \tag{3.83a,b}$$

$$\ddot{x}_2 + \tfrac{1}{4}x_2 + (\delta_1 + \cos t)x_1 + \delta_2 x_0 = 0, \tag{3.83c}$$

and so on. The general solution of equation (3.83a) is simply

$$x_0(t) = A\sin\left(\tfrac{1}{2}t\right) + B\cos\left(\tfrac{1}{2}t\right)$$

for arbitrary constants A and B; we may now proceed, collecting all terms proportional to A, and correspondingly to B, but it is far easier (and more usual) to treat these two sets of terms separately. Thus we select $A = 1$, $B = 0$, and $A = 0$, $B = 1$; this will generate two transitional curves: one associated with $x_0 = \sin(\frac{1}{2}t)$ and one with $x_0 = \cos(\frac{1}{2}t)$, which is the usual presentation adopted. Let us choose $x_0(t) = \sin(\frac{1}{2}t)$, then (3.83b) can be written

$$\ddot{x}_1 + \tfrac{1}{4}x_1 = -(\delta_1 + \cos t)\sin\left(\tfrac{1}{2}t\right) = -\delta_1\sin\left(\tfrac{1}{2}t\right) - \tfrac{1}{2}\left\{\sin\left(\tfrac{3}{2}t\right) - \sin\left(\tfrac{1}{2}t\right)\right\}$$

and a periodic solution for x_1 requires $\delta_1 = 1/2$ (because otherwise there would be a term proportional to $t\cos(\frac{1}{2}t)$); therefore we obtain

$$x_1(t) = C\sin\left(\tfrac{1}{2}t\right) + D\cos\left(\tfrac{1}{2}t\right) + \tfrac{1}{4}\sin\left(\tfrac{3}{2}t\right)$$

where C and D are arbitrary constants. Finally, from (3.83c), we have the equation

$$\begin{aligned}\ddot{x}_2 + \tfrac{1}{4}x_2 &= -(\delta_1 + \cos t)\left\{C\sin\left(\tfrac{1}{2}t\right) + D\cos\left(\tfrac{1}{2}t\right) + \tfrac{1}{4}\sin\left(\tfrac{3}{2}t\right)\right\} - \delta_2\sin\left(\tfrac{1}{2}t\right)\\ &= -\tfrac{1}{2}\left\{C\sin\left(\tfrac{1}{2}t\right) + D\cos\left(\tfrac{1}{2}t\right) + \tfrac{1}{4}\sin\left(\tfrac{3}{2}t\right)\right\} - \delta_2\sin\left(\tfrac{1}{2}t\right)\\ &\quad - \tfrac{1}{2}C\left\{\sin\left(\tfrac{3}{2}t\right) - \sin\left(\tfrac{1}{2}t\right)\right\} - \tfrac{1}{2}D\left\{\cos\left(\tfrac{3}{2}t\right) + \cos\left(\tfrac{1}{2}t\right)\right\}\\ &\quad - \tfrac{1}{8}\left\{\sin\left(\tfrac{5}{2}t\right) + \sin\left(\tfrac{1}{2}t\right)\right\}\end{aligned}$$

for which periodic solutions, $x_2(t)$, require $\delta_2 = -1/8$ (and $D = 0$). Thus, to this order, the transitional curve for $x_0(t) = \sin(\tfrac{1}{2}t)$ is

$$\delta(\varepsilon) \sim \tfrac{1}{4} + \tfrac{1}{2}\varepsilon - \tfrac{1}{8}\varepsilon^2 \quad (n = 1);$$

it is left as an exercise to show that, with the choice $x_0(t) = \cos(\tfrac{1}{2}t)$, then

$$\delta(\varepsilon) \sim \tfrac{1}{4} - \tfrac{1}{2}\varepsilon - \tfrac{1}{8}\varepsilon^2 \quad (n = 1).$$

In summary, we have the stability boundaries given by

$$\delta(\varepsilon) \sim -\tfrac{1}{2}\varepsilon^2 \quad (n = 0); \quad \delta(\varepsilon) \sim \tfrac{1}{4} \pm \tfrac{1}{2}\varepsilon - \tfrac{1}{8}\varepsilon^2 \quad (n = 1) \quad \text{as } \varepsilon \to 0.$$

Set as an exercise, the case $n = 2$ can be found in exercise Q3.15. A corresponding calculation to those described here, but now formulated in a way consistent with Floquet theory, is set in Q3.16. More information about Mathieu equations and functions, and their applications, is available in the excellent text: McLachlan (1964).

For our final discussion in this chapter, we will incorporate the idea introduced at the end of E3.2, namely, a 'correct' solution in the 'wrong place', but applied here to ordinary differential equations. We will describe the *method of strained coordinates*, which has a long history; it was used first, in an explicit way, by Poincaré (1892), but other authors had certainly been aware if the idea earlier, in one form or another. Some authors refer to this as the *PLK method* after Poincaré, Lighthill (1949) and Kuo (1953). The idea is exactly as mentioned above: the solution $y(x;\varepsilon) \sim y_0(x) + \varepsilon y_1(x)$ (say), which is not uniformly valid, is made so by writing

$$y(x;\varepsilon) \sim y_0\{\hat{x}(x;\varepsilon)\} + \varepsilon\hat{y}_1(\hat{x})$$

where $\hat{x}(x;\varepsilon)$ is a suitably chosen *strained coordinate*. Of course, only relatively special problems have solutions that possess this structure, but it is regarded as a significant improvement—over matched expansions—when it occurs. Indeed, we have met a problem of this type in Chapter 1: E1.1, our very first example. There we found that a straightforward asymptotic expansion led to

$$x(t;\varepsilon) \sim \sin t + \tfrac{1}{2}\varepsilon(t\cos t - \sin t)$$

(see equation (1.7)), but the use of the 'strained' coordinate $\tau = t\sqrt{1+\varepsilon}$ (see (1.8)) immediately removes the non-uniformity that is otherwise present as $t \to \infty$. Further, the leading term above ($\sim \sin t$) is essentially correct, but 'in the wrong place'; if t is replaced by τ, then $\sin \tau$ becomes a uniformly valid first approximation as $\tau \to \infty$. We will show how this method develops for a particular type of equation (first examined by Lighthill, 1949 & 1961; similar examples have been discussed by Carrier, 1953 & 1954).

E3.7 An ordinary differential equation with a strained coordinate asymptotic structure

We consider the problem

$$(x+\varepsilon y)\frac{\mathrm{d}y}{\mathrm{d}x} + q(x;\varepsilon)y = 0, \quad 0 \le x \le 1, \tag{3.84}$$

with

$$y(1;\ \varepsilon) = 1 \quad \text{and} \quad q(0;\ 0) > 0 \tag{3.85}$$

(and $q(x;\varepsilon)$ possesses the property that it can be expanded uniformly for $\forall x \in [0,\ 1]$ as $\varepsilon \to 0^+$); this is essentially the problem discussed by Lighthill (1949). The ideas are satisfactorily presented in a special case (which leads to a more transparent calculation); we choose to examine the problem with

$$q(x;\varepsilon) = \alpha + \beta x, \quad \alpha > 0,$$

where α and β are constants independent of ε. Thus (3.84) and (3.85) become

$$(x+\varepsilon y)y' + (\alpha+\beta x)y = 0, \quad 0 \le x \le 1, \quad \text{with} \quad y(1;\varepsilon) = 1 \tag{3.86}$$

We will start by seeking the conventional type of asymptotic solution, as $\varepsilon \to 0^+$, in the form

$$y(x;\varepsilon) \sim \sum_{n=0}^{\infty} \varepsilon^n y_n(x) \tag{3.87}$$

and then $y_0(x)$ satisfies

$$xy_0' + (\alpha+\beta x)y_0 = 0 \quad \text{or} \quad \frac{y_0'}{y_0} = -\left(\frac{\alpha}{x} + \beta\right)$$

which produces the general solution

$$y_0(x) = Ax^{-\alpha}\mathrm{e}^{-\beta x},$$

where A is an arbitrary constant. It is immediately clear that, with $\alpha > 0$, this solution is not defined on $x = 0$, although we may use the boundary condition on $x = 1$, which we choose to do, to give:

$$y_0(x) = x^{-\alpha} e^{\beta(1-x)}. \tag{3.88}$$

The behaviour of $y_1(x)$, as $x \to 0^+$, is discussed in Q3.18, where it is shown that the asymptotic expansion is not uniformly valid as $x \to 0^+$: it breaks down where $x = O(\varepsilon^{1/(1+\alpha)})$. Here, we will approach the problem of finding a solution by introducing a strained coordinate.

The strained coordinate, ξ, is defined by

$$x \sim \xi + \sum_{n=1}^{\infty} \varepsilon^n x_n(\xi), \quad \text{with} \quad x_n(1) = 0 \quad (n = 1, 2, \ldots), \tag{3.89}$$

which, if this expansion is uniformly valid on the domain in ξ (that corresponds to $x \in [0, 1]$), may be inverted to find $\xi(x; \varepsilon)$; note that, if (3.89) *is* uniformly valid, then $\xi \sim x$ for all x on the domain. The solution we seek is now written in terms of the strained coordinate as

$$y\{x(\xi; \varepsilon); \varepsilon\} \sim \sum_{n=0}^{\infty} \varepsilon^n Y_n(\xi), \tag{3.90}$$

and the reason for using (3.89), rather than $\xi = \xi(x; \varepsilon)$, becomes evident when we see that we transform of the original problem into functions of, and derivatives with respect to, only ξ. Thus, with

$$\frac{\mathrm{d}}{\mathrm{d}x} \equiv \frac{\mathrm{d}\xi}{\mathrm{d}x}\frac{\mathrm{d}}{\mathrm{d}\xi} \sim \left(1 + \sum_{n=1}^{\infty} \varepsilon^n x_n'(\xi)\right)^{-1} \frac{\mathrm{d}}{\mathrm{d}\xi},$$

the equation in (3.86) can be written

$$\begin{aligned}(\xi + \varepsilon x_1 + \varepsilon Y_0 + \cdots)(Y_0' + \varepsilon Y_1' + \cdots) \\ + (1 + \varepsilon x_1' + \cdots)\{\alpha + \beta(\xi + \varepsilon x_1 + \cdots)\}(Y_0 + \varepsilon Y_1 + \cdots) = 0\end{aligned} \tag{3.91}$$

where, as in our previous convention, '= 0' means zero to all orders in ε. From equation (3.91) we obtain

$$\xi Y_0' + (\alpha + \beta\xi) Y_0 = 0; \tag{3.92}$$

$$\xi Y_1' + (x_1 + Y_0) Y_0' + (\alpha + \beta\xi) Y_1 + \{(\alpha + \beta\xi) x_1' + \beta x_1\} Y_0 = 0, \tag{3.93}$$

and so on. Because we have defined (3.89) with each $x_n(1) = 0$, the boundary condition on $x = 1$ becomes simply

$$Y_0(1) = 1; \quad Y_n(1) = 0, \quad n = 1, 2, \ldots . \tag{3.94}$$

The solution of (3.98), with (3.94), produces exactly the solution obtained earlier, (3.88), but now expressed in terms of ξ rather than x:

$$Y_0(\xi) = \xi^{-\alpha} e^{\beta(1-\xi)}. \tag{3.95}$$

We now turn to the vexing issue of solving (3.93)—and it is vexing because this is *one* equation in *two* unknown functions: $Y_1(\xi)$ and $x_1(\xi)$. How do we proceed?

The aim of this new technique is to obtain a uniformly valid asymptotic expansion—if that is at all possible—for $\xi \in [\xi_0(\varepsilon), 1]$, although we have yet to determine $\xi_0(\varepsilon)$ (which corresponds to $x = 0$). If there is to be any chance of success, then we must remove any terms that generate non-uniformities in the asymptotic expansion for $y\{\xi(x; \varepsilon); \varepsilon\}$; from Q3.18, it is clear that the only such term here is $y_0 y_0'$ i.e. $Y_0 Y_0'$. Thus we define $x_1(\xi)$ so as to remove this term from the equation for $Y_1(\xi)$; it is sufficient to remove such singular terms in any suitable manner, but if it is possible to choose $x_1(\xi)$ so as to leave an *homogeneous* equation for $Y_1(\xi)$, then this is the usual move. (Other choices produce different asymptotic representations of the same solution, but all equivalent to a given order in ε.) Here, therefore, we elect to write the equation for x_1 as

$$(\alpha + \beta\xi) Y_0 x_1' + (Y_0' + \beta Y_0) x_1 + Y_0 Y_0' = 0 \tag{3.96}$$

leaving

$$\xi Y_1' + (\alpha + \beta\xi) Y_1 = 0. \tag{3.97}$$

An immediate response to this procedure is to observe that the term that causes all the difficulty, $Y_0 Y_0'$, has now appeared as a forcing term in the equation for $x_1(\xi)$—so all we have succeeded in doing is moving the non-uniformity from one asymptotic expansion to another! As we shall see, this is indeed the case, *but* the non-uniformity in the expansion for the strained coordinate is *not* as severe as that in the expansion for y. Before we address this critically important issue, we may note, from (3.97), that

$$Y_1(\xi) = B\xi^{-\alpha} e^{-\beta\xi}$$

where B is an arbitrary constant; but from (3.94) we see that we require $B = 0$ and so $Y_1 \equiv 0$. Further, if this same strategy for selecting the equations for each x_n is adopted, then $Y_n(\xi) \equiv 0$ for every $n = 1, 2, \ldots$, and the *exact* solution, in terms of ξ, becomes

$$y\{\xi(x; \varepsilon); \varepsilon\} = \xi^{-\alpha} e^{\beta(1-\xi)}. \tag{3.98}$$

It is typical of these problems that it is not necessary to solve completely the equations for each $x_n(\xi)$; it is sufficient to examine the nature of the solutions as $\xi \to 0^+$. Thus we will employ the same approach as described in Q3.18. First, from (3.92), we substitute for Y_0' into (3.96) and then cancel $Y_0(\xi)$ ($\neq 0$ everywhere), to give

$$(\alpha + \beta\xi)x_1' - \frac{\alpha}{\xi}x_1 = \left(\frac{\alpha}{\xi} + \beta\right) Y_0$$

and then, for $\xi \to 0^+$, this can be written

$$x_1' - \frac{1}{\xi}x_1 \sim \xi^{-1-\alpha}e^{\beta}.$$

This is easily integrated to produce the solution

$$x_1(\xi) \sim -\frac{e^{\beta}}{1+\alpha}\xi^{-\alpha} \quad \text{as } \xi \to 0^+,$$

where the term involving the arbitrary constant of integration is suppressed because it is less singular than the term retained. Thus the asymptotic expansion (3.89) becomes

$$x \sim \xi - \varepsilon\frac{e^{\beta}}{1+\alpha}\xi^{-\alpha} \quad \text{as } \varepsilon \to 0^+ \quad \text{and then as } \xi \to 0^+, \tag{3.99}$$

which exhibits a breakdown where $\xi = O(\varepsilon^{1/(1+\alpha)})$, exactly as for (3.87) (see Q3.18)—just what we most feared! But there is a very important difference here: the original expansion broke down where $x = O(\varepsilon^{1/(1+\alpha)})$ and we still required a solution valid on $x = 0$; now we have a breakdown at $\xi = O(\varepsilon^{1/(1+\alpha)})$ but, because $x = 0$ corresponds to

$$\xi = \xi_0(\varepsilon) \sim \left(\frac{\varepsilon e^{\beta}}{1+\alpha}\right)^{1/(1+\alpha)} = O(\varepsilon^{1/(1+\alpha)}),$$

obtained directly from (3.99), we require ξ to be *no smaller* than $O(\varepsilon^{1/(1+\alpha)})$. Of course, the burning question now is: are we allowed to use (3.99) with $\xi = O(\varepsilon^{1/(1+\alpha)})$ and hence define $\xi_0(\varepsilon)$? The answer is quite surprising.

We have seen that

$$x_1(\xi) \sim A_1\xi^{-\alpha} \quad \text{as } \xi \to 0^+,$$

where A_1 is the constant $-e^{\beta}/(1+\alpha)$; it is fairly straightforward to show that

$$x_n(\xi) \sim A_n\xi^{-n\alpha} \quad \text{as } \xi \to 0^+, \quad n = 1, 2, \ldots,$$

where the A_n are constants bounded as $n \to \infty$. Thus the asymptotic expansion (3.89), for the strained coordinate, becomes

$$x \sim \xi + \sum_{n=1}^{\infty} \varepsilon^n A_n\xi^{-n\alpha} \quad \text{as } \varepsilon \to 0^+ \quad \text{and then as } \xi \to 0^+$$

and this series *converges* for

$$\varepsilon\xi^{-\alpha} < \gamma,$$

where γ is some constant independent of ε. Hence the expansion converges for $\xi > (\varepsilon/\gamma)^{1/\alpha} = \mathrm{O}(\varepsilon^{1/\alpha})$—but ξ is to be no smaller than $\xi_0(\varepsilon) = \mathrm{O}(\varepsilon^{1/(1+\alpha)})$, which is *larger* than $\mathrm{O}(\varepsilon^{1/\alpha})$. Thus the asymptotic expansion with $\xi \in [\xi_0(\varepsilon),\ 1]$ (which maps to $x \in [0,\ 1]$) is not just uniformly valid—it is convergent! This is an altogether unlooked-for bonus; the method of strained coordinates, in this example, has proved to be a very significant improvement on our standard matched-expansions approach.

Some further examples that make use of a strained coordinate are set as exercises Q3.19–3.25. Although our example, E3.7, has demonstrated, to the full, the advantages of the strained coordinate method, not all problems are quite this successful. Many ordinary differential equations of the type discussed in E3.7 do indeed possess convergent series for the coordinate—so the complete solution is no longer simply asymptotic—but other problems may produce a strained coordinate that is uniformly valid only (without being convergent). In the exercises, the question of convergence is not explored (but, of course, the interested and skilful reader may wish to investigate this aspect).

In this text, at this stage, we have introduced many of the ideas and techniques of singular perturbation theory, and have applied them—in the main—to ordinary and partial differential equations of various types. In the next chapter we present one further technique for solving singular perturbation problems. This is a method which subsumes most of what we have presented so far and is, probably, the single most powerful approach that we have available. When this has been completed, we will employ all our methods in the examination of a selection of examples taken from a number of different scientific fields.

FURTHER READING

A few regular perturbation problems that are described by partial differential equations are discussed in van Dyke (1964, 1975) and also in Hinch (1991). A discussion of wave propagation and breakdown, and especially with reference to supersonic flow past thin aerofoils, can be found in van Dyke (above), Kevorkian & Cole (1981, 1996) and in Holmes (1995). Any good text on compressible fluid mechanics will cover these ideas, and much more, for the interested reader; we can recommend Courant & Friedrichs (1967), Ward (1955), Miles (1959), Hayes & Probstein (1966) and Cox & Crabtree (1965), but there are many others to choose from. A nice collection of partial differential equations that exhibit boundary-layer behaviour are presented in Holmes (1995). Most of the standard texts that discuss more general aspects of singular perturbation theory include Mathieu's equation, and related problems, as examples; good, dedicated works on ordinary differential equations will give a broad and general background to the

Mathieu equation (such as Ince, 1956). A lot of detail, with analytical and numerical results, including many applications, can be found in McLachlan (1964).

The method of strained coordinates is described quite extensively in van Dyke (1964, 1975), Nayfeh (1973), Hinch (1991) and, with only slightly less emphasis, in Kevorkian & Cole (1981, 1996).

EXERCISES

Q3.1 *Flow past a distorted circle.* Find the third term ($\mathrm{O}(\varepsilon^2)$) in the asymptotic expansion, for $\varepsilon \to 0$, of the problem described by

$$\frac{\partial^2 \psi}{\partial r^2} + \frac{1}{r}\frac{\partial \psi}{\partial r} + \frac{1}{r^2}\frac{\partial^2 \psi}{\partial \theta^2} = 0$$

with $\psi/r \to U \sin\theta$ as $r \to \infty$, and $\psi\{a[1 + \varepsilon f(\theta)], \theta; \varepsilon\} = 0$, for $f(\theta) = \cos\theta$.

Hence write down the asymptotic solution to this order and observe that, formally at least, there is a breakdown where $r = \mathrm{O}(\varepsilon^{-1})$. Deduce that the solution in the new scaled region is *identical* (to the appropriate order) to that obtained for $r = \mathrm{O}(1)$, the only adjustment being the order in which the terms appear in the asymptotic expansion (and so the expansion can be regarded as regular). Use your results to find an approximation to the velocity components on the surface of the distorted circle.

Q3.2 *Weak shear flow past a circle.* Cf. Q3.1; now we consider a flow with constant, small vorticity past a circle. Let the flow at infinity be $U(1 + \varepsilon y) = U(1 + \varepsilon r \sin\theta)$, which has the constant vorticity $-\varepsilon U$ (in the z-direction); the problem is therefore to solve

$$\psi_{rr} + \frac{1}{r}\psi_r + \frac{1}{r^2}\psi_{\theta\theta} = \varepsilon U$$

with $(\psi/r - U\sin\theta)/r \to \frac{1}{4}\varepsilon U(1 - \cos 2\theta)$ as $r \to \infty$, and $\psi(a, \theta; \varepsilon) = 0$.

Seek a solution $\psi(r, \theta; \varepsilon) \sim \sum_{n=0}^{\infty} \varepsilon^n \psi_n(r, \theta)$, find the first two terms and, on the basis of this evidence, show that this constitutes a two-term, uniformly valid asymptotic expansion. Indeed, show that your two-term expansion is the *exact* solution of the problem.

Q3.3 *Potential function outside a distorted circle.* (This is equivalent to finding the potential outside a nearly circular, infinite cylinder.) We seek a solution, $\phi(r, \theta; \varepsilon)$, of the problem

$$\phi_{rr} + \frac{1}{r}\phi_r + \frac{1}{r^2}\phi_{\theta\theta} = 0, \quad r \geq r_0(\theta; \varepsilon),$$

with $\quad \phi\{r_0(\theta; \varepsilon), \theta; \varepsilon\} = 1 \quad$ and $\quad \phi - \ln r \to 1 \quad$ as $r \to \infty$, where $r_0(\theta;\varepsilon) = 1 + \varepsilon f(\theta)$. Write $\phi(r,\theta;\varepsilon) \sim \sum_{n=0}^{\infty} \varepsilon^n \phi_n(r,\theta)$, find $\phi_0(r,\theta)$ and then determine $\phi_1(r,\theta)$ in the case $f(\theta) = \sin\theta$. On the basis of this evidence, confirm that you have a two-term, uniformly valid asymptotic expansion in $r \geq r_0$.

Q3.4 *The classical (model) Boussinesq equations for water waves.* These equations are written as the coupled pair

$$u_t + \varepsilon u u_x + \eta_x = \frac{1}{2}\varepsilon\lambda u_{xxt}; \quad \eta_t + [(1+\varepsilon\eta)u]_x = \frac{1}{6}\varepsilon\lambda u_{xxx},$$

where $u(x,t;\varepsilon)$ is the horizontal velocity component in the flow, $\eta(x,t;\varepsilon)$ the surface displacement i.e. the surface wave, and λ is a constant independent of ε. Find the first terms in the asymptotic expansions $q(x,t;\varepsilon) \sim \sum_{n=0}^{\infty} \varepsilon^n q_n(x,t)$ ($q \equiv u, \eta$) for $x = \mathrm{O}(1)$, $t = \mathrm{O}(1)$, as $\varepsilon \to 0$ (the near-field). Then introduce the far-field variables: $\xi = x - t$, $\tau = \varepsilon t$, for right-running waves, and hence find the equations defining the leading order; show that the equation for u takes the *form*

$$U_\tau + 6UU_\xi + U_{\xi\xi\xi} = 0,$$

the Korteweg-de Vries equation.
Show that a solution of this equation is the *solitary wave*

$$U = 2\kappa^2 \mathrm{sech}^2\{\kappa(\xi - 4\kappa^2\tau)\},$$

where κ is a free parameter.

Q3.5 *Long, small-amplitude waves with dissipation.* A model for the propagation of *long* waves, with some contribution from dissipation (damping), is

$$c_t + \varepsilon u c_x + (\hat{c}_0 + \varepsilon c)u_x = 0; \quad u_t + \varepsilon u u_x + (\hat{c}_0 + \varepsilon c)c_x = \varepsilon\mu u_{xx},$$

where $\hat{c}_0$ and μ are positive constants independent of ε. Follow the same procedure as in Q3.4 (near-field then far-field, although here the right-going characteristic will be $\xi = x - \hat{c}_0 t$). Show that, in the far-field, the leading term, for u (say), satisfies an equation of the *form*

$$U_\tau + UU_\xi = U_{\xi\xi},$$

the Burgers equation. Show that this equation has a steady-state shock-profile solution

$$U = \alpha[1 - \tanh\{k(\xi - C\tau)\}],$$

for suitable constants α, C and k (> 0), which should be identified. (This solution is usually called the Taylor shock profile; Taylor, 1910.)

Q3.6 *A nonlinear wave equation.* A wave is described by the equation

$$\left(\frac{\partial}{\partial t}+\varepsilon u\frac{\partial}{\partial x}\right)^2 u-\frac{\partial^2 u}{\partial x^2}=\varepsilon\frac{\partial^4 u}{\partial x^4},$$

where $\varepsilon \to 0$. Seek a solution (for right-running waves only) in the form

$$u(x, t; \varepsilon) \sim \sum_{n=0}^{\infty} \varepsilon^n u_n(x, t), \quad x = \mathrm{O}(1), \quad t = \mathrm{O}(1),$$

find the first two terms and demonstrate the existence of a breakdown when $t = \mathrm{O}(\varepsilon^{-1})$. Now introduce $\xi = x - t$, $\tau = \varepsilon t$, and show that the leading term in the expansion valid for $\tau = \mathrm{O}(1)$ ($u \sim U_0$) satisfies the equation

$$2U_{0\tau\xi} + 2U_0 U_{0\xi\xi} + U_{0\xi}^2 + U_{0\xi\xi\xi\xi} = 0. \qquad (*)$$

This calculation is now extended: define

$$\xi = x - t + \varepsilon f(x + t, \tau) + \mathrm{o}(\varepsilon); \quad \eta = x + t + \varepsilon g(x - t, \tau) + \mathrm{o}(\varepsilon)$$

and then determine f and g so that

$$U = U_0(\xi, \tau) + V_0(\eta, \tau) + \mathrm{o}(\varepsilon),$$

where U_0 satisfies (*) and V_0 satisfies the corresponding equation for left-going waves. (You may assume that both f and g possess Taylor expansions about $x + t$ and $x - t$, respectively.)

Q3.7 *A multi-wave speed equation.* A particular wave profile, $u(x, t; \varepsilon)$, with $u \to 0$ as $x \to \infty$, is described by the equation

$$\left[\frac{\partial}{\partial t}+(c_1+\varepsilon u)\frac{\partial}{\partial x}+\varepsilon u_x\right]\left[\frac{\partial}{\partial t}+(c_2+\varepsilon u)\frac{\partial}{\partial x}\right]u+\varepsilon\left[\frac{\partial}{\partial t}+(c+\sqrt{\varepsilon}u)\frac{\partial}{\partial x}\right]u=0,$$

where c_1, c_2 and c are constants (independent of ε). Show that, if $c_1 < c < c_2$, then on the time scale $\mathrm{O}(\varepsilon^{-1})$, the wave moving at speed c_1, and the wave moving at speed c_2, each decay exponentially (in time), to leading order as $\varepsilon \to 0^+$. Show, also, that on the time scale $\mathrm{O}(\varepsilon^{-2})$, the wave moving at speed c has diffused a distance $\mathrm{O}(\varepsilon^{-3/2})$ about the wave front and, to leading order, it satisfies a Burgers equation (see Q3.5).

Q3.8 *Water waves with weak nonlinearity, damping and dispersion.* The propagation of a one-dimensional wave on the surface of water can be modelled by the equations

$$u_t + \varepsilon u u_x + \eta_x = \varepsilon\left(u_{xx} + u_{xxt}\right); \quad \eta_t + \left[(1 + \varepsilon\eta)u\right]_x = 0,$$

where $u(x, t; \varepsilon)$ is the horizontal velocity component in the flow, and $\eta(x, t; \varepsilon)$ is the surface wave; cf. Q3.4 Find the first terms in the near-field expansions of u and η, as $\varepsilon \to 0^+$, and then obtain the equation for the leading term in η, valid in the far-field ($t = \mathrm{O}(\varepsilon^{-1})$). You should consider only right-going waves. (The equation that you obtain here is a Korteweg-de Vries-Burgers (KVB) equation; see Johnson, 1997.)

Q3.9 *Supersonic, thin-aerofoil theory: characteristic approach.* The characteristics for equation (3.22) can be defined by the equation $dy/dx = \tan(\theta + \nu)$, where θ is the streamline direction (so that $\tan\theta = \varepsilon\phi_y/(1 + \varepsilon\phi_x)$) and ν is the inclination of the characteristic relative to the streamline (so that $\tan\nu = 1/\sqrt{M^2 - 1}$, where M is the *local* Mach Number). Show that

$$M^2 = \frac{1 + \varepsilon^2\left(\phi_x^2 + \phi_y^2\right)}{\left(1/M_0^2\right) + \frac{1}{2}(\gamma - 1)\left[\varepsilon^2\left(\phi_y^2 - \phi_x^2\right) - 2\varepsilon\phi_x\right]}$$

and hence deduce that, on the characteristics,

$$\frac{dy}{dx} \sim \frac{\dfrac{1}{\beta}\left[1 - \dfrac{\varepsilon}{\beta^2}M_0^2\phi_x\left(1 + \dfrac{\gamma - 1}{2}M_0^2\right)\right] + \varepsilon\phi_y}{1 - \dfrac{1}{\beta}\varepsilon\phi_y}.$$

Finally, since to leading order $\phi_x \sim -\phi_y/\beta$, show that

$$\frac{dy}{dx} \sim \frac{1}{\beta}\left[1 - \frac{\varepsilon(\gamma + 1)M_0^4}{2\beta^2}\phi_x\right],$$

and confirm that this is recovered from equation (3.39).

Q3.10 *Thin aerofoil in a transonic flow.* Show that the asymptotic expansion (3.35) is not uniformly valid as $M_0 \to 1$.

(a) Set $M_0 = 1$, write $\hat{y} = \delta(\varepsilon)y$ and $\phi = \Delta(\varepsilon)\Phi$, and hence deduce that a scaling consistent with equations (3.22) and (3.24) is $\delta = \varepsilon^{1/3}$, $\Delta = \varepsilon^{-1/3}$, and that Φ then satisfies, to leading order,

$$\Phi_{\hat{y}\hat{y}} = (1 + \gamma)\Phi_x\Phi_{xx}.$$

(b) Given that $M_0 = 1 + \lambda\mu(\varepsilon)$, use the scaling in (a) to show that there is a distinguished limit in which $\mu(\varepsilon) = \varepsilon^{2/3}$; what now is the equation for Φ, to leading order?

Q3.11 *Thin aerofoil in a hypersonic flow.* See (3.35); show that this expansion breaks down as $M_0 \to \infty$ when $\varepsilon M_0 = \mathrm{O}(1)$. Introduce $y = \varepsilon Y$, leave x unscaled and write $\phi = \delta(\varepsilon)\Phi(x, Y; \varepsilon)$; show that terms from both the left-hand and right-hand sides of equation (3.22) are of the same order in the case $\varepsilon M_0 = \mathrm{O}(1)$, for a particular choice of $\delta(\varepsilon)$. What is the resulting leading-order equation for Φ? [This result can only give a flavour of how things might proceed because, with $M_0 \to \infty$, strong shock waves most certainly appear and then a potential function does not exist. To investigate this properly, we need to return to the original governing equations, without the isentropic assumption.]

Q3.12 *Asymmetrical bending of a pre-stressed annular plate.* The lateral displacement, $w(r, \theta; \varepsilon)$, of a plate is described by the equation (written in non-dimensional variables)

$$\varepsilon^2 \nabla^4 w - \nabla^2 w = 0 \quad \text{where} \quad \nabla^2 \equiv \frac{\partial^2}{\partial r^2} + \frac{1}{r}\frac{\partial}{\partial r} + \frac{1}{r^2}\frac{\partial^2}{\partial \theta^2},$$

and $\varepsilon^2 \to 0$ corresponds to weak bending rigidity. The annular plate is defined by $0 < r_0 \le r \le 1, 0 \le \theta \le 2\pi$, with the boundary conditions

$$w = r_0 \cos\theta \quad \text{and} \quad \frac{\partial w}{\partial r} = \cos\theta \quad \text{at} \quad r = r_0, \quad 0 \le \theta \le 2\pi,$$

$$w = \frac{\partial w}{\partial r} = 0 \quad \text{at} \quad r = 1, \quad 0 \le \theta \le 2\pi$$

where r_0 (< 1) is a constant independent of ε. Seek a solution $w(r, \theta; \varepsilon) = u(r; \varepsilon)\cos\theta$, find the equation for u and then find the first two terms in each of the asymptotic expansions of $u(r; \varepsilon)$, as $\varepsilon \to 0^+$, valid away from the boundaries of the region, and in the two boundary layers (near $r = 1, r_0$). Match your expansions as necessary. [For more details, see Nayfeh, 1973.]

Q3.13 *A nonlinear elliptic equation.* The function $u(x, y; \varepsilon)$ satisfies the equation

$$\varepsilon(u_{xx} + uu_{yy}) + u(\mathrm{e}^{-y} + u_y) = 0$$

and it is defined in $0 \le y \le 1$, $x \ge 0$. The boundary conditions are

$$u(x, 0; \varepsilon) = 2 \quad \text{and} \quad u(x, 1; \varepsilon) = \mathrm{e}^{-1} + \mathrm{e}^{-x}, \quad x \ge 0,$$

$$u(0, y) = 1 + \mathrm{e}^{-y}, \quad 0 \le y \le 1.$$

Use the asymptotic sequence $\{\varepsilon^n\}$, $\varepsilon \to 0^+$, and hence obtain the first two terms in the asymptotic expansion valid away from the boundaries $x = 0$ and $y = 0$; this solution *is* valid on $y = 1$. Now find the first term only in the asymptotic expansion valid in the boundary layer near $y = 0$, having first found the size of the layer; match as necessary. Repeat this procedure for the boundary layer near $x = 0$ and show that, to leading order, no such layer is required. However, deduce that one is needed to accommodate the boundary condition at $\mathrm{O}(\varepsilon)$. Write the solution in this boundary layer as

$$u = 1 + e^{-y} + \varepsilon V$$

(written in appropriate variables) and formulate the problem for the leading term in the asymptotic expansion for V, but do not solve for V.

Q3.14 *The steady temperature distribution in a square plate.* The temperature, $T(x, y; \varepsilon)$, in a plate is described by the heat conduction equation (written in non-dimensional variables)

$$\varepsilon T_{xx} + T_{yy} + \alpha(x) T_x = 0, \quad \alpha(x) > 0,$$

where the plate is $0 \le x \le 1, 0 \le y \le 1$, with the temperature on the boundary given by

$$T(1, y; \varepsilon) = \sin \pi y \quad \text{and} \quad T(0, y; \varepsilon) = 0, \quad 0 \le y \le 1,$$

$$T(x, 0; \varepsilon) = T(x, 1; \varepsilon) = 0, \quad 0 \le x \le 1.$$

Seek the first term of a *composite* expansion by writing

$$T(x, y; \varepsilon) = t_0(x, y) + \tau_0(x, y) \exp[-h(x)/\varepsilon] + O(\varepsilon)$$

where $h(x) = \int_0^x \alpha(x')\, dx'$ is the relevant boundary-layer variable. Determine completely t_0 and τ_0, and then show that

$$T_x(0, y; \varepsilon) \sim \frac{\alpha(0)}{\varepsilon} \exp[-\pi^2 k] \sin \pi y,$$

where $k = \int_0^1 dx/\alpha(x)$.

Q3.15 *Mathieu's equation for $n = 2$.* See E3.6; find the asymptotic expansion of $\delta(\varepsilon)$, as far as the term $O(\varepsilon^2)$, in the case $n = 2$ (and there will be two versions of this, depending on the choice of either sin or cos).

Q3.16 *Mathieu's equation based on Floquet theory.* Write the Mathieu equation

$$\ddot{x} + (\delta + \varepsilon \cos t)x = 0$$

as an equation in $y(t; \varepsilon)$ by setting $x(t; \varepsilon) = e^{\alpha t} y(t; \varepsilon)$, where $\alpha = \alpha(\varepsilon)$ is a constant; $y(t; \varepsilon)$ is periodic with period 2π or 4π. (Note that the transitional curves are now $\alpha(\varepsilon) = 0$.) Seek a solution

$$y(t; \varepsilon) \sim \sum_{m=0}^{\infty} \varepsilon^m y_m(t), \quad \delta(\varepsilon) \sim \frac{n^2}{4} + \sum_{m=1}^{\infty} \varepsilon^m \delta_m, \quad \alpha(\varepsilon) \sim \sum_{m=1}^{\infty} \varepsilon^m \alpha_m$$

in the case $n = 1$; because we now have $\alpha \ne 0$, the solution will be that which is valid *near* the transitional curves. Show that

$$\alpha(\varepsilon) = \pm \frac{1}{2} \varepsilon \sqrt{1 - 4\delta_1^2} + O(\varepsilon^2); \quad \delta(\varepsilon) \sim \frac{1}{4} + \varepsilon \delta_1,$$

where δ_1 is a free parameter.

Q3.17 *A particular Hill equation.* (Hill's equation is a generalisation of the Mathieu equation.) Consider the equation

$$\ddot{x} + (\delta + \varepsilon \cos t + \varepsilon\lambda \cos 3t)x = 0, \quad t \geq 0,$$

where λ is a constant independent of ε; seek a solution

$$x(t;\varepsilon) \sim \sum_{n=0}^{\infty} \varepsilon^n x_n(t), \quad \delta(\varepsilon) \sim 1 + \sum_{n=1}^{\infty} \varepsilon^n \delta_n,$$

where the δ_n are independent of ε. Impose the condition that x_0, x_1 and x_2 are to be periodic; what condition(s) must δ_1 and δ_2 satisfy?

Q3.18 *Matched expansion applied to E3.7.* Consider

$$(x + \varepsilon y)y' + (\alpha + \beta x)y = 0, \quad 0 \leq x \leq 1, \quad y(1;\varepsilon) = 1$$

with $\alpha > 0$ (and α, β constants independent of ε). Find the equation for the second term of the expansion, $y_1(x)$, and from this deduce that

$$y_1(x) \sim -\frac{\alpha e^{2\beta}}{1+\alpha} x^{-1-2\alpha} \quad \text{as } x \to 0^+;$$

this can be done by first approximating the equation for y_1 *before* integrating it—see the method leading to equation (3.96). Hence show that the asymptotic expansion for $y(x;\varepsilon)$ breaks down where $x = O(\varepsilon^{1/(1+\alpha)})$; rescale x and y in the neighbourhood of $x = 0$, and then find and solve the equation describing the dominant term in this region (matching as necessary). What is the behaviour of $y(0;\varepsilon)$, as $\varepsilon \to 0^+$, based on your solution?

Q3.19 *A strained-coordinate problem I.* (This is a problem introduced by Carrier, 1953.) Find an asymptotic solution of

$$(x^2 + \varepsilon y)y' + y = 0, \quad 0 \leq x \leq 1, \quad y(1;\varepsilon) = \mathrm{e},$$

as $\varepsilon \to 0^+$, in the form

$$y\{x(\xi;\varepsilon);\varepsilon\} \sim \sum_{n=0}^{\infty} \varepsilon^n Y_n(\xi) \quad \text{with} \quad x \sim \xi + \sum_{n=1}^{\infty} \varepsilon^n x_n(\xi).$$

Find $Y_0(\xi)$ and $x_1(\xi)$; use your solution to find the dominant behaviour of $\xi_0(\varepsilon)$ where $\xi = \xi_0(\varepsilon) \Leftrightarrow x = 0$. (You may assume that the asymptotic expansion of the coordinate is uniformly valid on $[\xi_0(\varepsilon), 1]$.)

Q3.20 *A strained-coordinate problem II.* See Q3.19; follow the same procedure for the problem

$$(x + \varepsilon y)y' + y + \varepsilon y^2 = 0, \quad 0 \leq x \leq 1, \quad y(1;\varepsilon) = 1$$

as $\varepsilon \to 0^+$. Use your results to show that $y(0;\varepsilon) \sim \sqrt{2/\varepsilon}$ as $\varepsilon \to 0^+$.

Q3.21 *A strained-coordinate problem III.* See Q3.20; follow this same procedure for

$$(x - \varepsilon y)y' + xy = e^{-x}, \quad 0 \le x \le 1, \quad y(1; \varepsilon) = e^{-1}$$

as $\varepsilon \to 0^+$. Show that $y(0; \varepsilon) \sim -\xi_0(\varepsilon)/\varepsilon$, as $\varepsilon \to 0^+$, where ξ_0 is the appropriate solution of the equation $\xi_0 + \varepsilon \ln \xi_0 = 0$ (which does possess one real root).

Q3.22 *A strained-coordinate problem IV.* See Q3.19; follow this same procedure for the problem

$$(x + \varepsilon y)y' + y = x, \quad 0 \le x \le 1, \quad y(1; \varepsilon) = 1$$

as $\varepsilon \to 0^+$. (You may observe that this problem can be solved exactly.)

Q3.23 *A strained coordinate problem V.* See Q3.19; follow this same procedure for the problem

$$(x^2 + \varepsilon y)y' + y^2 = 0, \quad 0 \le x \le 1, \quad y(1; \varepsilon) = -1/2$$

as $\varepsilon \to 0^+$. If the boundary condition had been $y(1; \varepsilon) = 1$, with the same domain, briefly investigate the nature of this new problem.

Q3.24 *Duffing's equation.* The equation for the motion of a simple pendulum, without the approximation for small angles of swing, takes the form

$$\ddot{x} + k \sin x = 0, \quad x = x(t), \quad k > 0 \text{ (constant)}.$$

If x is small, and we retain terms as far as x^3, we obtain an equation like

$$\ddot{x} + x - \varepsilon x^3 = 0;$$

this is Duffing's equation (Duffing, 1918) which was introduced to improve the approximation for the simple pendulum (without the complications of working with $\sin x$). Seek a solution of this equation, for $t \ge 0$ and

$$x(0; \varepsilon) = 1 \quad \text{and} \quad \dot{x}(0; \varepsilon) = 0,$$

by using a strained-coordinate formulation:

$$x\{t(\tau; \varepsilon); \varepsilon\} \sim \sum_{n=0}^{\infty} \varepsilon^n x_n(\tau), \quad \tau \sim \left(1 + \sum_{n=1}^{\infty} \varepsilon^n \omega_n\right) t$$

where the ω_n are constants. Determine the solution as far as terms in ε^2, choosing each ω_n in order to ensure that the solution is periodic.

Q3.25 *Weakly nonlinear wave propagation.* A wave motion is described by the equation

$$u_{tt} - (1 + \varepsilon u_x)u_{xx} = 0, \quad t \geq 0, \quad -\infty < x < \infty,$$

with $\quad u(x, 0; \varepsilon) = f(x), \quad u_t(x, 0; \varepsilon) = -f'(x), \quad -\infty < x < \infty$

(which ensures, certainly to leading order, that we have only a right-going wave). Seek a solution in the form

$$u \sim \sum_{n=0}^{\infty} \varepsilon^n u_n(\xi), \quad x - t \sim \xi + \left(\sum_{n=1}^{\infty} \varepsilon^n \omega_n(\xi) \right) t$$

and hence find the solution correct at $O(\varepsilon)$.

4. THE METHOD OF MULTIPLE SCALES

The final stage in our presentation of the essential tools that constitute singular perturbation theory is to provide a description of the *method of multiple scales*, arguably the most important and powerful technique at our disposal. The idea, as the title implies, is to introduce a number of different scales, each one (measured in terms of the small parameter) associated with some property of the solution. For example, one scale might be that which governs an underlying oscillation and another the scale on which the amplitude evolves (as in *amplitude modulation*). Indeed, this type of problem is the most natural one with which to start; we will explore a particularly simple example and use this as a vehicle to present the salient features of the method. However, before we embark on this, one word of warning: this process necessarily transforms all differential equations into partial differential equations—even ordinary differential equations! This could well cause some anxiety, but the comforting news is that the underlying mathematical problem is no more difficult to solve. So, for example, an ordinary differential equation, subjected to this procedure, involves an integration method that is essentially unaltered; the only adjustment is simply that arbitrary constants become arbitrary functions of all the other variables.

4.1 NEARLY LINEAR OSCILLATIONS

We will show how these ideas emerge in this class of relatively simple problems; indeed, we start with an example for which an exact solution exists. Let us consider the linear,

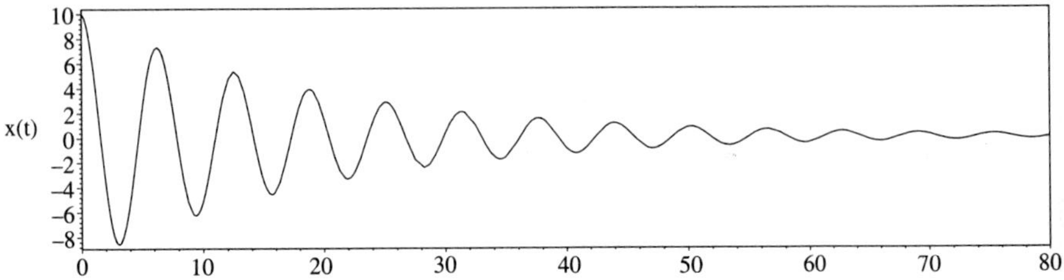

(a)

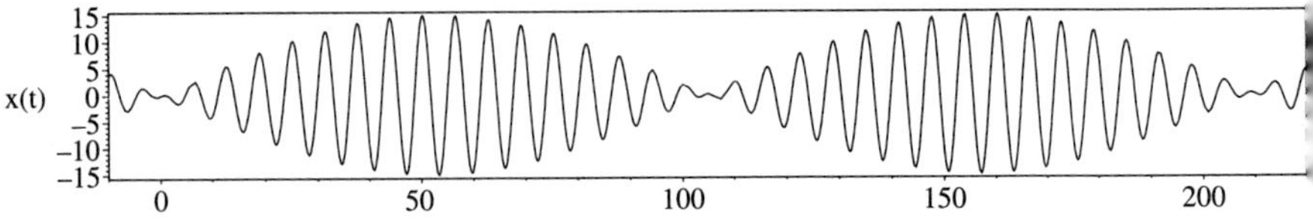

(b)

Figure 6. (a) Upper figure is a plot of the function $y = 10e^{-\varepsilon x}\cos x$ for $0 \le x \le 80$, with $\varepsilon = 0.05$. (b) Lower figure is a plot of the function $y = 15\sin(\varepsilon x)\cos x$ for $-10 \le x \le 220$, with $\varepsilon = 0.03$.

damped oscillator which is governed by

$$\ddot{x} + 2\varepsilon\dot{x} + x = 0, \quad t \ge 0, \tag{4.1}$$

with

$$x(0;\varepsilon) = 0 \quad \text{and} \quad \dot{x}(0;\varepsilon) = 1,$$

where $\varepsilon \to 0^+$. This problem can be solved exactly (and having the solution available will help initiate the discussion); the solution is

$$x_e(t;\varepsilon) = e^{-\varepsilon t}\frac{\sin\left(t\sqrt{1-\varepsilon^2}\right)}{\sqrt{1-\varepsilon^2}} \tag{4.2}$$

where the 'e' subscript denotes 'exact solution'. This solution represents an oscillation, with a fixed period, and with an amplitude which decays exponentially, albeit slowly. (This type of solution is depicted in figure 6a, and another function with a different modulation is shown in figure 6b.) Now this solution, (4.2), has three important characteristics: first, it is an oscillation controlled by $T = t\sqrt{1-\varepsilon^2}$ (usually called the *fast* scale); second, the amplitude decays slowly according to $\tau = \varepsilon t$ (usually called the *slow* scale); third, even if we express the solution in terms of T and τ, it will still require an asymptotic expansion, as $\varepsilon \to 0^+$, by virtue of the factor $\sqrt{1-\varepsilon^2}$ in the denominator. Any construction of an asymptotic solution directly from (4.1) must accommodate all these elements.

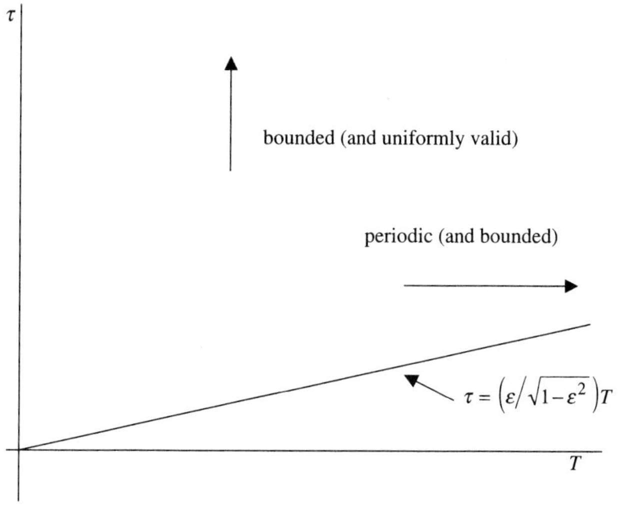

Figure 7. Sketch of the (T, τ)-plane, with the line $\tau = \left(\varepsilon/\sqrt{1-\varepsilon^2}\right) T$, $T \geq 0$, included.

The appearance of two time scales in (4.2) is quite clear: T (fast) and τ (slow), so we could write the solution as

$$x = e^{-\tau}\frac{\sin T}{\sqrt{1-\varepsilon^2}}, \quad T = t\sqrt{1-\varepsilon^2}, \quad \tau = \varepsilon t.$$

The underlying idea in the method of multiple scales is to formulate the *original* problem in terms of these two scales from the outset and then to treat $x(t;\varepsilon) \equiv X(T, \tau; \varepsilon)$—a function of *two* variables; this will lead to a *partial* differential equation for X. Clearly, T and τ are *not* independent variables—they are both proportional to t—so we have, apparently, a significant mathematical inconsistency. How, therefore, do we proceed with any confidence? The method and philosophy are surprisingly straightforward.

We seek an asymptotic solution for $X(T, \tau; \varepsilon)$, as $\varepsilon \to 0^+$, as a function with its domain in 2-space; this is certainly more general than in the original formulation. The aim is to obtain a uniformly valid expansion in $T \geq 0$ and $\tau \geq 0$. This will, typically, require us to invoke periodicity (and boundedness) in T, and boundedness (and uniformity) in τ. If we are able to construct such an asymptotic solution, it will be valid *throughout* the quadrant $T \geq 0$, $\tau \geq 0$, in (T, τ)-space. Because the solution is valid in this region, it will be valid along any and every path that we may wish to follow in this region; in particular, it will be valid along the line $\tau = (\varepsilon/\sqrt{1-\varepsilon^2})T$ which is the statement that T and τ are suitably related to t. This important interpretation is represented in figure 7, and this idea is at the heart of all multiple-scale techniques. We will now apply this method to (4.1), presented as a formal example.

E4.1 A linear, damped oscillator

Given

$$\ddot{x} + 2\varepsilon\dot{x} + x = 0, \quad t \geq 0, \tag{4.3}$$

with

$$x(0;\varepsilon) = 0 \quad \text{and} \quad \dot{x}(0;\varepsilon) = 1, \tag{4.4}$$

we introduce

$$\tau = \varepsilon t \quad \text{and} \quad T \sim \left(1 + \sum_{n=1}^{\infty} \varepsilon^n \omega_n\right) t, \tag{4.5}$$

where the ω_n are constants, and write

$$x(t;\varepsilon) \equiv X(T,\tau;\varepsilon) \sim \sum_{n=0}^{\infty} \varepsilon^n X_n(T,\tau) \tag{4.6}$$

where each X_n is to be periodic in T and, we hope, uniformly valid as $T \to \infty$ and as $\tau \to \infty$. (Note that the choice of T, in (4.5), follows exactly the pattern of a strained coordinate; cf. Q3.24.) From (4.5) we obtain the operator identity

$$\frac{\mathrm{d}}{\mathrm{d}t} \equiv \left(1 + \sum_{n=1}^{\infty} \varepsilon^n \omega_n\right) \frac{\partial}{\partial T} + \varepsilon \frac{\partial}{\partial \tau}$$

and so equation (4.3) becomes

$$\left(1 + \sum_{n=1}^{\infty} \varepsilon^n \omega_n\right)^2 X_{TT} + 2\varepsilon \left(1 + \sum_{n=1}^{\infty} \varepsilon^n \omega_n\right) X_{T\tau} + \varepsilon^2 X_{\tau\tau}$$
$$+ 2\varepsilon \left\{ \left(1 + \sum_{n=1}^{\infty} \varepsilon^n \omega_n\right) X_T + \varepsilon X_\tau \right\} + X = 0,$$

where, as usual, '$= 0$' means zero to all orders in ε. When we insert (4.6) for X, we obtain the sequence of equations

$$X_{0TT} + X_0 = 0; \quad X_{1TT} + X_1 + 2\omega_1 X_{0TT} + 2X_{0T\tau} + 2X_{0T} = 0; \tag{4.7a,b}$$

$$X_{2TT} + X_2 + (2\omega_2 + \omega_1^2) X_{0TT} + 2\omega_1 (X_{1TT} + X_{0T\tau}) + 2X_{1T\tau} + X_{0\tau\tau}$$
$$+ 2(X_{1T} + \omega_1 X_{0T} + X_{0\tau}) = 0, \tag{4.7c}$$

and so on. The initial conditions, (4.4), become

$$X(0, 0; \varepsilon) = 0; \quad \left(1 + \sum_{n=1}^{\infty} \varepsilon^n \omega_n\right) X_T(0, 0; \varepsilon) + \varepsilon X_\tau(0, 0; \varepsilon) = 1,$$

which give

$$X_n(0, 0) = 0, \quad n = 0, 1, 2, \ldots; \quad X_{0T}(0, 0) = 1, \quad X_{1T}(0, 0) + \omega_1 X_{0T}(0, 0) + X_{0\tau}(0, 0) = 0, \tag{4.8}$$

et cetera. It is now a very straightforward exercise to solve each of these problems, in sequence.

The general solution of (4.7a) is

$$X_0(T, \tau) = A_0(\tau) \sin T + B_0(\tau) \cos T,$$

although it is rather more convenient to write this as

$$X_0(T, \tau) = A_0(\tau) \sin[T + \phi_0(\tau)], \tag{4.9}$$

where $A_0(\tau)$ and $\phi_0(\tau)$ are arbitrary functions; we require, from (4.8),

$$A_0(0) \sin[\phi_0(0)] = 0 \quad \text{and} \quad A_0(0) \cos[\phi_0(0)] = 1,$$

which we satisfy by selecting

$$A_0(0) = 1, \quad \phi_0(0) = 0. \tag{4.10}$$

The equation for $X_1(T, \tau)$, (4.7b), can now be written

$$X_{1TT} + X_1 = 2\omega_1 A_0 \sin\theta_0 - 2\left\{A_0' \cos\theta_0 - A_0\phi_0' \sin\theta_0\right\} - 2A_0 \cos\theta_0,$$

where, for simplicity, we have written $\theta_0 = T + \phi_0(\tau)$. But for $X_1(T, \tau)$ to be periodic in T, all terms on the right-hand side, in $\sin\theta_0$ and $\cos\theta_0$, must vanish (for otherwise we will generate particular integrals for X_1 like $T\sin[T + \phi_0(\tau)]$); this removal of *secular* (i.e. non-periodic) terms requires

$$A_0' + A_0 = 0; \quad (\omega_1 + \phi_0')A_0 = 0. \tag{4.11a,b}$$

The solution of (4.11a) and then (4.11b), with (4.10), gives directly

$$A_0(\tau) = \mathrm{e}^{-\tau} \quad \text{and} \quad \phi_0(\tau) = -\omega_1\tau,$$

leaving

$$X_1(T, \tau) = A_1(\tau)\sin[T + \phi_1(\tau)]. \tag{4.12}$$

Thus, at this stage, we have the asymptotic solution

$$X(T, \tau; \varepsilon) \sim \mathrm{e}^{-\tau}\sin(T - \omega_1\tau), \tag{4.13}$$

but $A_1(\tau)$, $\phi_1(\tau)$ *and* ω_1 are, as yet, unknown. Before we proceed to examine the equation for X_2, it is instructive to note, in (4.13), that the oscillatory component of the solution depends on

$$T - \omega_1\tau \sim (1 + \varepsilon\omega_1)t - \omega_1\varepsilon t = t.$$

Thus, at this order, ω_1 vanishes identically and so we may just as well set $\omega_1 = 0$; this is the usual simplification that is adopted in these problems. The reason for this redundancy readily becomes clear: in the definition of the fast scale, the term $\varepsilon\omega_1 t$ could be written $\omega_1\tau$ and then subsumed into the general τ-dependence of the solution. Thus to set $\omega_1 = 0$ from the outset is permitted.

The equation for X_2, from (4.7c) and with $\omega_1 = 0$, becomes

$$X_{2TT} + X_2 = (2\omega_2 A_0 - 2A_0' - A_0'')\sin T - 2(A_1'\cos\theta_1 - A_1\phi_1'\sin\theta_1) - 2A_1\cos\theta_1,$$

where $\theta_1 = T + \phi_1(\tau)$; to make plain all the terms ($\sin T$, $\cos T$) which generate secular behaviour, we further write this as

$$X_{2TT} + X_2 = (2\omega_2 A_0 - 2A_0' - A_0'')\sin T - 2(A_1 + A_1')(\cos T\cos\phi_1 - \sin T\sin\phi_1)$$
$$+ 2A_1\phi_1'(\sin T\cos\phi_1 + \cos T\sin\phi_1).$$

Thus $X_2(T, \tau)$ is periodic in T if

$$A_0'' + 2A_0' - 2\omega_2 A_0 - 2(A_1' + A_1)\sin\phi_1 - 2A_1\phi_1'\cos\phi_1 = 0; \tag{4.14}$$

$$(A_1' + A_1)\cos\phi_1 - A_1\phi_1'\sin\phi_1 = 0, \tag{4.15}$$

with initial conditions (from (4.48))

$$A_1(0)\sin[\phi_1(0)] = 0; \quad A_1(0)\cos[\phi_1(0)] = 0$$

and so

$$A_1(0) = 0 \quad \text{with} \quad \phi_1(0) \text{ arbitrary}. \tag{4.16}$$

Equation (4.15) can be written

$$(A_1 \cos\phi_1)' + A_1 \cos\phi_1 = 0 \quad \text{i.e. } A_1(\tau)\cos[\phi_1(\tau)] = C_1 \mathrm{e}^{-\tau}$$

where C_1 is an arbitrary constant; but (4.16) then requires

$$A_1(\tau)\cos[\phi_1(\tau)] = 0. \tag{4.17}$$

Equation (4.14) can be written in a similar fashion:

$$(A_1 \sin\phi_1)' + A_1 \sin\phi_1 = \tfrac{1}{2}A_0'' + A_0' - \omega_2 A_0 = -\left(\tfrac{1}{2} + \omega_2\right)\mathrm{e}^{-\tau}$$

and it follows that the solution for $A_1 \sin\phi_1$ will contain a term $\tau\mathrm{e}^{-\tau}$ unless $\omega_2 = -1/2$. We must make this choice in order to avoid a non-uniformity as $\tau \to \infty$ (because this second term in the asymptotic expansion will grow like τ relative to the first). Thus we are left with

$$A_1(\tau)\sin[\phi_1(\tau)] = 0$$

and hence, noting (4.17), we have $A_1(\tau) \equiv 0$ (and the arbitrary $\phi_1(0)$ is now unimportant). Thus we have the solution

$$X(T, \tau; \varepsilon) = \mathrm{e}^{-\tau}\sin T + \mathrm{O}(\varepsilon^2)$$

where $T \sim (1 - \frac{1}{2}\varepsilon^2)t$; this should be compared with the expansion of the exact solution, (4.2).

We have used this example to introduce and illustrate all the essential features of the technique. It should be clear that the transformation from an ordinary to a partial differential equation does not introduce any undue complications in the method of solution. We do see that we must impose periodicity and uniformity at each order, and that this produces conditions that uniquely describe the solution at the previous order. Indeed, the removal of terms that generate *secularities* is fundamental to the approach. Further, as we have seen, suitable freedom in the choice of the fast scale enables non-uniformities also to be removed. The only remaining question, at least in the context of ordinary differential equations, is how the method fares when the equation cannot be solved exactly (so we have no simple guide to the form of solution, as we did above). We explore this aspect *via* another example.

E4.2 A Duffing equation with damping

We consider the problem

$$\ddot{x} + x + \varepsilon(2\lambda\dot{x} - x^3) = 0, \quad t \geq 0, \tag{4.18}$$

with

$$x(0;\varepsilon) = 1, \quad \dot{x}(0;\varepsilon) = 0, \tag{4.19}$$

where $\varepsilon \to 0^+$ and λ (> 0) is a constant independent of ε; cf. Q3.24. We seek an asymptotic solution in the form

$$x(t;\varepsilon) \equiv X(T,\tau;\varepsilon) \sim \sum_{n=0}^{\infty} \varepsilon^n X_n(T,\tau)$$

where
$$T \sim \left(1 + \sum_{n=2}^{\infty} \varepsilon^n \omega_n\right) t, \quad \tau = \varepsilon t,$$

and note that we have omitted the term $\varepsilon\omega_1 t$ in accordance with our earlier observation. Equation (4.18) becomes

$$\left(1 + \sum_{n=2}^{\infty} \varepsilon^n \omega_n\right)^2 X_{TT} + 2\varepsilon\left(1 + \sum_{n=2}^{\infty} \varepsilon^n \omega_n\right) X_{T\tau} + \varepsilon^2 X_{\tau\tau} + X$$
$$+ 2\varepsilon\lambda\left\{\left(1 + \sum_{n=2}^{\infty} \varepsilon^n \omega_n\right) X_T + \varepsilon X_\tau\right\} - \varepsilon X^3 = 0$$

which gives the sequence of equations

$$X_{0TT} + X_0 = 0; \quad X_{1TT} + X_1 + 2X_{0T\tau} + 2\lambda X_{0T} - X_0^3 = 0; \tag{4.20a,b}$$

$$X_{2TT} + X_2 + 2\omega_2 X_{0TT} + 2X_{1T\tau} + X_{0\tau\tau} + 2\lambda(X_{1T} + X_{0\tau}) - 3X_0^2 X_1 = 0, \tag{4.20c}$$

and so on. The initial conditions, (4.19), become

$$X_0(0,0) = 1; \quad X_n(0,0) = 0 \quad \text{for } n = 1,\ 2,\ 3\ldots; \tag{4.21}$$

$$0 = X_{0T}(0,0) = X_{1T}(0,0) + X_{0\tau}(0,0) = X_{2T}(0,0) + \omega_2 X_{0T}(0,0) + X_{1T}(0,0) \tag{4.22}$$

as far as terms $O(\varepsilon^2)$. The solution to (4.20a), with (4.21) and (4.22), is

$$X_0(T,\tau) = A_0(\tau)\cos[T + \phi_0(\tau)] \quad \text{with} \quad A_0(0) = 1 \quad \text{and} \quad \phi_0(0) = 0. \tag{4.23}$$

The problem for $X_1(T,\tau)$ can be written

$$\begin{aligned} X_{1TT} + X_1 &= 2(A_0' \sin\theta_0 + A_0\phi_0' \cos\theta_0) + 2\lambda A_0 \sin\theta_0 - A_0^3 \cos^3\theta_0 \\ &= 2(A_0' + \lambda A_0)\sin\theta_0 + (2A_0\phi_0' - \tfrac{3}{4}A_0^3)\cos\theta_0 - \tfrac{1}{4}A_0^3\cos 3\theta_0, \end{aligned} \tag{4.24}$$

where $\theta_0 = T + \phi_0(\tau)$ with

$$X_1(0,0) = 0 \quad \text{and} \quad X_{1T}(0,0) = -A_0'(0). \tag{4.25}$$

The function $X_1(T, \tau)$ is periodic in T only if

$$A_0' + \lambda A_0 = 0 \quad \text{and} \quad \phi_0' = \tfrac{3}{8} A_0^2;$$

thus, with (4.23), we obtain

$$A_0(\tau) = \mathrm{e}^{-\lambda\tau}; \quad \phi_0(\tau) = \frac{3}{16\lambda}(1 - \mathrm{e}^{-2\lambda\tau}). \tag{4.26}$$

(Note that, if we allow $\lambda \to 0$, then $A_0 \to 1$ and $\phi_0 \to 3\tau/8$, which corresponds to the results obtained in Q3.24; the change of sign in ϕ_0 is because here the leading term involves *cos* rather than *sin*.) This leaves the solution of (4.24) as

$$X_1(T, \tau) = A_1(\tau)\cos[T + \phi_1(\tau)] + \tfrac{1}{32}\mathrm{e}^{-3\lambda\tau}\cos[3\{T + \phi_0(\tau)\}]$$

with $\quad A_1(0)\cos[\phi_1(0)] = -\frac{1}{32} \quad$ and $\quad A_1(0)\sin[\phi_1(0)] = -\lambda.$

The analysis of equation (4.20c), for $X_2(T, \tau)$, follows the same pattern (and see also E4.1), but here the details are considerably more involved; we will not pursue this calculation any further (for we learn nothing of significance, other than to show that $\omega_2 = -1/(2\lambda^2)$, which is left as an exercise). The solution, to this order, is therefore

$$x \sim \mathrm{e}^{-\lambda\tau}\cos\left[t + \tfrac{3}{16\lambda}(1 - \mathrm{e}^{-2\lambda\tau})\right] + \varepsilon\left\{A_1(\tau)\cos[t + \phi_1(\tau)] + \tfrac{1}{32}\mathrm{e}^{-3\lambda\tau}\cos[3t + \tfrac{9}{16\lambda}(1 - \mathrm{e}^{-2\lambda\tau})]\right\}.$$

Other examples based on small adjustments to the equation for a linear oscillator can be found in exercises Q4.1–4.9. It might be anticipated, in the context of oscillators governed by ordinary differential equations, that the method of multiple scales is successful only if the underlying problem is a *linear* oscillation i.e. controlled by an equation such as $\ddot{x} + x = 0$; this would be false. In an important extension of these techniques, Kuzmak (1959) showed that they work equally well when the oscillator is predominantly *nonlinear*. Of course, the fundamental oscillation will no longer be represented by functions like *sin* or *cos*, but by functions that are solutions of nonlinear equations e.g. the *Jacobian elliptic functions*.

4.2 NONLINEAR OSCILLATORS

Equations such as

$$\ddot{x} + x + x^3 = 0, \quad t \geq 0,$$

possess solutions that can be expressed in terms of *sn*, *cn* or *dn*, for example. (In the Appendix we present all the basic information about these functions that is necessary for the calculations that we describe.) We discuss an example, taken from Kuzmak (1959), which exemplifies this technique.

E4.3 A nonlinear oscillator with slowly varying coefficients

We consider the problem

$$\ddot{x} + a(\varepsilon t)x + b(\varepsilon t)x^3 = 0, \quad t \geq 0, \tag{4.27}$$

for $\varepsilon \to 0^+$, together with some suitable initial data; this is a Duffing equation with slowly varying coefficients. The equation clearly implies that the slow scale should be $\tau = \varepsilon t$, but what do we use for the fast scale? Here, we define a general form of fast scale, T, by

$$\frac{dT}{dt} = \theta(\tau), \tag{4.28}$$

where $\theta(\tau)$ is to be determined, and the period of the oscillation in T is defined to be a *constant*—an essential requirement in the application of multiple scales in this problem. (In some more involved problems, it might be necessary to introduce $\theta(\tau;\varepsilon) \sim \sum_{n=0}^{\infty} \varepsilon^n \theta_n(\tau)$.) Equation (4.27) is transformed according to

$$\frac{d}{dt} \equiv \theta(\tau)\frac{\partial}{\partial T} + \varepsilon\frac{\partial}{\partial \tau},$$

which gives, with $x(t;\varepsilon) \equiv X(T, \tau; \varepsilon)$,

$$\theta^2 X_{TT} + 2\varepsilon\theta X_{T\tau} + \varepsilon\theta' X_T + \varepsilon^2 X_{\tau\tau} + a(\tau)X + b(\tau)X^3 = 0 \tag{4.29}$$

and we seek a solution

$$X(T, \tau; \varepsilon) \sim \sum_{n=0}^{\infty} \varepsilon^n X_n(T, \tau) \tag{4.30}$$

where each X_n is periodic in T. (The boundedness and uniform validity, that we aim for as $\tau \to \infty$, will depend on the particular $a(\tau)$, $b(\tau)$; we will assume that these functions allow this.) Now (4.30) in (4.29) yields the sequence of equations

$$\theta^2 X_{0TT} + a X_0 + b X_0^3 = 0; \tag{4.31}$$

$$\theta^2 X_{1TT} + a X_1 + 3b X_0^2 X_1 + 2\theta X_{0T\tau} + \theta' X_{0T} = 0, \tag{4.32}$$

as far as $O(\varepsilon)$. Note that, although the first equation—for $X_0(T, \tau)$—is necessarily nonlinear, all equations thereafter are *linear*.

An exact solution of equation (4.31) is

$$X_0(T,\tau) = A_0(\tau)\mathrm{sn}[4K(m)(T-T_0); m(\tau)] \tag{4.33}$$

where
$$16(1+m)\theta^2K^2 = a \quad \text{and} \quad A_0^2 b + \left(\frac{2m}{1+m}\right) a = 0. \tag{4.34a,b}$$

(The confirmation of this, using the properties given in the Appendix, is left as an exercise.) Here, T_0 is a constant (to ensure that the period in T is a constant which, with the chosen form (4.33), is 1). Given $a(\tau)$ and $b(\tau)$, equations (4.34a,b) provide two equations for the three unknown functions: $A_0(\tau)$, $m(\tau)$ and $\theta(\tau)$. A third equation is obtained by imposing periodicity on $X_1(T,\tau)$.

The task of solving (4.32) is not as difficult as it might appear; the important manoeuvre is to write

$$X_1(T,\tau) = f(T,\tau)X_{0T}(T,\tau) \tag{4.35}$$

and then to solve for $f(T,\tau)$. Equation (4.32), with (4.35), gives

$$\theta^2(f\,X_{0TTT} + 2f_T X_{0TT} + f_{TT}X_{0T}) + af\,X_{0T} + 3bf\,X_0^2 X_{0T} + 2\theta\,X_{0T\tau} + \theta' X_{0T} = 0; \tag{4.36}$$

but from (4.31) we also have that

$$\theta^2 X_{0TTT} + a\,X_{0T} + 3b\,X_0^2 X_{0T} = 0$$

and so (4.36) becomes, after multiplication by X_{0T},

$$\theta^2\left(f_T X_{0T}^2\right)_T = -\left(2\theta\,X_{0T}X_{0T\tau} + \theta' X_{0T}^2\right).$$

Now, for $X_1(T,\tau)$ to be periodic with period 1, $f(T,\tau)$ must be similarly periodic (because X_{0T} is, in (4.35)). Thus we must have, for any T,

$$\int_T^{T+1}\left(2\theta\,X_{0T}X_{0T\tau} + \theta' X_{0T}^2\right)\mathrm{d}T = \int_0^1\left(2\theta\,X_{0T}X_{0T\tau} + \theta' X_{0T}^2\right)\mathrm{d}T = 0$$

which can be written

$$\frac{\mathrm{d}}{\mathrm{d}\tau}\left(\int_0^1 \theta X_{0T}^2\,\mathrm{d}T\right) = 0 \quad \text{or} \quad \theta\int_0^1 X_{0T}^2\,\mathrm{d}T = \text{constant}, \tag{4.37}$$

which is our third relation. When (4.33) is used in (4.37), the integration is possible (but this does require some additional skills with, and knowledge of, elliptic functions and integrals; see e.g. Byrd & Friedman, 1971), to give

$$\theta^2\frac{A_0^2K(m)}{m}\left[(1+m)E(m) - (1-m)K(m)\right] = \text{constant}; \tag{4.38}$$

initial data will determine both this constant and T_0. Then, for known and suitable $a(\tau)$ and $b(\tau)$, equations (4.34a,b) and (4.38) enable the complete description of the first term in the asymptotic expansion (4.30).

This example has demonstrated that we are not restricted to nearly-linear oscillations, although we must accept that mathematical intricacies, and the required mathematical skills, are rather more extensive here than in the two previous examples. In addition, problems of this type, because they are strongly nonlinear, often force us to address other difficulties: we have used a periodicity condition, (4.37), but this fails if the solution is not periodic—and this can happen. If the solution evolves so that $m \to 1$, then the periodicity is lost because the period becomes infinite in this limit. In this situation, it is necessary to match the solution for $m = 1$ to the periodic solution approximated as $m \to 1$; an example of this procedure can be found in Johnson (1970). Some additional material related to this topic is available in Q4.10 & 4.11.

4.3 APPLICATIONS TO CLASSICAL ORDINARY DIFFERENTIAL EQUATIONS

The method of multiple scales is particularly useful in the analysis of certain types of ordinary differential equation which incorporate a suitable small parameter. We will discuss three such problems, the first of which we have already encountered: the Mathieu equation (§3.4 and E3.6). The next two involve a discussion of a particular class of problems—associated with the presence or absence of turning points (see §2.8)—with a solution-technique usually referred to as *WKB* (or, sometimes, *WKBJ*); we will write more of this later.

The Mathieu equation, discussed in §3.4, is

$$\ddot{x} + (\delta + \varepsilon \cos t)x = 0$$

to which we will apply the method of multiple scales. However, before we undertake this, we need to know what the fast and slow scales should be; this requires a little care. Let us consider the equation with $\delta = \omega^2$ (fixed independent of ε):

$$\ddot{x} + \omega^2 x = -\varepsilon x \cos t \quad \text{with} \quad x(t;\varepsilon) \sim \sum_{n=0}^{\infty} \varepsilon^n x_n(t),$$

then we could select $x_0 = \cos(\omega t)$. The equation for x_1 becomes

$$\ddot{x}_1 + \omega^2 x_1 = -\cos t \cos(\omega t) = -\tfrac{1}{2}\left[\cos\{(\omega+1)t\} + \cos\{(\omega-1)t\}\right].$$

In general, we find that each $x_n(t)$ has a particular integral proportional to $\cos[(\omega \pm n)t]$, *unless* $\omega \pm n = \pm\omega$ and then we have particular integrals that grow in t. This condition will occur for $\omega = n/2$ and so the *critical* values of δ are $\delta = n^2/4$ ($n = 0,\ 1,\ 2\ \ldots$). From these points on the ε-axis will emanate the transitional curves, in the (ε,δ)-plane, which separate purely oscillatory from exponentially growing solutions; see §3.4. Further, for any given n, the asymptotic expansion will take the form

$$x(t;\varepsilon) \sim \cos(\tfrac{n}{2}t) + \cdots\cdots + \varepsilon^n t \times (\text{trigonometric terms})$$

and so we should then use the slow scale $\tau = \varepsilon^n t$. We note that the fast scale can be taken simply as t. With these points in mind, we consider an example in some detail.

E4.4 Mathieu's equation ($n = 1$) for $\varepsilon \to 0$

We use the method of multiple scales for the equation

$$\ddot{x} + [\delta(\varepsilon) + \varepsilon \cos t]x = 0, \quad t \geq 0, \tag{4.39}$$

with $\delta(0) = 1/4$; thus we have the case $n = 1$ and so we introduce the scales

$$T = t \quad \text{and} \quad \tau = \varepsilon t.$$

Equation (4.39), with $x(t;\varepsilon) \equiv X(T, \tau; \varepsilon)$, then becomes

$$X_{TT} + 2\varepsilon X_{T\tau} + \varepsilon^2 X_{\tau\tau} + \left(\tfrac{1}{4} + \sum_{n=1}^{\infty} \varepsilon^n \delta_n + \varepsilon \cos T\right) X = 0,$$

where δ has been expanded in the usual way, and then we seek a solution

$$X(T, \tau; \varepsilon) \sim \sum_{n=0}^{\infty} \varepsilon^n X_n(T, \tau); \tag{4.40}$$

thus we generate the sequence of equations

$$X_{0TT} + \tfrac{1}{4} X_0 = 0; \quad X_{1TT} + \tfrac{1}{4} X_1 + 2X_{0T\tau} + (\delta_1 + \cos T)X_0 = 0, \tag{4.41a,b}$$

and so on.

Equation (4.41a) has the general solution

$$X_0(T, \tau) = A_0(\tau)\sin(\tfrac{1}{2}T) + B_0(\tau)\cos(\tfrac{1}{2}T)$$

and then (4.41b) can be written

$$\begin{aligned} X_{1TT} + \tfrac{1}{4} X_1 &= -2X_{0T\tau} - (\delta_1 + \cos T)X_0 \\ &= -\left[A_0' \cos(\tfrac{1}{2}T) - B_0' \sin(\tfrac{1}{2}T)\right] - \delta_1\left[A_0 \sin(\tfrac{1}{2}T) + B_0 \cos(\tfrac{1}{2}T)\right] \\ &\qquad - \tfrac{1}{2} A_0\left[\sin(\tfrac{3}{2}T) - \sin(\tfrac{1}{2}T)\right] - \tfrac{1}{2} B_0\left[\cos(\tfrac{3}{2}T) + \cos(\tfrac{1}{2}T)\right]. \end{aligned}$$

Thus a solution for X_1 which is periodic in T requires

$$A_0' + \delta_1 B_0 + \tfrac{1}{2} B_0 = 0; \quad B_0' - \delta_1 A_0 + \tfrac{1}{2} A_0 = 0, \tag{4.42}$$

which possess the general solution

$$A_0(\tau) = \alpha_0 e^{\lambda\tau}; \quad B_0(\tau) = \beta_0 e^{\lambda\tau}$$

where $\lambda = \pm\sqrt{\frac{1}{4} - \delta_1^2}$ and α_0, β_0 are arbitrary—possibly complex which then requires the complex conjugate to be included—constants. We see that

(a) if $|\delta_1| > \frac{1}{2}$ then $A_0(\tau)$ and $B_0(\tau)$ are oscillatory and so $X_0(T, \tau)$ is oscillatory in both T and τ;
(b) if $|\delta_1| < \frac{1}{2}$ then there exists a solution which grows exponentially;
(c) if $\delta_1 = \pm\frac{1}{2}$ then (see (4.42)) one of $A_0(\tau)$ or $B_0(\tau)$ is constant and the other grows linearly.

The method of multiple scales has enabled us, albeit in the limit $\varepsilon \to 0^+$, to describe all the essential features of solutions of Mathieu's equation, and how these change as the parameters select different positions and regions in (ε,δ)-space. The corresponding problem for $n = 2$ is discussed in Q4.12, and related exercises are given in Q4.13 & 4.14.

We now turn to an important class of problems that are exemplified by the equation (cf. Q2.24)

$$y'' + a(x;\varepsilon)y = 0, \quad x \in D, \tag{4.43}$$

where $a(x;\varepsilon)$ is given. In the simplest problem of this type, $a(x;\varepsilon)$ takes one sign throughout the given domain (D) i.e. $a > 0$ (oscillatory) or $a < 0$ (exponential). A more involved situation arises if a changes sign in the domain: a turning-point problem (see Q2.24). The intention here is to examine the solution of the equation in the case $a(x;\varepsilon) = \alpha(\varepsilon x)$, so that, for $\varepsilon \to 0^+$, the coefficient is slowly varying. We use the method of multiple scales to analyse this problem and hence give a presentation of the technique usually referred to as 'WKB'. (This is after Wentzel, 1926; Kramers, 1926; Brillouin, 1926, although the essential idea can be traced back to Liouville and Green. Some authors extend the label to *WKBJ*, to include Jeffreys, 1924.) We will formulate an oscillatory problem ($\alpha > 0$) and use this example to describe the WKB approach.

E4.5 WKB method for a slowly-varying oscillation

We consider

$$\ddot{x} + \omega^2(\varepsilon t)x = 0, \quad t \geq 0, \tag{4.44}$$

with $\varepsilon \to 0^+$, $\omega > 0$ for $t \geq 0$ and with suitable initial conditions. The most natural and convenient formulation of the multiple-scale problem follows E4.3: we introduce

$$\tau = \varepsilon t \quad \text{and} \quad \frac{\mathrm{d}T}{\mathrm{d}t} = \Omega(\tau;\varepsilon)$$

and seek a solution

$$x(t;\varepsilon) \equiv X(T,\tau;\varepsilon) \sim \sum_{n=0}^{\infty} \varepsilon^n X_n(T,\tau). \tag{4.45}$$

Equation (4.44) becomes

$$\Omega^2 X_{TT} + 2\varepsilon\Omega X_{T\tau} + \varepsilon\Omega' X_T + \varepsilon^2 X_{\tau\tau} + \omega^2 X = 0$$

and, in this first exercise, we will not expand Ω (but see Q4.15). Thus (4.45) leads to the sequence of equations

$$\Omega^2 X_{0TT} + \omega^2 X_0 = 0; \quad \Omega^2 X_{1TT} + \omega^2 X_1 + 2\Omega X_{0T\tau} + \Omega' X_{0T} = 0, \tag{4.46a,b}$$

for the $\mathrm{O}(1)$ and $\mathrm{O}(\varepsilon)$ problems.

To solve (4.46a), we first select $\Omega = \omega$, and then obtain the general solution

$$X_0(T,\tau) = A_0(\tau)\cos[T + \phi_0(\tau)],$$

where A_0 and ϕ_0 are arbitrary functions. (Other choices of Ω lead to a formulation in terms of ω/Ω, but then the period will depend on τ, which leads to additional non-uniformities—see E4.6—unless ω/Ω is constant; we have made the simplest choice of this constant.) The equation for X_1 now becomes

$$\begin{aligned}\omega^2(X_{1TT} + X_1) &= -2\omega X_{0T\tau} - \omega' X_{0T}\\ &= 2\omega(A_0' \sin\theta_0 + A_0\phi_0' \cos\theta_0) + A_0\omega' \sin\theta_0\end{aligned}$$

where we have written $\theta_0 = T + \phi_0(\tau)$. The solution for X_1 is periodic in T, i.e. in θ_0, if

$$2\omega A_0' + \omega' A_0 = 0; \quad A_0\phi_0' = 0$$

and so

$$\omega A_0^2 = \text{constant and } \phi_0 = \text{constant}.$$

So, for example, if we are given the initial data $x(0;\varepsilon) = 1$, $\dot{x}(0;\varepsilon) = 0$, then we have

$$A_0(0) = 1, \quad \phi_0(0) = 0 \quad \text{i.e. } \omega A_0^2 = \omega(0) \quad \text{and} \quad \phi_0(\tau) \equiv 0$$

and a solution (with $T = 0$ at $\tau = 0$, say), expressed in terms of τ, is

$$x \equiv X \sim \sqrt{\frac{\omega(0)}{\omega(\tau)}} \cos\left\{\frac{1}{\varepsilon}\int_0^{\tau} \omega(\tau')\mathrm{d}\tau'\right\}. \tag{4.47}$$

This describes a *fast* oscillation (by virtue of the factor ε^{-1}) with a *slow* evolution of the amplitude; these are the salient features of a WKB(J) solution. (Note that a property of this solution is $\omega A_0^2 =$ constant, which is usually called 'action'; typically, energy $(E) \propto \omega^2 A_0^2$ and so E/ω (action) is conserved—not the energy itself.)

The problem of finding higher-order terms in the WKB solution is addressed in Q4.15, and the corresponding problem with $a(x;\varepsilon) = \alpha(\varepsilon x) < 0$ is discussed in Q4.16, and an interesting associated problem is discussed in Q4.17. We now consider the case of a *turning point.*

It is apparent that the solution (4.47) is not valid if $\omega(\tau) = 0$, which is the case at a turning point. In (4.43), we will write $a(x;\varepsilon) = \varepsilon x f(\varepsilon x)$, with $f(\varepsilon x) > 0$ throughout the domain D, and analytic (to the extent that $f(\varepsilon x)$ may be written as a uniformly valid asymptotic expansion, as $\varepsilon \to 0^+$, for $\forall x \in D$). This choice of $a(x;\varepsilon)$ has a single (simple) turning point at $x = 0$; a turning point elsewhere can always be moved to $x = 0$ by a suitable origin shift. The intention is to find a solution valid near the turning point and then, away from this region, use the WKB method in $x < 0$ and in $x > 0$. Thus the turning-point solution is to be inserted between the two WKB solutions and, presumably, matched appropriately. We will present all these ideas, using the method of multiple scales valid as $\varepsilon \to 0^+$, in the following example.

E4.6 A turning-point problem

We consider

$$y'' + \varepsilon x f(\varepsilon x) y = 0, \qquad -x_0 \le x \le x_1, \tag{4.48}$$

where both x_0 and x_1 are positive, O(1) constants, $f(\varepsilon x) > 0$ and analytic for $\forall x \in [-x_0, x_1]$, and $\varepsilon \to 0^+$. The turning point is at $x = 0$, and the first issue is to decide what scales to use in the neighbourhood of this point; this has already been addressed in §2.8. Let us write $\varepsilon x = \delta(\varepsilon) Z$ (and any scaling on y is redundant, in so far as the governing equation is concerned, because the equation, (4.48), is linear) to give

$$\left(\frac{\varepsilon}{\delta}\right)^2 \frac{\mathrm{d}^2 y}{\mathrm{d}Z^2} + \delta Z f(\delta Z) y = 0$$

and so a balance of terms is possible if $\delta = \varepsilon^{2/3}$ i.e. $x = \varepsilon^{-1/3} Z$—a fast scale. The slow scale is simply $X = \varepsilon x$. However, a more convenient choice of the fast variable (cf. §2.7) is

$$Z = \varepsilon^{1/3} x h(\varepsilon x) = \varepsilon^{-2/3} X h(X) \tag{4.49}$$

where $h(X) > 0$ is to be determined; then we have the operator identity

$$\frac{\mathrm{d}}{\mathrm{d}x} \equiv \varepsilon \frac{\partial}{\partial X} + \varepsilon^{1/3}[Xh(X)]' \frac{\partial}{\partial Z}. \tag{4.50}$$

Finally, before we use this to transform equation (4.48), we need to decide how to replace x in the coefficient $\varepsilon x f(\varepsilon x)$ in equation (4.48): should we use Z or X or both? Now the important property of the fast scale, (4.49), is that it is zero at the turning point; thus we elect to write any part of a coefficient which has this same property in terms of Z, and otherwise use X. With this in mind, we use (4.49) and (4.50) in (4.48) and then, with $y(x;\varepsilon) \equiv Y(Z, X; \varepsilon)$, we obtain

$$g^2 Y_{ZZ} + \varepsilon^{2/3}(2g\, Y_{ZX} + g' Y_Z) + \varepsilon^{4/3} Y_{XX} + Z\frac{f(X)}{h(X)} Y = 0,$$

where, for simplicity, we have written

$$g(X) = [Xh(X)]' = h(X) + Xh'(X) \quad (\neq 0 \text{ on } X = 0). \tag{4.51}$$

We seek a (bounded) solution in the form

$$Y(Z, X; \varepsilon) \sim \sum_{n=0}^{\infty} \varepsilon^{2n/3} Y_n(Z, X)$$

and hence we obtain the sequence of equations

$$g^2 Y_{0ZZ} + Z\frac{f}{h} Y_0 = 0; \quad g^2 Y_{1ZZ} + Z\frac{f}{h} Y_1 + 2g\, Y_{0ZX} + g' Y_{0Z} = 0, \tag{4.52a,b}$$

and so on.

The bounded solution of (4.52a) can be expressed in terms of the *Airy function*, Ai (see e.g. Abramowitz & Stegun, 1964). At this stage, however, we have not yet made a suitable choice for $h(X)$; let us choose

$$g^2 = f/h \tag{4.53}$$

and then we have

$$Y_0(Z, X) = A_0(X)\mathrm{Ai}(Z).$$

This solution is oscillatory for $Z > 0$ and exponentially decaying for $Z < 0$; in particular

$$\mathrm{Ai}(Z) \sim \frac{1}{\sqrt{\pi}} Z^{-1/4} \sin\left(\tfrac{2}{3} Z^{3/2} + \tfrac{\pi}{4}\right) \quad \text{as } Z \to +\infty \tag{4.54a}$$

and

$$\mathrm{Ai}(Z) \sim \frac{1}{2\sqrt{\pi}} (-Z)^{-1/4} \exp\left[-\tfrac{2}{3}(-Z)^{3/2}\right] \quad \text{as } Z \to -\infty. \tag{4.54b}$$

The equation for Y_1 now becomes

$$Y_{1ZZ} + ZY_1 = -\frac{2}{g} A_0' \mathrm{Ai}' - \frac{g'}{g^2} A_0 \mathrm{Ai}',$$

and a particular integral of this equation is necessarily proportional to $Z\mathrm{Ai}(Z)$, which immediately leads to a non-uniformity in $Y \sim Y_0 + \varepsilon^{2/3} Y_1$ as $|Z| \to \infty$. Thus we must select

$$\frac{g'}{g^2} A_0 + \frac{2}{g} A_0' = 0 \quad \text{or} \quad g A_0^2 = \text{constant};$$

it is left as an exercise to show that, if we had written $Y_{0ZZ} + k(X) Y_0 = 0$, $k = f/(hg^2)$, then another non-uniformity would be present unless $k = $ constant, and we have already set $k = 1$. We alluded to this difficulty at the end of E4.5. Finally, the leading-order solution will be completely determined once we have found $h(X)$ (introduced in (4.49)).

From (4.53) and (4.51), we have the equation

$$[h + Xh']^2 h = f;$$

we consider the case $X > 0$, then it is convenient to write

$$G(X) = Xh(X) \quad \text{so that} \quad G(G')^2 = Xf(X).$$

This is

$$G'\sqrt{G} = \pm\sqrt{Xf(X)}$$

and so

$$\tfrac{2}{3} G^{3/2} = \pm \int^X \sqrt{X' f(X')}\, \mathrm{d}X'$$

which gives (with the appropriate choice of sign)

$$G(X) = Xh(X) = \left[\tfrac{3}{2} \int_0^X \sqrt{X' f(X')}\, \mathrm{d}X'\right]^{2/3} \tag{4.55}$$

with $X > 0$ and $h(X) > 0$. (It follows directly from this result that

$$h(X) \to [f(0)]^{1/3} \quad \text{as } X \to 0^+.)$$

The case $X < 0$, where the corresponding choice is $G(X) = -Xh(X)$, is left as an exercise; you should find that $h(x) \to [f(0)]^{1/3}$ as $X \to 0^-$: a continuous $h(X)$ can be defined.

In order to complete the calculation, we require the solution in $X > 0$ and in $X < 0$, away from the neighbourhood of the turning point. Following E4.5 we have, for $X > 0$ and writing $\omega = \sqrt{Xf(X)}$,

$$y \sim \alpha \left[\frac{X_1 f(X_1)}{Xf(X)} \right]^{1/4} \sin \left\{ \varepsilon^{-1} \int_X^{X_1} \sqrt{X' f(X')}\, dX' + \theta_0 \right\} \tag{4.56a}$$

where $X_1 = \varepsilon x_1$, and α and θ_0 are arbitrary constants (because we will not impose any particular conditions at $x = x_1$). For $X < 0$ (see Q4.16) we obtain

$$y \sim \beta \left[\frac{-X_0 f(X_0)}{Xf(X)} \right]^{1/4} \exp \left\{ -\varepsilon^{-1} \int_{-X_0}^{X} \sqrt{-X' f(X')}\, dX' \right\} \tag{4.56b}$$

for a bounded solution (as $\varepsilon \to 0^+$) and with $y(-x_0; \varepsilon) = y(-X_0/\varepsilon; \varepsilon) = \beta$. (Remember that, since the original equation, (4.48), is second order, only two boundary conditions may be independently assigned.)

The solution in the neighbourhood of the turning point is

$$y \sim Y_0 = \frac{\gamma}{\sqrt{G'(X)}} \mathrm{Ai}(Z) \tag{4.57}$$

where $G(X)$ is given by (4.55) and γ is an arbitrary constant. The final task is therefore to match (4.57) with (4.56a,b). First, in $Z < 0$, $X < 0$, we have (from (4.57) and (4.54b))

$$\begin{aligned} y &\sim \gamma \left(\frac{h}{f} \right)^{1/4} \frac{1}{2\sqrt{\pi}} \left[-\varepsilon^{-2/3} Xh(X) \right]^{-1/4} \exp \left\{ -\frac{2}{3} \left[-\varepsilon^{-2/3} Xh(X) \right]^{3/2} \right\} \\ &= \frac{\varepsilon^{-1/6} \gamma}{2\sqrt{\pi}} [-Xf(X)]^{-1/4} \exp \left\{ -\frac{2}{3} \varepsilon^{-1} [-Xh(X)]^{3/2} \right\}. \end{aligned}$$

From (4.56b) we have

$$y \sim \beta \left[\frac{X_0 h(X_0)}{-Xh(X)} \right]^{1/4} \exp \left\{ -\frac{2}{3} \varepsilon^{-1} \sqrt{f(0)} (-X)^{3/2} - \varepsilon^{-1} \int_{-X_0}^{0} \sqrt{-Xf(X)}\, dX \right\}$$

where we have used (4.55) (and note that $h^{3/2} \to \sqrt{f(0)}$); thus matching occurs if we choose

$$\frac{\varepsilon^{-1/6}\gamma}{2\sqrt{\pi}} = \beta[-X_0 f(X_0)]^{1/4} \exp\left\{-\varepsilon^{-1}\int_{-X_0}^{0} \sqrt{-Xf(X)}\,\mathrm{d}X\right\}.$$

For $Z > 0$, $X > 0$, from (4.57) and (4.54a), we have

$$\begin{aligned} y &\sim \gamma\left(\frac{h}{f}\right)^{1/4} \frac{1}{\sqrt{\pi}}\left[\varepsilon^{-2/3} Xh(X)\right]^{-1/4} \sin\left\{\frac{2}{3}\left[\varepsilon^{-2/3} Xh(X)\right]^{3/2} + \frac{\pi}{4}\right\} \\ &= \frac{\varepsilon^{-1/6}\gamma}{\sqrt{\pi}}[Xf(X)]^{-1/4} \sin\left\{\frac{2}{3}\varepsilon^{-1}[Xh(X)]^{3/2} + \frac{\pi}{4}\right\}; \end{aligned}$$

from (4.56a) we obtain

$$y \sim \alpha\left[\frac{X_1 f(X_1)}{Xf(X)}\right]^{1/4} \sin\left\{-\frac{2}{3}\varepsilon^{-1}\sqrt{f(0)}\,X^{3/2} + \varepsilon^{-1}\int_0^{X_1} \sqrt{Xf(X)}\,\mathrm{d}X + \theta_0\right\}$$

which also matches if we choose

$$\frac{\varepsilon^{-1/6}\gamma}{\sqrt{\pi}} = -\alpha\,[X_1 f(X_1)]^{1/4} \quad \text{and} \quad \theta_0 + \varepsilon^{-1}\int_0^{X_1} \sqrt{Xf(X)}\,\mathrm{d}X = -\frac{\pi}{4}.$$

The matching conditions are usually called, in this context, *connection formulae*: they 'connect' the solutions on either side of the turning point i.e. the relation between α and β. Here, we have three relations between the four constants α, β, γ and θ_0, so only one is free; that only one occurs here is because, of the two (independent) boundary conditions that we may prescribe, one has been fixed by seeking a bounded solution in $X < 0$ i.e. the exponentially growing solution has already been excluded.

A number of other examples of turning-point problems are offered in the exercises; see Q4.18–4.21. This completes all that we will write about the routine applications to ordinary differential equations; we now take a brief look at how these same techniques are relevant to the study of partial differential equations.

4.4 APPLICATIONS TO PARTIAL DIFFERENTIAL EQUATIONS

It is no accident that we will discuss partial differential equations which are associated with wave propagation; this type of equation is analogous to oscillatory solutions of ordinary differential equations. (These two categories of equations are the most natural vehicles for the method of multiple scales, although others are certainly possible.) In particular we will start with an equation that has become a classical example of its type: Bretherton's model equation for the weak, nonlinear interaction of dispersive waves (Bretherton, 1964).

E4.7 Bretherton's equation

The equation that we will discuss is

$$\phi_{tt} + \phi_{xx} + \phi_{xxxx} + \phi = \varepsilon\phi^3, \tag{4.58}$$

where $\phi(x, t; \varepsilon)$ is defined in $t \geq 0$ and $-\infty < x < \infty$; we will further assume that we have suitable initial data for the type of solution that we seek. The aim is to produce a solution, *via* the method of multiple scales, for $\varepsilon \to 0^+$. First we observe that, with $\varepsilon = 0$, there is a solution

$$\phi = a\cos(kx - \omega t) \tag{4.59}$$

where, given k (the *wave number*), $\omega(k)$ (the *frequency*) is defined by the *dispersion relation*:

$$\omega^2 = k^4 - k^2 + 1 = (k^2 - 1)^2 + k^2 > 0. \tag{4.60}$$

The presence of the parameter, ε, together with a naïve asymptotic solution (generating terms proportional to εt or εx), suggests that we must expect changes on the slow scales εx and εt. The fast scale is defined in much the same way that we adopted for E4.3 and E4.5; thus we write

$$X = \varepsilon x, \quad T = \varepsilon t, \quad \frac{\partial\theta}{\partial x} = k(X, T; \varepsilon), \quad \frac{\partial\theta}{\partial t} = -\omega(X, T; \varepsilon). \tag{4.61}$$

The solution now sits in a domain in 3-space, defined by $T \geq 0$, $-\infty < X < \infty$, $-\infty < \theta < \infty$. A solution described by these variables will have the property that both the wave number and the frequency slowly evolve. (Note that the correct form of the solution with $\varepsilon = 0$ is recovered if $k =$ constant and $\omega =$ constant.) The retention of the parameter in the definitions of k and ω allows us to treat these functions as asymptotic expansions, if that is useful and relevant; often this is unnecessary.

From (4.61) we have the operator identities

$$\frac{\partial}{\partial x} \equiv k\frac{\partial}{\partial\theta} + \varepsilon\frac{\partial}{\partial X}, \quad \frac{\partial}{\partial t} \equiv -\omega\frac{\partial}{\partial\theta} + \varepsilon\frac{\partial}{\partial T}$$

and also

$$\frac{\partial^2}{\partial t^2} \equiv \omega^2\frac{\partial^2}{\partial\theta^2} - 2\varepsilon\omega\frac{\partial^2}{\partial\theta\partial T} - \varepsilon\omega_T\frac{\partial}{\partial\theta} + \varepsilon^2\frac{\partial^2}{\partial T^2},$$

etc., as far as $\partial^4/\partial x^4$. It is sufficient, for the results that we present here, to transform equation (4.58) but retain terms no smaller than $O(\varepsilon)$; thus, with

$\phi(x, t; \varepsilon) \equiv \Phi(\theta, X, T; \varepsilon)$ we obtain

$$\omega^2\Phi_{\theta\theta} - 2\varepsilon\omega\Phi_{\theta T} - \varepsilon\omega_T\Phi_\theta + k^2\Phi_{\theta\theta} + 2\varepsilon k\Phi_{\theta X} + \varepsilon k_X\Phi_\theta$$
$$+ k^4\Phi_{\theta\theta\theta\theta} + 4\varepsilon k^3\Phi_{\theta\theta\theta X} + 6\varepsilon k^2 k_X\Phi_{\theta\theta\theta} + \Phi - \varepsilon\Phi^3 = \mathrm{O}(\varepsilon^2). \tag{4.62}$$

Before we proceed with the details, we should make an important observation. The transform, (4.61), defining the fast scale θ (usually called the *phase* in these problems), implies a consistency condition that must exist if θ is a twice-differentiable function, namely

$$\frac{\partial^2\theta}{\partial x\partial t} = \varepsilon\frac{\partial k}{\partial T} = -\varepsilon\frac{\partial\omega}{\partial X} \quad \text{i.e. } k_T + \omega_X = 0. \tag{4.63}$$

this additional equation is called the *conservation of waves* (or of *wave crests*), and it arises quite naturally from an elementary argument. Consider (one-dimensional) waves entering and leaving the region $[x_0, x]$; the number of waves, per unit time, crossing x_0 into the region is given as ω_0 and the number leaving, across x, we will write as ω. The total number of waves (wave crests) between x_0 and x is $\int_{x_0}^{x} k\mathrm{d}x$; if the number of waves does not change (which is what is typically observed, even if they change shape) then

$$\frac{\partial}{\partial t}\int_{x_0}^{x} k\mathrm{d}x = \omega_0 - \omega$$

or, upon allowing differentiation with respect to x,

$$k_t + \omega_x = 0,$$

which immediately recovers (4.63) if the dependence on (x, t) is *via* (X, T).

Returning to equation (4.62), we seek a solution

$$\Phi(\theta, X, T; \varepsilon) \sim \sum_{n=0}^{\infty} \varepsilon^n\Phi_n(\theta, X, T)$$

with
$$k(X, T; \varepsilon) \sim \sum_{n=0}^{\infty} \varepsilon^n k_n(X, T), \quad \omega(X, T; \varepsilon) \sim \sum_{n=0}^{\infty} \varepsilon^n\omega_n(X, T), \tag{4.64}$$

and then we obtain the sequence of equations

$$\omega_0^2\Phi_{0\theta\theta} + k_0^2\Phi_{0\theta\theta} + k_0^4\Phi_{0\theta\theta\theta\theta} + \Phi_0 = 0; \tag{4.65}$$

$$\omega_0^2\Phi_{1\theta\theta} + k_0^2\Phi_{1\theta\theta} + k_0^4\Phi_{1\theta\theta\theta\theta} + \Phi_1 + 2\omega_0\omega_1\Phi_{0\theta\theta} + 2k_0k_1\Phi_{0\theta\theta} + 4k_0^3k_1\Phi_{0\theta\theta\theta\theta} - 2\omega_0\Phi_{0\theta T}$$
$$- \omega_{0T}\Phi_{0\theta} + 2k_0\Phi_{0\theta X} + k_{0X}\Phi_{0\theta} + 4k_0^3\Phi_{0\theta\theta\theta X} + 6k_0^2k_{0X}\Phi_{0\theta\theta\theta} - \Phi_0^3 = 0, \tag{4.66}$$

and so on. We will take the solution of (4.65) to be (cf. (4.59))

$$\Phi_0(\theta, X, T) = A_0(X, T)\cos\theta \tag{4.67}$$

and this requires

$$\omega_0^2 = k_0^4 - k_0^2 + 1. \tag{4.68}$$

The equation for Φ_1, (4.66), will include terms generated by Φ_0 which involve $\sin\theta$ or $\cos\theta$—and all these must be removed if Φ_1 is to be periodic. The terms in $\cos\theta$ give

$$-2\omega_0\omega_1 A_0 - 2k_0 k_1 A_0 + 4k_0^3 k_1 A_0 - \tfrac{3}{4}A_0^3 = 0; \tag{4.69}$$

and those in $\sin\theta$:

$$2\omega_0 A_{0T} + \omega_{0T} A_0 - 2k_0 A_{0X} - k_{0X} A_0 + 6k_0^3 k_{0X} A_0 + 4k_0^3 A_{0X} = 0, \tag{4.70}$$

leaving

$$\omega_0^2 \Phi_{1\theta\theta} + k_0^2 \Phi_{1\theta\theta} + k_0^4 \Phi_{1\theta\theta\theta\theta} + \Phi_1 = \tfrac{1}{4}A_0^3 \cos 3\theta. \tag{4.71}$$

The two equations, (4.69) and (4.70), which ensure the removal of secular terms, look rather daunting, but quite a lot can be done with them. However, we first need to introduce two familiar properties of a propagating wave. One is the speed at which the underlying wave—the *carrier* wave—travels, usually called the *phase speed*. This is defined as the speed at which lines $\theta =$ constant move i.e. lines such that

$$\frac{\mathrm{d}}{\mathrm{d}t}[\theta(X, T; \varepsilon)] = 0 \quad \text{or} \quad \frac{\mathrm{d}}{\mathrm{d}T}[\theta(X, T; \varepsilon)] = 0$$

so

$$\theta_X \frac{\mathrm{d}X}{\mathrm{d}T} + \theta_T = 0 \quad \text{or} \quad \frac{\mathrm{d}X}{\mathrm{d}T} = c = \frac{\omega}{k}.$$

The other—and for us, the far more significant—property is the speed at which the energy propagates (and therefore, for example, the speed at which the amplitude modulation moves); this is defined as $\mathrm{d}\omega/\mathrm{d}k = c_g$, the *group speed*. From our result in (4.68), we have

$$c_g = \frac{\mathrm{d}\omega}{\mathrm{d}k} \sim \frac{\mathrm{d}\omega_0}{\mathrm{d}k_0} = c_{g0} \quad \text{where} \quad \omega_0 \frac{\mathrm{d}\omega_0}{\mathrm{d}k_0} = 2k_0^3 - k_0. \tag{4.72}$$

Equation (4.70), after multiplication by A_0, can be written as

$$\frac{\partial}{\partial T}\left(\omega_0 A_0^2\right) + \frac{\partial}{\partial X}\left[\left(2k_0^3 - k_0\right) A_0^2\right] = 0$$

or

$$\frac{\partial}{\partial T}\left(\omega_0 A_0^2\right) + c_{g0}\frac{\partial}{\partial X}\left(\omega_0 A_0^2\right) = -\omega_0 A_0^2 \frac{\partial c_{g0}}{\partial X},$$

which describes the property $\omega_0 A_0^2$ (the *wave action*; see E4.5) propagating at the group speed, c_{g0}, and evolving by virtue of the non-zero right-hand side of this equation. Similarly, equation (4.69) can be written first as

$$-\omega_1 + c_{g0} k_1 = \frac{3}{8}\frac{A_0^2}{\omega_0}$$

and if we elect to write $\theta \sim \sum_{n=0}^{\infty} \varepsilon^n \theta_n$, then (4.61) and (4.64) allow us to interpret $k_1 = \theta_{1x}$ and $\omega_1 = -\theta_{1t}$, and then we obtain

$$\theta_{1t} + c_{g0}\theta_{1x} = \frac{3}{8}\frac{A_0^2}{\omega_0}.$$

Thus the correction to the phase, θ_1, also propagates at the group speed and evolves. Finally, from (4.63), and noting that (4.68) (or (4.72)) implies $\omega_0 = \omega_0(k_0)$, we obtain

$$k_{0T} + \frac{\mathrm{d}\omega_0}{\mathrm{d}k_0} k_{0X} = 0 \quad \text{or} \quad k_{0T} + c_{g0} k_{0X} = 0 :$$

the wave number (and, correspondingly, the frequency) propagate unchanged at the group speed. (The initial data will include the specification of $k(X, 0; \varepsilon)$.)

This example has demonstrated that the method of multiple scales can be used to analyse appropriate partial differential equations, even if we had to tease out the details by introducing, in particular, the group speed. A related exercise can be found in Q4.22.

In our analysis of Bretherton's equation, we worked—not surprisingly—with real-valued functions throughout i.e. $\cos\theta$. There is, sometimes, an advantage in working within a complex-valued framework e.g. $A\mathrm{e}^{\mathrm{i}\theta}+$ complex conjugate. We will use this approach in the next example, but also take note of the general structure that is evident in E4.7.

E4.8 A nonlinear wave equation: the NLS equation

A wave profile, $u(x, t; \varepsilon)$, satisfies the equation

$$u_{tt} - u_{xx} + u = \varepsilon\left(u_x^2 + u^2\right), \tag{4.73}$$

where $\varepsilon \to 0^+$. In this problem we seek a solution which depends on $x - ct$ (here, c is the phase speed), on $x - c_g t$ (c_g is the group speed) and on t (time). We expect that the dependence on $x - c_g t$ is *slow* i.e. in the form $\varepsilon(x - c_g t)$, but we will allow an even slower time dependence; we define

$$\theta = x - ct, \quad X = \varepsilon(x - c_g t), \quad T = \varepsilon^2 t, \tag{4.74}$$

although we have yet to determine c and c_g. With $u(x, t; \varepsilon) \equiv U(\theta, X, T; \varepsilon)$ and (4.74), equation (4.73) becomes

$$c^2 U_{\theta\theta} + 2\varepsilon c c_g U_{\theta X} + \varepsilon^2 c_g^2 U_{XX} - 2\varepsilon^2 c U_{\theta T} - (U_{\theta\theta} + 2\varepsilon U_{\theta X} + \varepsilon^2 U_{XX}) + U$$
$$= \varepsilon \left\{ U_\theta^2 + 2\varepsilon U_\theta U_X + U^2 \right\} + \mathrm{O}(\varepsilon^3) \tag{4.75}$$

and we seek a solution

$$U(\theta, X, T; \varepsilon) \sim \sum_{n=0}^{\infty} \varepsilon^n U_n(\theta, X, T) \quad \text{with} \quad U_n = \sum_{m=0}^{n+1} A_{nm}(X, T) \mathrm{e}^{\mathrm{i}mk\theta} + \mathrm{cc} \tag{4.76a,b}$$

with $A_{00} \equiv 0$; 'cc' denotes the complex conjugate. This solution represents a primary harmonic wave ($A_{01}\mathrm{e}^{\mathrm{i}k\theta}$), together with appropriate higher harmonics (the number of which depends on ε^n i.e. on the hierarchy of nonlinear interactions). The wave number of the primary wave, k, is given and real; θ, X and T are real (of course), but each A_{nm} is complex-valued. We will assume that c and c_g are independent of ε.

First we use (4.76a) in (4.75) to give the sequence of equations

$$(c^2 - 1)U_{0\theta\theta} + U_0 = 0; \quad (c^2 - 1)U_{1\theta\theta} + U_1 + 2(cc_g - 1)U_{0\theta X} = U_{0\theta}^2 + U_0^2; \tag{4.77a,b}$$

$$(c^2 - 1)U_{2\theta\theta} + U_2 + 2(cc_g - 1)U_{1\theta X} + (c_g^2 - 1)U_{0XX}$$
$$= 2(U_{0\theta}U_{1\theta} + U_0 U_1 + U_{0\theta}U_{0X}) \tag{4.77c}$$

as far as the $\mathrm{O}(\varepsilon^2)$ terms. The solution of (4.77a) is given in the form

$$U_0 = A_{01}(X, T)\mathrm{e}^{\mathrm{i}k\theta} + \mathrm{cc}$$

(see (4.76b)), and so we require

$$c^2 = 1 + k^{-2} \tag{4.78}$$

which defines the phase speed; at this stage A_{01} is unknown. Equation (4.77b), for U_1, now becomes

$$(c^2 - 1)U_{1\theta\theta} + U_1 = (1 + k^2)\,|A_{01}|^2 + \left\{(1 - k^2)A_{01}^2 \mathrm{e}^{2\mathrm{i}k\theta} - 2(cc_g - 1)\mathrm{i}kA_{01X}\mathrm{e}^{\mathrm{i}k\theta} + \mathrm{cc}\right\}$$

with the given form of solution

$$U_1 = A_{10} + A_{11}e^{ik\theta} + A_{12}e^{2ik\theta} + \text{cc}$$

which requires

$$A_{10} = (1+k^2)\,|A_{01}|^2, \quad c\,c_g = 1, \quad A_{12} = -\tfrac{1}{3}(1-k^2)A_{01}^2$$

with $A_{11}(X,T)$ (and A_{01}) still arbitrary. (We note that $c_g = 1/c$ and that, with $c = \omega/k$, we have $\omega^2 = 1+k^2$ so $d\omega/dk = k/\omega = 1/c = c_g$: the classical result for group speed.)

The final stage is to find the solution—or, at least, the relevant part of the solution—of equation (4.77c). The usual aim in such calculations is to find, completely, the first term in (we hope) a uniformly valid asymptotic expansion. In this case we have yet to find $A_{01}(X,T)$ (although we know both $c(k)$ and $c_g(k)$); the determination of A_{01} arises from the terms $e^{ik\theta}$ in equation (4.77c). We will find just this one term; the rest of the solution for U_2 is left as an exercise, the essential requirement being to check that there are no inconsistencies that appear as U_2 is determined. These terms, $e^{ik\theta}$, in (4.77c) give the equation

$$\begin{aligned}2(c\,c_g - 1)ik\,A_{11X} + (c_g^2 - 1)A_{01XX} - 2c\,ik\,A_{01T}\\ = 4k^2\bar{A}_{01}A_{12} + 2(A_{12}\bar{A}_{01} + A_{01}A_{10} + A_{01}\bar{A}_{10})\end{aligned}$$

where the over-bar denotes the complex conjugate. When the earlier results are incorporated here, we find the equation for A_{01}:

$$-2ikc\,A_{01T} + \left(c_g^2 - 1\right)A_{01XX} = \tfrac{2}{3}(2k^4 + 5k^2 + 5)A_{01}\,|A_{01}|^2. \tag{4.79}$$

This equation, (4.79), is a *Nonlinear Schrödinger (NLS) equation*, another of the extremely important exactly-integrable equations within the framework of *soliton theory*; see Drazin & Johnson (1992). Thus we have a complete description of the first term in the asymptotic expansion (4.76a,b):

$$u \equiv U \sim A_{01}(X,T)e^{ik\theta} + \text{cc} = A_{01}\{\varepsilon(x - c_g t), \varepsilon^2 t\}e^{ik(x-ct)} + \text{cc} \tag{4.80}$$

where $A_{01}(X,T)$ is a solution of (4.78) and both $c(k)$ and $c_g(k)$ are known.

We should comment that, because of the particular form of solution that we have constructed in this example, it is appropriate only for certain types of initial data. Thus, from (4.80), we see that, at $t = 0$, we must have an initial wave-profile that is predominantly a harmonic wave, but one that admits a slow amplitude modulation i.e.

$$u(x,0;\varepsilon) \sim A_{01}(\varepsilon x, 0)e^{ikx} + \text{cc}.$$

Any initial conditions that do not conform to this pattern would require a different asymptotic, and multi-scale, structure. Other examples that correspond to E4.8 are set as exercises Q4.23–4.26; see also Q4.27.

Before we describe one last general application of the method of multiple scales—perhaps a rather surprising one—we note a particular limitation on the method.

4.5 A LIMITATION ON THE USE OF THE METHOD OF MULTIPLE SCALES

The foregoing examples that have shown how to generate asymptotic solutions of partial differential equations appear reasonably routine and highly successful. However, there is an underlying problem that is not immediately evident and which cannot be ignored. In the context of wave propagation, which is the most common application of this technique to partial differential equations, we encounter difficulties if the predominant solution is a *non-dispersive* wave i.e. waves with different wave number (k) all travel at the *same* speed. To see how this difficulty can arise, we will examine an example which is close to that introduced in E4.8.

E4.9 Dispersive/non-dispersive wave propagation

We consider the equation (cf. (4.73))

$$u_{tt} - u_{xx} + \lambda u = \varepsilon u^2 \tag{4.81}$$

where λ is a given constant and $\varepsilon \to 0^+$; we seek a solution in the form $u(x, t; \varepsilon) \equiv U(\theta, X, T; \varepsilon)$ where

$$\theta = x - ct, \quad X = \varepsilon(x - c_g t), \quad T = \varepsilon^2 t.$$

The equation then becomes

$$c^2 U_{\theta\theta} + 2\varepsilon c c_g U_{\theta X} + \varepsilon^2 c_g^2 U_{XX} - 2\varepsilon^2 c U_{\theta T} - (U_{\theta\theta} + 2\varepsilon U_{\theta X} + \varepsilon^2 U_{XX}) + \lambda U = \varepsilon U^2$$

and we look for a solution periodic in θ, in the form of a harmonic wave:

$$U(\theta, X, T; \varepsilon) \sim \sum_{n=0}^{\infty} \varepsilon^n U_n(\theta, X, T) \quad \text{with} \quad U_n = \sum_{m=0}^{n+1} A_{nm}(X, T)\mathrm{e}^{\mathrm{i}mk\theta} + \mathrm{cc}$$

where $A_{00} \equiv 0$; all this follows the procedure described in E4.8.

Here, we find that

$$U_0 = A_{01}(X, T)\mathrm{e}^{\mathrm{i}k\theta} + \mathrm{cc} \quad \text{with} \quad c^2 = 1 + \lambda/k^2, \tag{4.82}$$

but A_{01} is yet to be determined. At the next order, the equation for U_1 can be written

$$(c^2 - 1)U_{1\theta\theta} + \lambda U_1 = |A_{01}|^2 + \left\{A_{01}^2 \mathrm{e}^{2\mathrm{i}k\theta} - 2(cc_g - 1)\mathrm{i}k A_{01X}\mathrm{e}^{\mathrm{i}k\theta} + \mathrm{cc}\right\} \tag{4.83}$$

and so we require

$$A_{10} = \frac{1}{\lambda} |A_{01}|^2, \quad cc_g = 1, \quad A_{12} = -\frac{1}{3\lambda} A_{01}^2.$$

It is immediately evident that we have a non-uniformity as $\lambda \to 0$. From (4.82) we see that $\lambda = 0$ corresponds to a non-dispersive wave i.e. $c = \pm 1$ for all wave numbers, k. When we set $\lambda = 0$ in (4.83), and use $c^2 = 1$, no solution exists, although for any $\lambda \neq 0$ no complications arise and we may proceed.

This failure is not to be regarded as fatal: for the assumed initial data (the harmonic wave), or any other reasonable initial conditions, an asymptotic solution (with $\lambda = 0$) can be found by the method of strained coordinates. For a wave problem, as we have seen (e.g. Q3.25), this simply requires a suitable representation of the characteristic variables. Nevertheless, in the most extreme cases, when the method of multiple scales is still deemed to be the best approach, the resulting asymptotic solution may not be *uniformly valid* as T (or τ in our discussion of ordinary differential equations) $\to \infty$. Typically, the multiple-scale solution will be valid for T (or τ) no larger than O(1), but this is usually an improvement on the validity of a straightforward asymptotic expansion.

4.6 BOUNDARY-LAYER PROBLEMS

The examples that we have discussed so far involve, usually with a rather straightforward physical interpretation, the slow evolution or development of an underlying solution. Boundary-layer problems (see §2.6–2.8), on the other hand, might appear not to possess this structure. Such problems have different—but matched—solutions away from, and near to, a boundary. However, the solution of such problems expressed as a *composite expansion* (see §1.10) exhibits precisely the multiple-scale structure: a fast scale which describes the solution in the boundary layer, and a slow scale describing the solution elsewhere. We will demonstrate the details of this procedure by considering again our standard boundary-layer-type problem given in equation (2.63) (and see also (1.16)).

E4.10 A boundary-layer problem

We consider

$$\varepsilon y'' + (1+\varepsilon)y' + y = 0, \quad 0 \le x \le 1, \tag{4.84}$$

with $y(0;\varepsilon) = \alpha$, $y(1;\varepsilon) = \beta$, where α and β are constants independent of ε, $\alpha \neq \beta e$ and $\varepsilon \to 0^+$. The boundary-layer variable for this problem (see (2.66)) is $X = x/\varepsilon$, and so we write $y(x;\varepsilon) \equiv Y(x, X;\varepsilon)$ which leads to equation (4.84) written in the form

$$Y_{XX} + 2\varepsilon Y_{xX} + \varepsilon^2 Y_{xx} + (1+\varepsilon)(Y_X + \varepsilon Y_x) + \varepsilon Y = 0 \tag{4.85}$$

with

$$Y(0, 0; \varepsilon) = \alpha \quad \text{and} \quad Y(1, \varepsilon^{-1}; \varepsilon) = \beta. \tag{4.86}$$

Note the appearance of the evaluation on $X = \varepsilon^{-1}$, which may cause some anxiety; we will now address this issue. To see the consequences of this, and before we write down the complete asymptotic expansion, it is instructive first to solve for $Y \sim Y_0(x, X)$ i.e.

$$Y_{0XX} + Y_{0X} = 0 \quad \text{so} \quad Y_0(x, X) = A(x) + B(x)\mathrm{e}^{-X}$$

with the boundary conditions

$$Y_0(0, 0) = A(0) + B(0) = \alpha; \quad Y_0(1, \varepsilon^{-1}) = A(1) + B(1)\mathrm{e}^{-1/\varepsilon} = \beta.$$

This second condition involves $\mathrm{e}^{-1/\varepsilon}$, and to accommodate such terms the asymptotic expansion must include them; thus we seek an asymptotic solution

$$Y(x, X; \varepsilon) \sim \sum_{m=0}^{\infty} \varepsilon^{-m/\varepsilon} \sum_{n=0}^{\infty} \varepsilon^n Y_{mn}(x, X) \tag{4.87}$$

and evaluation on $X = \varepsilon^{-1}$ is now no longer an embarrassment.

The sequence of equations, generated by using (4.87) in (4.85), starts

$$Y_{00XX} + Y_{00X} = 0; \quad Y_{01XX} + Y_{01X} + 2Y_{00xX} + Y_{00X} + Y_{00} = 0; \tag{4.88a,b}$$

$$Y_{10XX} + Y_{10X} = 0. \tag{4.88c}$$

The general solution of (4.88a) is

$$Y_{00}(x, X) = A_0(x) + B_0(x)\mathrm{e}^{-X}$$

and then we may write (4.88b) as

$$Y_{01XX} + Y_{01X} = B_0'\mathrm{e}^{-X} - (A_0' + A_0)$$

which itself has the general solution

$$Y_{01}(x, X) = -(A_0' + A_0)X - B_0' X e^{-X} + A_1(x) + B_1(x)e^{-X}.$$

We seek a solution which is uniformly valid for $0 \le x \le 1$ and $0 \le X \le \varepsilon^{-1}$, but this latter implies $X \to \infty$ (as $\varepsilon \to 0^+$); thus we require, for uniformity,

$$A_0' + A_0 = 0 \quad \text{and} \quad B_0' = 0.$$

The boundary conditions, at this order, give

$$A_0(0) + B_0(0) = \alpha, \quad A_0(1) = \beta$$

and so we have the solutions

$$A_0(x) = \beta\,\mathrm{e}^{1-x}, \quad B_0(x) = \alpha - \beta\,\mathrm{e}$$

which completely determines the first term of the asymptotic expansion:

$$y(x;\varepsilon) \sim \beta\,\mathrm{e}^{1-x} + (\alpha - \beta\,\mathrm{e})\,\mathrm{e}^{-x/\varepsilon}$$

as given in equation (2.76). The other terms in the expansion follow directly by imposing uniformity at each order; the terms Y_{mn}, for $m \geq 1$, are used to remove the exponentially small contributions $\mathrm{e}^{-m/\varepsilon}$ that appear in the boundary condition on $x = 1$.

The method of multiple scales is therefore equally valid, and beneficial, for the analysis of boundary-layer problems. However, the example that we have presented is particularly straightforward (and, of course, we have available the exact solution). We conclude this chapter by applying the method to a more testing example (taken from Q2.17(a)).

E4.11 A nonlinear boundary-layer problem

We consider the problem

$$\varepsilon y'' + (1+x)^2 y' - y^2 = 0, \quad 0 \geq x \geq 1, \tag{4.89}$$

with $y(0;\varepsilon) = y(1;\varepsilon) = 2$, for $\varepsilon \to 0^+$. The slow scale is clearly x, but for the fast scale we will use the most general formulation of the boundary-layer variable; see §2.7. Thus we introduce

$$X = \frac{1}{\varepsilon}\int_0^x (1+x')^2\mathrm{d}x' = \frac{1}{3\varepsilon}[(1+x)^3 - 1]$$

and then we write $y(x;\varepsilon) \equiv Y(x, X;\varepsilon)$ so that equation (4.89) becomes

$$(1+x)^4(Y_{XX} + Y_X) + \varepsilon\{(1+x)^2(Y_x + 2Y_{xX}) + 2(1+x)Y_X - Y^2\} + \varepsilon^2 Y_{xx} = 0 \tag{4.90}$$

with

$$Y(0, 0;\varepsilon) = Y\left(1, \tfrac{7}{3}\varepsilon^{-1};\varepsilon\right) = 2.$$

The terms that are exponentially small on $x = 1$, as $\varepsilon \to 0^+$, are $\mathrm{e}^{-7m/3\varepsilon}$ and so we seek a solution

$$Y(x, X;\varepsilon) \sim \sum_{m=0}^{\infty} e^{-7m/3\varepsilon} \sum_{n=0}^{\infty} \varepsilon^n Y_{mn}(x, X). \tag{4.91}$$

The first two equations in the sequence, obtained by using (4.91) in (4.90), are

$$Y_{00XX} + Y_{00X} = 0; \tag{4.92a}$$

$$Y_{01XX} + Y_{01X} + \frac{1}{(1+x)^4}\left\{(1+x)^2(Y_{00x} + 2Y_{00xX}) + 2(1+x)Y_{00X} - Y_{00}^2\right\} = 0, \tag{4.92b}$$

and (4.92a) has the general solution

$$Y_{00}(x, X) = A_0(x) + B_0(x)\mathrm{e}^{-X}, \tag{4.93}$$

with

$$A_0(0) + B_0(0) = A_0(1) = 2. \tag{4.94}$$

Equation (4.92b) can then be written

$$\begin{aligned} Y_{01XX} + Y_{01X} = &-\frac{1}{(1+x)^2}(A_0' + B_0'\mathrm{e}^{-X}) + \frac{2}{(1+x)^3}B_0\mathrm{e}^{-X} \\ &- \frac{2}{(1+x)^2}B_0'\mathrm{e}^{-X} - \frac{1}{(1+x)^4}\left(A_0^2 + 2A_0B_0e^{-X} + B_0^2\mathrm{e}^{-2X}\right) \end{aligned}$$

which has the general solution

$$\begin{aligned} Y_{01}(x, X) = &-\frac{1}{(1+x)^4}\{(1+x)^2B_0' + 2(1+x)B_0 + 2A_0B_0\}X\mathrm{e}^{-X} \\ &- \frac{1}{(1+x)^4}\left\{(1+x)^2A_0' - A_0^2\right\}X - \frac{1}{2(1+x)^4}B_0^2\mathrm{e}^{-2X} + A_1(x) + B_1(x)\mathrm{e}^{-X}. \end{aligned}$$

A solution that is uniformly valid as $X \to \infty$ requires

$$(1+x)^2A_0' - A_0^2 = 0; \quad (1+x)^2B_0' + 2(1+x)B_0 + 2A_0B_0 = 0$$

and so, incorporating the boundary conditions, (4.94), we readily obtain

$$A_0(x) = 1 + x, \quad B_0(x) = (1+x)^{-4}.$$

Combining this with (4.93), we have the first term of a uniformly valid representation of the solution:

$$y(x; \varepsilon) \sim 1 + x + \frac{1}{(1+x)^4}\exp\{-[(1+x)^3 - 1]/3\varepsilon\},$$

which should be compared with the solution obtained in Q2.17(a).

Now that we have seen the method of multiple scales applied to boundary-layer problems, it should be evident that this provides the simplest and most direct approach to the solution of this type of problem. Not only do we avoid the need to match—although, of course, the correct selection of fast and slow variables is essential—but we also generate a composite expansion directly, which may be used as the basis for numerical or graphical representations of the solution. Further examples are given in exercises Q4.28–4.35.

This concludes our presentation of the method of multiple scales, and is the last technique that we shall describe. In the final chapter, the plan is to work through a number of examples taken from various branches of the mathematical, physical and related sciences, grouped by subject area. These, we hope, will show how our various techniques are relevant and important. It is to be hoped that those readers with interests in particular fields will find something to excite their curiosity and to point the way to the solution of problems that might otherwise appear intractable.

FURTHER READING

Most of the texts that we have mentioned earlier discuss the method of multiple scales, to a greater or a lesser extent. Two texts, in particular, give a good overview of the subject: Nayfeh (1973), in which a number of problems are investigated in many different ways (including variants of the method of multiple scales), and Kevorkian & Cole (1996) which provides an up-to-date and wide-ranging discussion. The applications to ordinary differential equations are nicely presented in both Smith (1985) and O'Malley (1991), where a lot of technical detail is included, as well as a careful discussion of asymptotic correctness. Holmes (1995) provides an excellent account of both the method of multiple scales and the WKB method. This latter is also discussed in Wasow (1965) and in Eckhaus (1979). Finally, an excellent introduction to the rôle of asymptotic methods in the analysis of oscillations (mainly those that are nonlinear) can be found in Bogoliubov & Mitropolsky (1961)—an older text, but a classic that can be highly recommended (even though it does not possess an index!).

EXERCISES

Q4.1 *Nearly linear oscillator I.* A weakly nonlinear oscillator is described by the equation

$$\ddot{x} + x + \varepsilon (x - a\dot{x})^3 = 0, \quad t \geq 0,$$

where $a\,(> 0)$ is a constant (independent of ε) and $\varepsilon \to 0^+$; the initial conditions are

$$x(0;\varepsilon) = 1, \quad \dot{x}(0;\varepsilon) = 0.$$

Use the method of multiple scales to find, completely, the first term in a uniformly valid asymptotic expansion. (It is sufficient to use the time scales $T = t$ and $\tau = \varepsilon t$.) Explain why your solution fails if $a < 0$.

Q4.2 *Nearly linear oscillator II.* See Q4.1; repeat this for the problem

$$\ddot{x} + [1 + \varepsilon \sin(\varepsilon t)]x + \varepsilon(\dot{x})^3 = 0, \quad t \geq 0,$$

with
$$x(0; \varepsilon) = 1, \quad \dot{x}(0; \varepsilon) = 0.$$

Q4.3 *Nearly linear oscillator III.* See Q4.1; repeat this for the problem

$$(\dot{x})^2 + [1 + 2\varepsilon f(\varepsilon t)]x^2 = \alpha^2, \quad t \geq 0,$$

where α (> 0) is a constant (independent of ε) and $f(\varepsilon t)$ is an integrable function; the initial condition is $x(0; \varepsilon) = 0$. What condition must $f(\varepsilon t)$ satisfy if, on the basis of the evidence of your two-term expansion, the asymptotic expansion is to be uniformly valid?
[Hint: remember that the general solution of

$$(\dot{x})^2 + \beta^2 x^2 = \alpha^2 \quad \text{is} \quad x(t) = \frac{\alpha}{\beta} \sin(\beta t + \gamma),$$

where γ is an arbitrary constant.]

Q4.4 *A nearly linear oscillator with forcing.* See Q4.1; repeat this for the problem

$$\ddot{x} + x + \varepsilon x^3 = \alpha \cos \omega t, \quad t \geq 0,$$

where α and ω are constants, both independent of ε. Find the general form of the first term, given that $\omega \neq 0, \pm 1, \pm 3, \pm 1/3$. Explain the consequences of (a) $\omega = 0$; (b) $\omega = \pm 1$. Also, write down the equations defining the slow evolution of the solution in the cases: (c) $\omega = \pm 3$; (d) $\omega = \pm 1/3$. (See Q4.7, 4.8 for more details.)

Q4.5 *A slowly varying linear oscillator.* A damped, linear oscillator is described by the equation

$$\ddot{x} + \varepsilon f(\varepsilon t)\dot{x} + f^2(\varepsilon t)x = 0, \quad t \geq 0,$$

with $\varepsilon \to 0^+$ and $f(\varepsilon t) > 0$. Obtain the complete description of the (general) first term of a uniformly valid asymptotic expansion, for which you should use the fast scale defined by $\mathrm{d}T/\mathrm{d}t = \theta(\varepsilon t)$ (and the slow scale is $\tau = \varepsilon t$).

Q4.6 *A coupled oscillatory system.* An oscillation is described by the pair of equations

$$\dot{u} + v = -\varepsilon u; \quad \dot{v} - u = \varepsilon v^3, \quad t \geq 0,$$

with $\varepsilon \to 0^+$, and the initial conditions are $u(0; \varepsilon) = 1$, $v(0; \varepsilon) = 0$. Introduce $T = \omega t$, $\omega \sim 1 + \sum_{n=2}^{\infty} \varepsilon^n \omega_n$, and $\tau = \varepsilon t$, use the method of multiple scales and find, completely, the first term of a uniformly valid asymptotic expansions.

Also, from the equations that define the third term in the expansions, show that the solution is uniformly valid (as $\tau \to \infty$) only if $\omega_2 = 1/8$.

Q4.7 *Forcing near resonance I.* Consider an oscillation described by a Duffing equation with (weak) forcing:

$$\ddot{x} + x + \varepsilon x^3 = \varepsilon \cos[\Omega(\varepsilon)t], \quad t \geq 0,$$

with $\varepsilon \to 0^+$, $\Omega(\varepsilon)$ is the given frequency of the forcing and the initial conditions are $x(0;\varepsilon) = 1$, $\dot{x}(0;\varepsilon) = 0$. Given that $\Omega(\varepsilon) = 1 + \varepsilon\Omega_1$ (where $\Omega_1 \neq 0$ is a constant independent of ε), find the equation which describes completely the first term of a uniformly valid asymptotic expansion. [Hint: write the forcing term as $\cos(T + \Omega_1\tau)$, where T/τ are the fast/slow scales, respectively.]

Q4.8 *Forcing near resonance II.* See Q4.7; repeat this for $\Omega(\varepsilon) = 3 + \varepsilon\Omega_1$ in the equation

$$\ddot{x} + x + \varepsilon x^3 = \alpha \cos[\Omega(\varepsilon)t]$$

(see Q4.4), where α is a constant independent of ε (and note the appearance of *subharmonics*).

Q4.9 *Failure of the method of multiple scales.* An oscillator is described by the equation

$$\ddot{x} + x = \varepsilon x^2, \quad t \geq 0,$$

with $\varepsilon \to 0^+$. Introduce $T = \omega t$ and $\tau = \varepsilon t$ (with $\omega \sim 1 + \sum_{n=2}^{\infty} \varepsilon^n \omega_n$) and analyse as far as the term at $O(\varepsilon^2)$, which ensures the complete description of the solution as far as $O(\varepsilon)$. Show that a uniformly valid solution cannot be obtained using this approach. (You may wish to investigate why this happens by examining the energy integral for the motion.)

Q4.10 *Nonlinear oscillation I.* A (fully) nonlinear oscillation is described by the equation

$$\ddot{x} + x + x^3 = \varepsilon(\dot{x} + x^2), \quad t \geq 0,$$

with $\varepsilon \to 0^+$. Show that a solution with $\varepsilon = 0$ is $x = a\,\mathrm{cn}[4K(m)t; m]$ for a suitable relation between a and m. Now use the method of multiple scales, with the scales T (where $\mathrm{d}T/\mathrm{d}t = \theta(\tau)$) and $\tau = \varepsilon t$, to find the first term of an asymptotic expansion which is periodic in T. (The periodicity condition should be written as an integral, but this does not need to be evaluated.)

Q4.11 *Nonlinear oscillation II.* See Q4.10; follow this same procedure for the equation

$$\ddot{x} + x(x - 1) = \varepsilon\dot{x}, \quad t \geq 0, \quad \varepsilon > 0,$$

where a solution with $\varepsilon = 0$ can be written $x = a + b\,\mathrm{cn}^2[4K(m)t; m]$ for suitable a, b and m.

Q4.12 *Mathieu's equation for n = 2.* See E4.4; consider Mathieu's equation

$$\ddot{x} + (\delta + \varepsilon \cos t)x = 0, \quad t \geq 0,$$

with $\delta \sim 1 + \sum_{n=1}^{\infty} \varepsilon^n \delta_n$ (which is the case $n = 2$). Introduce $T = t$ and $\tau = \varepsilon^2 t$, show that $\delta_1 = 0$ and find the equation for the $O(\varepsilon^2)$ term in the asymptotic expansion for x; from this, deduce that the exponent which describes the amplitude modulation (cf. E4.4) is

$$\lambda = \pm\tfrac{1}{2}\sqrt{\left(\tfrac{5}{12} - \delta_2\right)\left(\tfrac{1}{12} + \delta_2\right)}.$$

What is the nature of the solution of this Mathieu equation, for various δ_2? (These results should be compared with those obtained in Q3.15.)

Q4.13 *A particular Hill equation.* See Q4.12; follow this same procedure for the Hill equation

$$\ddot{x} + (\delta + \varepsilon \cos t + \varepsilon\lambda \cos 3t)x = 0, \quad t \geq 0,$$

where λ is a fixed constant (independent of ε) and $\delta \sim 1 + \sum_{n=1}^{\infty} \varepsilon^n \delta_n$, for $\varepsilon \to 0^+$. Show that $\delta_1 = 0$ and that the leading term is periodic in T and τ only if

$$(\delta_2 - \delta_+)(\delta_2 - \delta_-) > 0$$

where $\delta_\pm(\lambda)$ are to be identified. (These results should be compared with those obtained in Q3.17.)

Q4.14 *Mathieu's equation away from critical δ.* See Q4.12; consider Mathieu's equation, but now with δ away from the critical values: set $\delta = \Omega^2$ ($\neq n^2/4$, $n = 0, 1, 2, \ldots$) and fixed independent of ε. Introduce $T = \omega t$, with $\omega \sim \Omega + \sum_{n=2}^{\infty} \varepsilon^n \omega_n$, and $\tau = \varepsilon t$, and find the solution correct at $O(\varepsilon)$. (You should note the singularities, for various Ω, that are evident here.)

Q4.15 *WKB: higher-order terms.* See E4.5; consider the equation

$$\ddot{x} + \omega^2(\varepsilon t)x = 0, \quad t \geq 0,$$

with $\varepsilon \to 0^+$, introduce $\tau = \varepsilon t$ and write

$$x(t;\varepsilon) = A(\tau;\varepsilon)\exp\left\{ \mathrm{i}\int_0^t \sigma(\varepsilon t;\varepsilon)\mathrm{d}t \right\} + \text{cc}.$$

Show that $$2\sigma A' + A\sigma' = 0; \quad (\omega^2 - \sigma^2)A + \varepsilon^2 A'' = 0$$

and then expand both $A \sim \sum_{n=0}^{\infty} \varepsilon^{2n} A_n(\tau)$, $\sigma \sim \sum_{n=0}^{\infty} \varepsilon^{2n}\sigma_n(\tau)$. Determine A_1 and σ_1 in terms of $\omega(\tau)$ and the constant $k = \sigma A^2$ (which is independent of ε). (This procedure is a very neat way to obtain higher-order terms in the WKB approach.)

Q4.16 *WKB (exponential case).* Consider the equation

$$\frac{d^2y}{dx^2} - \omega^2(\varepsilon x)y = 0, \quad x \in D,$$

where $\omega > 0$ for $\forall x \in D$ and $\varepsilon \to 0^+$. Introduce $X = \varepsilon x$ and $d\theta/dx = \Omega(X;\varepsilon)$ and hence find, completely, the first term of a uniformly valid asymptotic expansion (which will be the counterpart of equation (4.47)).

Q4.17 *Eigenvalues.* Use the method of multiple scales, in the WKB form, to find the leading approximation to the eigenvalues (λ) of the problem

$$\frac{d^2y}{dx^2} + \lambda a(\varepsilon x)y = 0, \quad 0 \le x \le 1, \quad \varepsilon > 0,$$

with $y(0;\varepsilon) = y(1;\varepsilon) = 0$ and $a > 0$ for $\forall x \in [0, 1]$. (You should introduce $X = \varepsilon x$ and $d\theta/dx = \Omega(X;\varepsilon)$.) Evaluate your results, explicitly, for the cases: (a) $a(X) = e^{-X}$; (b) $a(X) = 1 + X$. [Written in the form $d^2y/dX^2 + (\lambda/\varepsilon^2)a(X)y = 0$, it is evident that this is the problem of finding approximations to the large eigenvalues.]

Q4.18 *A turning-point problem I.* Consider

$$y'' + \varepsilon^{-2}(1-x)f(x)y = 0, \quad x_0 \le x \le x_1,$$

where $x_0 < 1 < x_1$ with $f > 0$ and analytic throughout the given domain. Show that the relevant scaling (cf. §2.7) in the neighbourhood of the turning point (at $x = 1$) is $x - 1 = O(\varepsilon^{2/3})$ and then introduce the more useful fast scale $X = \varepsilon^{-2/3}(1-x)h(x)$and use x as the slow scale. Show that $y(x;\varepsilon) \equiv Y(x, X;\varepsilon) \sim A_0(x)\mathrm{Ai}(X)$ where $[(1-x)h(x)]' A_0^2 =$ constant and $\mathrm{Ai}(X)$ is the (bounded) Airy function (a solution of $Y'' + XY = 0$). Determine $h(x)$ and write down the first term of the asymptotic expansion of $Y(x, X;\varepsilon)$.

Q4.19 *A turning-point problem II.* Show that the equation

$$\varepsilon^2 y'' + 2\varepsilon y' + (1 + x - x^2)y = 0, \quad x_0 \le x \le x_1,$$

where $x_0 < 0$ and $x_1 > 1$, has turning points at $x = 0$ and at $x = 1$. [Hint: write $y = e^{-x/\varepsilon}Y$.] Use the WKB approach (for $\varepsilon \to 0^+$) to find the first term in each of the asymptotic expansions valid in $x < 0$, $0 < x < 1$, $x > 1$. Also write down the leading term in the asymptotic expansions valid near $x = 0$ and near $x = 1$.

Q4.20 *A higher-order turning-point.* Show that the equation

$$\varepsilon^5 y'' + x^3 f(x)y = 0, \quad x_0 \le x \le x_1,$$

with $x_0 < 0 < x_1$, and $f > 0$ throughout the given domain, has a turning point at $x = 0$. Find the equation that, on an appropriate fast scale (X), describes the solution near $x = 0$ (as $\varepsilon \to 0^+$). Show that this equation has solutions which can be written in terms of *Bessel functions*: $X^{1/2}J_{\pm\nu}(\alpha X^{1/(2\nu)})$ for suitable α and ν.

Q4.21 *Schrödinger's equation for high energy.* The time-independent, one-dimensional Schrödinger equation for a simple-harmonic-oscillator potential can be written

$$\frac{d^2\psi}{dx^2} + (E - x^2)\psi = 0, \quad -\infty < x < \infty,$$

where E (= constant) is the total energy. Let us write $E = \varepsilon^{-2}$ ($\varepsilon > 0$) and define $z = \varepsilon x$, to give

$$\frac{d^2\psi}{dz^2} + \varepsilon^{-4}(1 - z^2)\psi = 0;$$

this equation has turning points at $z = \pm 1$ and we require exponentially decaying solutions as $|z| \to \infty$ (and oscillatory solutions exist for $-1 < z < 1$). Find the leading term in each of the regions, match (and thus develop appropriate connection formulae) and show that the eigenvalues (E) satisfy $E = \varepsilon^{-2} \sim 2n - 1$, where n is a large integer. (This problem can be solved exactly, using *Hermite functions*; it turns out that our asymptotic evaluation of E is exact for all n.)

Q4.22 *A weakly nonlinear wave.* A wave is described by the equation

$$\phi_t + \phi_{xxx} = \varepsilon\phi^2\phi_x, \quad -\infty < x < \infty, \quad t \geq 0,$$

where $\varepsilon \to 0^+$. Introduce $X = \varepsilon x$, $T = \varepsilon t$, $\theta_x = k(X, T)$, $\theta_t = -\omega(X, T)$ and derive the equations that completely describe the leading-order solution

$$\phi(x, t; \varepsilon) \equiv \Phi(\theta, X, T; \varepsilon) \sim A_0(X, T)e^{i\theta} + cc$$

which is uniformly valid. (Do not solve your equation for $A_0(X, T)$.)

Q4.23 *A weakly nonlinear wave: NLS I.* A wave is described by the equation

$$u_{tt} - u_{xx} + u = \varepsilon(uu_{xx} - u^2), \quad -\infty < x < \infty, \quad t \geq 0,$$

with $\varepsilon \to 0^+$. Introduce $\theta = x - ct$, $X = \varepsilon(x - c_g t)$ and $T = \varepsilon^2 t$ (where $c(k)$ and $c_g(k)$ are independent of ε) and seek a solution

$$u(x, t; \varepsilon) \equiv U(\theta, X, T; \varepsilon) \sim \sum_{n=0}^{\infty} \varepsilon^n \sum_{m=0}^{n+1} A_{nm}(X, T)e^{imk\theta} + cc$$

where $A_{00} \equiv 0$. Determine $c(k)$ and $c_g(k)$ (and confirm that c_g satisfies the usual condition for the group speed) and find the equation for $A_{01}(X, T)$.

Q4.24 *A weakly nonlinear wave: NLS II.* See Q4.23; repeat all this for the equation

$$u_{tt} - u_{xx} + u = \varepsilon(u^2 + uu_x).$$

Q4.25 *A weakly nonlinear wave: NLS III.* See Q4.23; repeat all this for the equation

$$u_{tt} - u_{xx} - u = \varepsilon\left(u_x^2 - uu_{xx}\right).$$

Q4.26 *KdV→NLS.* Consider the Korteweg-de Vries (KdV) equation

$$2u_t + 3uu_x + \frac{\lambda}{3}u_{xxx} = 0, \quad -\infty < x < \infty, \quad t \geq 0,$$

where $\lambda \to \infty$ is a parameter. Introduce $\theta = x + c\lambda t$, $X = \lambda^{-1}(x + c_g\lambda t)$ and $T = \lambda^{-1}t$ (and here $c(k)$ and $c_g(k)$ will be *corrections* to the original c and c_g because the given KdV equation has already been written in a suitable moving frame). Seek a solution

$$u(x, t; \lambda) \equiv U(\theta, X, T; \lambda) \sim \sum_{n=0}^{\infty} \lambda^{-n} \sum_{m=0}^{n+1} A_{nm}(X, T)E^m + \text{cc}$$

where $E = \exp(ik\theta)$; find $c(k)$, $c_g(k)$ and the equation for $A_{01}(X, T)$. (This example demonstrates that an underlying structure of the KdV equation is an NLS (Nonlinear Schrödinger) equation; indeed, it can be shown that, in the context of water waves, for example, the relevant NLS equation for that problem matches to this NLS equation—see Johnson, 1997.)

Q4.27 *Ray theory.* A wave (moving in two dimensions) slowly evolves, on the scale ε, so that $\theta_x = k(X, Y, T)$, $\theta_y = \ell(X, Y, T)$, $\theta_t = -\omega(X, Y, T)$ where $X = \varepsilon x$, $Y = \varepsilon y$, $T = \varepsilon t$; show that
(a) $\nabla\omega + \partial\mathbf{k}/\partial T = \mathbf{0}$ where $\nabla \equiv \left(\frac{\partial}{\partial X}, \frac{\partial}{\partial Y}\right)$ and $\mathbf{k} \equiv (\mathrm{k}, \ell)$;
(b) $\theta_x^2 + \theta_y^2 = |\mathbf{k}|^2$ (the *eikonal equation*);
(c) $k_Y - \ell_X = 0$ (so the vector $\mathbf{k}$ is 'irrotational').
(Given that the energy in the wave motion is $E(X, Y, T)$, it can be shown that $\frac{\partial}{\partial T}\left(\frac{E}{\omega}\right) + \nabla \cdot \left(\mathbf{c_g}\frac{E}{\omega}\right) = 0$, where $\mathbf{c_g} \equiv \left(\frac{\partial\omega}{\partial k}, \frac{\partial\omega}{\partial\ell}\right)$. All this is the basis for *ray theory*, or the *theory of geometrical optics*, which is used to describe the properties of waves that move through a slowly changing environment.)

Q4.28 *Boundary-layer problem I.* Use the method of multiple scales to find, completely, the first term of a uniformly valid asymptotic expansion of the solution of

$$\varepsilon y'' + (1 + \varepsilon x)y' - 3y = 3x, \quad 0 \leq x \leq 1,$$

where $y(0; \varepsilon) = y(1; \varepsilon) = 2$, for $\varepsilon \to 0^+$.

Q4.29 *Boundary-layer problem II.* See Q4.28; repeat this for the problem

$$\varepsilon y'' - (1 + 2x)y' + y = x, \quad 0 \le x \le 1,$$

with $y(0; \varepsilon) = 1,\ y(1; \varepsilon) = 0$.

Q4.30 *Boundary-layer problem III.* See Q4.28; repeat this for the problem

$$\varepsilon y'' + (1 + 2x)y' + 2y = 0, \quad 0 \le x \le 1,$$

with $y(0; \varepsilon) = 4,\ y(1; \varepsilon) = 1$.

Q4.31 *Boundary-layer problem IV.* See Q4.28; repeat this for the problem

$$\varepsilon y'' + (1 + x)^2 y' - y^2 = 0, \quad 0 \le x \le 1,$$

with $y(0; \varepsilon) = 0,\ y(1; \varepsilon) = 2/3$.

Q4.32 *Boundary-layer problem V.* See Q4.28; repeat this for the problem

$$\varepsilon y'' + y' + y^2 = 0, \quad 0 \le x \le 1,$$

with $y(0; \varepsilon) = 0,\ y(1; \varepsilon) = 1/2$.

Q4.33 *Heat transfer problem.* See Q4.28; repeat this for the heat transfer problem (as given in Q2.30)

$$\varepsilon T'' + x(T' - T) = 0, \quad 0 \le x \le 1,$$

with $T(0; \varepsilon) = T_0,\ T(1; \varepsilon) = T_1$. [Take care!]

Q4.34 *A more general boundary-layer problem.* See Q4.28; repeat this for the problem

$$\varepsilon y'' - y' = y^n, \quad 0 \le x \le 1,$$

where $n \ge 2$ is a given integer, with the boundary conditions $y(0; \varepsilon) = 1,\ y(1; \varepsilon) = 0$.

Q4.35 *Two boundary layers.* See Q4.28; repeat this for the problem

$$\varepsilon y''' - \left(\frac{1}{1 + 2x}\right) y' + y^2 = 0, \quad 0 \le x \le 1,$$

with $y(0; \varepsilon) = 0,\ y(1; \varepsilon) = 1,\ y'(1; \varepsilon) = 1/\sqrt{\varepsilon}$, but note that *two* fast scales are required here, to accommodate the two boundary layers—one near $x = 0$ and the other near $x = 1$.

5. SOME WORKED EXAMPLES ARISING FROM PHYSICAL PROBLEMS

In this final chapter, the aim is to present a number of worked examples where most of the details are given explicitly; what little is left undone may be completed by the interested reader (although no *formal* exercises are offered). Also, we will not dwell upon the purely technical aspects of finding the solution of a particular differential equation. These examples are taken from, or based on, texts and papers that introduce, describe, develop, explain and solve practical problems in various fields; references to appropriate source material will be included. Most have arisen—not surprisingly—from the physical sciences, but we have attempted to provide a fairly broad spread of topics. Each problem is described with sufficient detail (we hope) to enable it to be put into context, although it would be quite impossible to include all the background ideas for those altogether unfamiliar with the particular field. To this end, the problems are collected under various headings (such as 'mechanical & electrical systems, 'semiconductors' or 'chemical & biological reactions') and so the reader with particular interests might turn to specific ones first. Nevertheless, the hope is that every problem is accessible, as an example in singular perturbation theory, to those who have followed this (or any other suitable) text. The technique adopted to construct the asymptotic solution will be mentioned, and a reference will be given to a relevant section or example from the earlier chapters of this text.

A number of the examples and exercises that have already been discussed have been taken from various important applications; in some cases, those presented in this chapter build on and expand these earlier problems. The reader should be aware, therefore, that the relevant calculations in the previous chapters may need to be rehearsed before embarking on some of the new material presented here. In each group of problems, every example will be labelled by a suitable name, and the full list of these will appear in the preamble to that group. The titles of the groups are: 5.1 Mechanical & electrical systems; 5.2 Celestial mechanics; 5.3 Physics of particles & light; 5.4 Semi- and superconductors; 5.5 Fluid mechanics; 5.6 Extreme thermal processes; 5.7 Chemical & biochemical reactions. These chosen headings are intended simply to provide a general guide to the reader; there is no doubt that some examples could be placed in a different group—or appear in more than one group. Further, many other examples could have been included (and the author apologises if your favourite has been omitted); the intention in a text such as this is to give only a flavour of what is possible. Nevertheless, it is hoped that sufficient information is available to encourage the interested researcher to appreciate the power of the techniques that we have described.

Although the physical basis for each problem will be outlined, the relevant non-dimensional, scaled equations will usually be the starting point for the analysis. There is little to be gained by presenting the original physical problem, in all its detail, together with the non-dimensionalisation, *et cetera*, if only because of the requirement, for example, to define all the physical variables in every problem. Further, the reasonable limitation on space also precludes this. The interested reader should be able to fill in the details, particularly with the aid of the original reference(s).

5.1 MECHANICAL & ELECTRICAL SYSTEMS

The examples collected under this heading are based on fairly simple mechanical or physical principles; more advanced and specific topics (such as celestial mechanics) which might have appeared in this group are considered separately. The examples to be discussed are: E5.1 Projectile motion with small drag; E5.2 Child's swing; E5.3 Meniscus on a circular tube; E5.4 Drilling by laser; E5.5 The van der Pol/Rayleigh oscillator; E5.6 A diode oscillator with a current pump; E5.7 A Klein-Gordon equation.

E5.1 Projectile motion with small drag

We consider a projectile which is moving in the two-dimensional (x, z)-plane under the action of gravity (which is constant in the negative z-direction) and of a drag force proportional to the square of the speed (and acting back along the local direction of motion). The non-dimensional equations are most conveniently written as

$$\frac{\mathrm{d}u}{\mathrm{d}t} = -\varepsilon u\sqrt{u^2 + w^2}; \quad \frac{\mathrm{d}w}{\mathrm{d}t} = -1 - \varepsilon w\sqrt{u^2 + w^2} \tag{5.1a,b}$$

where $(u, w) = \mathrm{d}/\mathrm{d}t(x, z)$ and the initial conditions are given as

$$u = \cos\alpha, \quad w = \sin\alpha, \quad x = z = 0 \quad \text{all at time } t = 0; \tag{5.2}$$

α is the angle of projection ($0 < \alpha < \pi/2$). The small parameter is ε (>0), and for motion such as that for a shot-put (see Mestre, 1991) its value is typically about 0.01. In these projectile problems, the main interest is in estimating the range (and maximum range), x, for a given vertical displacement (z), which may be zero; here, we find x where $z = h$. We seek a solution, following the straightforward procedure (which may produce a uniformly valid solution; cf. §2.3):

$$u(t;\varepsilon) \sim \sum_{n=0}^{\infty} \varepsilon^n u_n(t); \quad w(t;\varepsilon) \sim \sum_{n=0}^{\infty} \varepsilon^n w_n(t),$$

and then from (5.1a,b) we obtain

$$u_0' = 0 \quad \text{and} \quad w_0' = -1 \quad \text{i.e.} \quad u_0(t) = \cos\alpha, \quad w_0(t) = \sin\alpha - t.$$

At the next order, we have the equations

$$u_1' = -u_0\sqrt{u_0^2 + w_0^2} = -\cos\alpha\sqrt{1 - 2t\sin\alpha + t^2};$$

$$w_1' = -w_0\sqrt{u_0^2 + w_0^2} = -(\sin\alpha - t)\sqrt{1 - 2t\sin\alpha + t^2},$$

which give, after an integration and on using the initial conditions $u_1 = w_1 = 0$ on $t = 0$,

$$u_1(t) = -\cos\alpha\left\{\frac{1}{2}(t - \sin\alpha)\sqrt{1 - 2t\sin\alpha + t^2} + \frac{\cos^2\alpha}{2}\operatorname{arcsinh}\left(\frac{t - \sin\alpha}{\cos\alpha}\right)\right.$$
$$\left. + \frac{1}{2}\sin\alpha + \frac{\cos^2\alpha}{2}\operatorname{arcsinh}(\tan\alpha)\right\}; \tag{5.3}$$

$$w_1(t) = \frac{1}{3}\{(1 - 2t\sin\alpha + t^2)^{3/2} - 1\}. \tag{5.4}$$

Thus we have, for example,

$$w(t;\varepsilon) = \frac{dz}{dt} \sim \sin\alpha - t + \frac{\varepsilon}{3}\{(1 - 2t\sin\alpha + t^2)^{3/2} - 1\},$$

which remains valid only if t is smaller than $O(\varepsilon^{-1/2})$ and at this stage we do not know the domain $t \in [0, t_0]$ for which $z = h$ at $t = t_0$.

Finally, we integrate once again to find how the position of the object, (x, z), depends on time (t), on α and on ε:

$$x(t;\varepsilon) \sim t\cos\alpha + \varepsilon\left\{\frac{1}{6}(1 - S^3)\cos\alpha + \frac{1}{2}(S - 1)\cos^3\alpha - \frac{1}{4}t\sin 2\alpha\right.$$
$$\left. - \frac{1}{2}(t - \sin\alpha)(\cos^3\alpha)\ln\left(\frac{t - \sin\alpha + S}{1 - \sin\alpha}\right)\right\}; \tag{5.5}$$

$$z(t;\varepsilon) \sim t\sin\alpha - \frac{1}{2}t^2 + \varepsilon\left\{\frac{1}{12}\sin\alpha + \frac{1}{12}(t-\sin\alpha)S^3 - \frac{1}{3}t - \frac{1}{8}\sin\alpha.\cos^2\alpha\right.$$
$$\left. - \frac{1}{8}S(t-\sin\alpha)\cos^2\alpha - \frac{1}{8}\cos^4\alpha.\ln\left(\frac{t-\sin\alpha+S}{1-\sin\alpha}\right)\right\}, \quad (5.6)$$

where $S = \sqrt{1-2t\sin\alpha+t^2}$, each for $\varepsilon \to 0^+$.

Now we suppose that $h = \mathrm{O}(1)$ as $\varepsilon \to 0$, then the time at which this is attained ($t = t_0$) is also $\mathrm{O}(1)$, and hence our asymptotic expansions are valid for $\forall t \in [0, t_0]$. In particular, for $z = h$ and selecting the larger of the roots for $t_0(\varepsilon)$ i.e. further away, we find (from (5.6)), that

$$t_0(\varepsilon) \sim \sin\alpha + S_0 + \varepsilon\left\{\frac{1}{12}(1-2h)^{3/2} - \frac{1}{8}(1-2h)^{1/2}\cos^2\alpha - \frac{1}{2}\right.$$
$$\left. - \frac{1}{8}\frac{(2+\cos^2\alpha)\sin\alpha}{S_0} - \frac{\cos^4\alpha}{8S_0}\ln\left(\frac{S_0+\sqrt{1-2h}}{1-\sin\alpha}\right)\right\},$$

where $S_0 = \sqrt{\sin^2\alpha - 2h}$, provided that $h < \frac{1}{2}\sin^2\alpha$. (For the case of the shot-put application of this model, the landing point is below the projection point, so $h < 0$ and this condition is certainly satisfied.) Finally, using this asymptotic expansion for $t_0(\varepsilon)$ in (5.5), we find that the range is

$$x \sim (\sin\alpha + S_0)\cos\alpha + \varepsilon\left\{\frac{5}{8}(1-2h)^{1/2}\cos^3\alpha - \frac{1}{12}(1-2h)^{3/2}\cos\alpha - \frac{2}{3}\cos\alpha\right.$$
$$\left. - \frac{1}{4}S_0\sin 2\alpha + \frac{1}{8}\frac{\sin\alpha.\cos^3\alpha - \sin 2\alpha}{S_0} + \frac{1}{24}\frac{\left(3\cos^2\alpha - 8S_0^2\right)\cos^3\alpha}{S_0}\ln\left(\frac{S_0+\sqrt{1-2h}}{1-\sin\alpha}\right)\right\}$$

from which, for example, we can estimate the angle which maximises the range; this is left as an exercise.

This example has proved to be particularly straightforward; indeed, because of the specific application that we had in mind—the shot-put—all the asymptotic expansions are uniformly valid. On the other hand, if we had projected the object from the top of a high cliff, then we would encounter the problem of $h \to -\infty$ i.e. $t_0 \to \infty$, and then the validity of the original expansions would be in doubt. The expansions are not valid when $t = \mathrm{O}(\varepsilon^{-1/2})$, but we still have $u = \mathrm{O}(1)$ although $w = \mathrm{O}(\varepsilon^{-1/2})$; this investigation is also left as an exercise.

E5.2 Child's swing

We are all familiar with the child's swing, and the technique for increasing the arc (i.e. the amplitude) of the swing. The process of swinging the legs (coupled with a

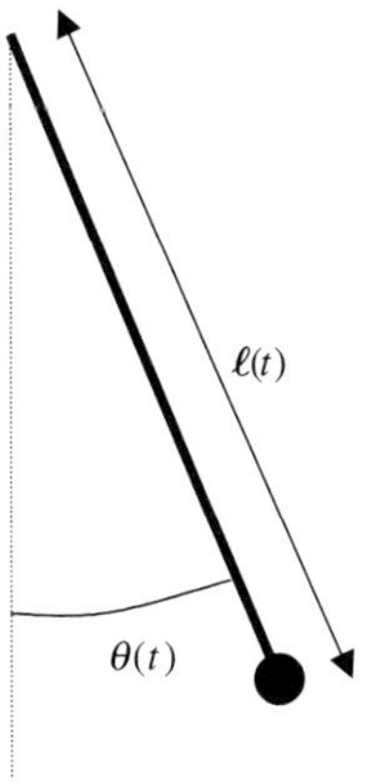

Figure 8. Pendulum of variable length, $\ell(t)$, swinging through the angle $\theta(t)$ (in a vertical plane).

small movement of the torso) causes the centre of gravity of the body to be raised and lowered periodically. This can be modelled by treating the swing as a pendulum which changes its length, $\ell(t)$, by a small amount; the model equation for this (in the absence of damping) is

$$\frac{\mathrm{d}^2\theta}{\mathrm{d}t^2} + \frac{2}{\ell}\frac{\mathrm{d}\ell}{\mathrm{d}t}\frac{\mathrm{d}\theta}{\mathrm{d}t} + \sin\theta = 0, \quad t \geq 0, \tag{5.7}$$

for $\theta(t)$, the angle of the swing, given $\ell(t)$; see figure 8. We choose to represent the child's movement on the swing by $\ell = \ell_0(1 + \varepsilon \sin \omega t)$ where ℓ_0 is a (positive) constant and ω is a constant frequency to be selected. We will further simplify the problem by analysing only the initial stages of the motion when θ is small, so we write $\sin\theta \sim \theta$. (For larger amplitudes, we must retain $\sin\theta$; this complicates the issue somewhat. A number of more general observations about this problem can be found in Holmes, 1995.)

Thus we approximate equation (5.7) as

$$\frac{\mathrm{d}^2\theta}{\mathrm{d}t^2} + \left(\frac{2\varepsilon\omega\cos\omega t}{1 + \varepsilon\sin\omega t}\right)\frac{\mathrm{d}\theta}{\mathrm{d}t} + \theta = 0, \tag{5.8}$$

which we will solve using the method of multiple scales (cf. E4.1).

We take the fast scale as $T = t$ (or, more generally, $T = \Omega t$, but there is no advantage in this, for we are led to the choice $\Omega = 1$) and, by virtue of the term in $\mathrm{d}\theta/\mathrm{d}t$, a slow scale $\tau = \varepsilon t$; thus we have the identity

$$\frac{\mathrm{d}}{\mathrm{d}t} \equiv \frac{\partial}{\partial T} + \varepsilon\frac{\partial}{\partial \tau}.$$

Equation (5.8), with $\theta(t;\varepsilon) \equiv \Theta(T,\tau;\varepsilon)$, becomes

$$\Theta_{TT} + 2\varepsilon\Theta_{T\tau} + \varepsilon^2\Theta_{\tau\tau} + \varepsilon\left(\frac{2\omega\cos\omega T}{1+\varepsilon\sin\omega T}\right)(\Theta_T + \varepsilon\Theta_\tau) + \Theta = 0, \tag{5.9}$$

and we seek a solution in the form

$$\Theta(T,\tau;\varepsilon) \sim \sum_{n=0}^{\infty} \varepsilon^n \Theta_n(T,\tau) \tag{5.10}$$

which is periodic in T. In this problem, we have yet to choose the frequency ω—it is what the child can control in order to increase the amplitude of the swing. Without any damping in the model, we must anticipate that we can find an ω which allows the amplitude to grow without bound. First, with (5.10) in (5.9), we obtain

$$\Theta_{0TT} + \Theta_0 = 0; \quad \Theta_{1TT} + \Theta_1 + 2\Theta_{0T\tau} + 2\omega(\cos\omega T)\Theta_{0T} = 0, \tag{5.11a,b}$$

and so on. Equation (5.11a) has the general solution

$$\Theta_0(T,\tau) = A_0(\tau)\sin\left(T + \phi_0(\tau)\right),$$

for arbitrary functions A_0 and ϕ_0, which leads to equation (5.11b) in the form

$$\Theta_{1TT} + \Theta_1 + 2[A_0'\cos(T+\phi_0) - A_0\phi_0'\sin(T+\phi_0)] + 2\omega A_0\cos(T+\phi_0).\cos\omega T = 0. \tag{5.12}$$

In order to make clear the forcing terms in equation (5.12), which may lead to secularities in T, we expand the last term to give

$$\Theta_{1TT} + \Theta_1 + 2[A_0'\cos(T+\phi_0) - A_0\phi_0'\sin(T+\phi_0)]$$
$$+ \omega A_0[\cos(\omega T + T + \phi_0) + \cos(\omega T - T - \phi_0)] = 0.$$

We see immediately that periodicity in T requires $\phi_0' = 0$ and so $\phi_0 =$ constant; we will, for simplicity, choose $\phi_0 = 0$. (We are not particularly concerned about how the motion is initiated, which would serve to select a particular value of ϕ_0.) Now if $(\omega \pm 1)T \neq \pm T$ (all sign combinations allowed), then we require $A_0' = 0$ for periodicity in T, and the amplitude remains constant: the arc of the swing is *not* increased. However, we are seeking that condition which will allow the amplitude to increase on the time scale τ; thus we expect that $|A_0(\tau)|$ will increase as τ increases. This is possible only if $A_0' \neq 0$ and this can arise here if $\omega \pm 1 = \pm 1$ (all sign combinations to be considered), and $\omega \neq 0$, so we may choose $\omega = \pm 2$ (and the sign of ω is immaterial). Thus with $\omega = \pm 2$, periodicity in T requires that

$$2A_0' \pm 2 = 0 \quad \text{i.e.} \quad A_0(\tau) = \mp\tau + \text{constant},$$

and the amplitude grows. We conclude, therefore, that the adjustments provided by the child on the swing must be at twice the frequency of the oscillation of the swing—which is what we learnt as children.

E5.3 Meniscus on a circular tube

The phenomenon of a liquid rising in a small-diameter tube that penetrates (vertically) the surface of the liquid is very familiar, as are the menisci that form inside and outside the tube. In this problem, we determine a first approximation to the shape of the surface (inside and outside) in the case when the surface tension dominates (or, equivalently, the tube is narrow). The basic model assumes that the mean curvature at the surface is proportional to the pressure difference across the surface (which is maintained by virtue of the surface tension). With the two principal curvatures of radii written as κ_1 and κ_2, then this assumption can be expressed as

$$\kappa_1 + \kappa_2 \propto z$$

where z is the vertical coordinate and the pressure difference is proportional to the (local) height of the liquid above the undisturbed level far away from the tube; this relation is usually referred to as *Laplace's formula*. In detail, written in non-dimensional form, this equation (for cylindrical symmetry) becomes

$$\frac{1}{r}\frac{\mathrm{d}}{\mathrm{d}r}\left\{\frac{r\,(\mathrm{d}z/\mathrm{d}r)}{\sqrt{1+(\mathrm{d}z/\mathrm{d}r)^2}}\right\} = \varepsilon z, \tag{5.13}$$

where the surface is $z = z(r\,;\varepsilon)$, r is the radial coordinate with $r = 0$ at the centre of the tube, and the tube wall (of infinitesimal thickness) is at $r = 1$; the liquid surface satisfies $z \to 0$ as $r \to \infty$ (see figure 9). The non-dimensional parameter ε is inversely proportional to the surface tension in the liquid and proportional to the square of the tube radius (and is usually called the *Bond number*). We will examine the problem of solving equation (5.13) for $\varepsilon \to 0^+$ with the boundary conditions

$$\frac{\mathrm{d}z}{\mathrm{d}r} = 0 \quad \text{at } r = 0;\ z \to 0 \text{ as } r \to \infty; \qquad \frac{\mathrm{d}z}{\mathrm{d}r} = \mp \cot\theta \quad \text{as } r \to 1^{\pm} \text{ (i.e. both inside and outside)}$$

where θ is the given contact angle between the meniscus and the tube (measured relative to the upward vertical side of the tube). For *wetting*, then we have $0 \le \theta < \pi/2$, which we will assume is the case for our liquid. We solve the interior ($0 \le r < 1$) and the exterior ($r > 1$) problems independently. The discussion that we present for the exterior problem is based on Lo (1983); another description of both the interior and exterior problems is given in Lagerstrom (1988). As we will see, this problem results in the construction of a uniformly valid expansion (interior), and a scaling and matching problem in the exterior (cf. §§2.4, 2.5).

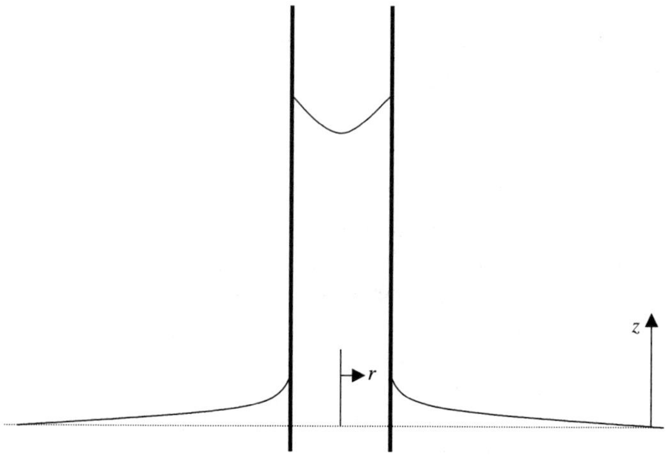

Figure 9. Circular tube (centre $r = 0$) penetrating the surface of a liquid whose undisturbed level is $z = 0$.

Interior problem

It is a familiar observation, at least for $0 < \theta < \pi/2$, that the narrower the tube then the higher the liquid rises in the tube; this suggests that the height of the liquid will increase as $\varepsilon \to 0$. When this is coupled with the property (the confirmation of which is left as an exercise) that no relevant solution of equation (5.13) exists if we ignore the term εz, we are led to write the solution in the form

$$z(r;\varepsilon) = \varepsilon^{-1}h(\varepsilon) + \varsigma(r;\varepsilon)$$

where $h(0) = \mathrm{O}(1)$ (and $h(0) > 0$). Now we seek an asymptotic solution

$$h(\varepsilon) \sim \sum_{n=0}^{\infty} \varepsilon^n h_n; \quad \varsigma(r;\varepsilon) \sim \sum_{n=0}^{\infty} \varepsilon^n \varsigma_n(r)$$

and so the leading-order problem becomes

$$\frac{1}{r}\frac{\mathrm{d}}{\mathrm{d}r}\left\{\frac{r(\mathrm{d}\varsigma_0/\mathrm{d}r)}{\sqrt{1+(\mathrm{d}\varsigma_0/\mathrm{d}r)^2}}\right\} = h_0 \tag{5.14}$$

with

$$\varsigma_0(0) = 0; \quad \varsigma_0'(0) = 0; \quad \varsigma_0' \to \cot\theta \quad \text{as } r \to 1^-. \tag{5.15a,b,c}$$

(The prime denotes the derivative with respect to r.) One integration of (5.14) gives directly that

$$\frac{r\varsigma_0'}{\sqrt{1+(\varsigma_0')^2}} = \frac{1}{2}h_0 r^2 + \text{constant}$$

and then (5.15b) requires that the constant of integration be zero. The condition (5.15c) now shows that

$$\tfrac{1}{2}h_0 = \cos\theta, \quad 0 \le \theta < \pi/2$$

which defines the height of the column of the liquid (measured at the centre-line of the tube). The solution for $\varsigma_0(r)$ can be written down immediately; it is conveniently expressed in terms of a parameter as

$$\varsigma_0 = \frac{2}{h_0}(1 - \cos\phi), \quad r = \frac{2}{h_0}\sin\phi, \quad 0 \le \phi \le \frac{\pi}{2} - \theta,$$

which satisfies condition (5.15a). We have found the first approximation to the height of the liquid at $r = 0$ (namely, $(2\cos\theta)/\varepsilon$) and the shape of the surface of the liquid inside the tube: a section of a spherical shell. Further terms in the asymptotic expansions can be found quite routinely; these expansions are uniformly valid for $0 \le r \le 1$.

Exterior problem

Finding the solution for the shape of the surface outside the tube is technically a more demanding exercise. First, the deviation of the surface from its level ($z = 0$) at infinity is observed to be not particularly large—as compared with what happens inside. This suggests that we attempt to solve equation (5.13) directly, subject to

$$z'(r;\varepsilon) \to -\cot\theta \quad \text{as } r \to 1^+; \quad z \to 0 \quad \text{and} \quad z' \to 0 \quad \text{as } r \to \infty. \tag{5.16a,b}$$

Let us write

$$z(r;\varepsilon) = z_0(r) + o(1) \quad \text{as } \varepsilon \to 0^+,$$

so that we are not committing ourselves, at this stage, to the size of the next term in the asymptotic expansion; in fact, as we shall see, logarithmic terms arise, although we will not pursue the details here. The equation for $z_0(r)$ is simply

$$\frac{1}{r}\frac{\mathrm{d}}{\mathrm{d}r}\left\{\frac{r z_0'}{\sqrt{1+(z_0')^2}}\right\} = 0 \quad \text{or} \quad \frac{r z_0'}{\sqrt{1+(z_0')^2}} = A \tag{5.17}$$

where the arbitrary constant A is determined from (5.16a) as $A = -\cos\theta$. One further integration of (5.17) then yields

$$z_0(r) = -(\cos\theta)\ln\left(r\sec\theta + \sqrt{r^2\sec^2\theta - 1}\right) + B, \tag{5.18}$$

where B is a second arbitrary constant. The complications alluded to earlier are now evident: $z_0(r) \sim -(\cos\theta)\ln r$ as $r \to \infty$, which can never accommodate the conditions at infinity, (5.16b), for any choice of B.

Now any solution of equation (5.13) which admits (5.16b) must balance contributions from each side of the equation; because the difficulties in (5.18) arise as $r \to \infty$, and this is where these other boundary conditions are to be applied, we write $r = R/\delta$ where $\delta(\varepsilon) \to 0^+$ as $\varepsilon \to 0^+$. Whether z also needs to be scaled is unclear at this stage; let us therefore write $z(R/\delta;\varepsilon) \equiv \Delta(\varepsilon)\,Z(R;\varepsilon)$ and then (5.13) becomes

$$\frac{\delta^2}{R}\frac{\mathrm{d}}{\mathrm{d}R}\left\{\frac{RZ'}{\sqrt{1+\delta^2\Delta^2 Z'^2}}\right\} = \varepsilon Z$$

and so, provided that $\delta\Delta \to 0$ as $\varepsilon \to 0$, we must select $\delta^2 = \mathrm{O}(\varepsilon)$. We choose $\delta = \sqrt{\varepsilon}$ and hence obtain the equation

$$\frac{1}{R}\frac{\mathrm{d}}{\mathrm{d}R}\left\{\frac{RZ'}{\sqrt{1+\varepsilon\Delta^2 Z'^2}}\right\} = Z \tag{5.19}$$

valid far away from the tube. The solution of this equation is to satisfy the boundary conditions at infinity and also to match to the solution valid for $r = \mathrm{O}(1)$ i.e. to (5.18). We seek a solution of equation (5.19) in the form

$$Z(R;\varepsilon) = Z_0(R) + \mathrm{o}(1) \quad \text{as } \varepsilon \to 0^+$$

where

$$\frac{1}{R}\frac{\mathrm{d}}{\mathrm{d}R}\left(R\frac{\mathrm{d}Z_0}{\mathrm{d}R}\right) = Z_0$$

which has solutions $\mathrm{K}_0(R)$ and $\mathrm{I}_0(R)$, the modified Bessel functions; but I_0 grows exponentially as $R \to \infty$ (and K_0 decays), so we select the solution

$$Z_0(R) = C\mathrm{K}_0(R) \tag{5.20}$$

where C is an arbitrary constant. Now this modified Bessel function has the properties:

$$\mathrm{K}_0(R) \sim -\ln R \quad \text{as } R \to 0; \qquad \mathrm{K}_0(R) \sim \sqrt{\frac{\pi}{2R}}\,\mathrm{e}^{-R} \quad \text{as } R \to \infty,$$

so the conditions at infinity, (5.16b), are satisfied.

Finally, we match the first term valid for $r = \mathrm{O}(\varepsilon^{-1/2})$, (5.20), to the first term valid for $r = \mathrm{O}(1)$, (5.18). The latter gives

$$z_0(R/\sqrt{\varepsilon}) = -(\cos\theta)\ln\left(\frac{R\sec\theta}{\sqrt{\varepsilon}} + \sqrt{\frac{R^2\sec^2\theta}{\varepsilon} - 1}\right) + B$$

and so

$$z_0 \sim -\left\{\ln R - \tfrac{1}{2}\ln\varepsilon + \ln 2 - \ln(\cos\theta)\right\}\cos\theta + B \quad \text{as } \varepsilon \to 0^+,\ R = \mathrm{O}(1), \tag{5.21}$$

where we have retained the logarithmic term in ε (as we have learnt previously is necessary; see §1.9). From (5.20) we obtain

$$z \sim -\Delta C(\ln r + \tfrac{1}{2}\ln\varepsilon) \quad \text{as } \varepsilon \to 0^+,\ r = \mathrm{O}(1),$$

or, reverting to the R-variable, simply

$$z \sim -\Delta C \ln R. \tag{5.22}$$

For (5.21) and (5.22) to match, we choose

$$C = \cos\theta \text{ (and } \Delta = 1); \quad B = -\left(\tfrac{1}{2}\ln\varepsilon + \ln(\cos\theta) - \ln 2\right)\cos\theta.$$

We note that the left-hand side of equation (5.13) allows z to be shifted by a constant, which at this order is B, and this contains a term $\ln\varepsilon$. The asymptotic procedure for the full equation may proceed provided εz (the right-hand side) $\to 0$ for $r = \mathrm{O}(1)$, and this is the case even in the presence of this logarithmic term, since $\varepsilon\ln\varepsilon \to 0$ as $\varepsilon \to 0^+$. Of course, the appearance of this term indicates the need to include logarithmic terms throughout the asymptotic expansions.

This completes the description of the exterior problem so far as we are concerned here; much more detail can be found in Lo (1983).

E5.4 Drilling by laser

This problem is a one-dimensional model for the process of drilling through a thick block of material using a laser. The laser heats the material until it vaporises, and we assume that the vapour is continuously removed; the essential character of this problem is therefore one of heat transfer at a boundary—the bottom of the drill hole—which is moving. In suitable non-dimensional variables, the temperature relative to ambient conditions satisfies the classical heat conduction equation

$$\frac{\partial T}{\partial t} = \frac{\partial^2 T}{\partial x^2}, \quad t > 0,\ 0 < x < \infty, \tag{5.23}$$

with $$T(x, 0; \varepsilon) = 0;\ T(x, t; \varepsilon) \to 0 \quad \text{as } x \to \infty \tag{5.24a,b}$$

(which describes the initial state and the condition at infinity, respectively), and

$$T(X(t;\varepsilon), t; \varepsilon) = 1, \tag{5.25}$$

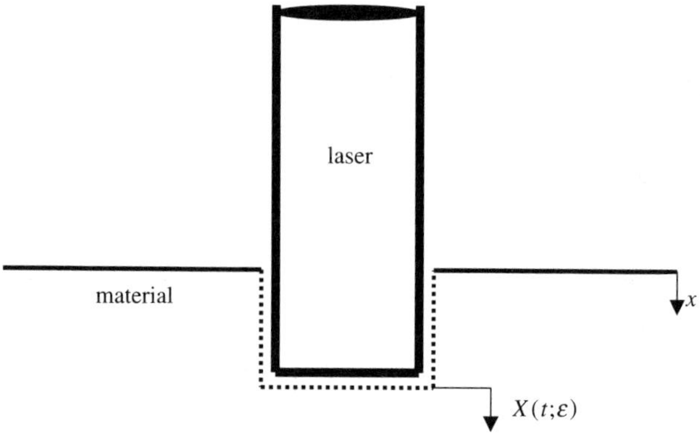

Figure 10. Sketch of a laser drilling through a block of material; the bottom of the drill hole moves according to $x = X(t;\varepsilon)$.

the vaporisation condition at the bottom of the drill hole at $x = X(t;\varepsilon)$; see figure 10. The speed of the drilling process is controlled by

$$\frac{\mathrm{d}X}{\mathrm{d}t} = 1 + \varepsilon \frac{\partial T}{\partial x}(X(t;\varepsilon), t; \varepsilon) \quad \text{with} \quad X(0;\varepsilon) = 0, \tag{5.26}$$

which is based on *Fourier's law* applied at the bottom of the hole. This set, (5.23)–(5.26), is an example of a *Stefan problem*; see Crank (1984) for more details (and important additional references) about this problem and other moving-boundary, heat-transfer examples. (A discussion of this particular problem can also be found in Andrews & McLone, 1976, and, in outline, in Fulford & Broadbridge, 2002.)

Our intention is to seek a solution of this set of equations for $\varepsilon \to 0^+$. For many common metals, ε is fairly small (about 0.2); small ε signifies that rather more (latent) heat than heat content is required to vaporise the material, once it has reached the vaporisation temperature. We shall approach the solution by seeking a straightforward expansion in powers of ε, but we will need to take care over the evaluation on the moving boundary. We will find that the resulting solution is not uniformly valid as $t \to 0$, and the way forward requires a careful examination of what is happening in the early stages of the heating process.

We write

$$T(x,t;\varepsilon) \sim \sum_{n=0}^{\infty} \varepsilon^n T_n(x,t); \quad X(t;\varepsilon) \sim \sum_{n=0}^{\infty} \varepsilon^n X_n(t) \tag{5.27a,b}$$

so that equations (5.25) and (5.26) yield

$$T_0(X_0(t),t) = 1, \quad T_1(X_0(t),t) + X_1(t)T_{0x}(X_0(t),t) = 0, \text{ etc.,} \tag{5.28a,b}$$

and

$$\dot{X}_0 = 1, \quad \dot{X}_1 = T_{0x}(X_0,t), \text{ etc.,} \tag{5.29a,b}$$

respectively, where all the evaluations on $x = X$ have been mapped to $x = X_0$ by allowing Taylor expansions about $x = X_0$ (and constructed for $\varepsilon \to 0$). The subscripts in x denote partial derivatives and the over-dot is the time derivative. The leading-order problem is then described by the equations

$$\frac{\partial T_0}{\partial t} = \frac{\partial^2 T_0}{\partial x^2}$$

with $T_0(x, 0) = 0$; $T_0(x, t) \to 0$ as $x \to \infty$; $T_0(X_0(t), t) = 1$ where $\dot{X}_0 = 1$.

Thus $X_0(t) = t$ and then the complete solution for $T_0(x, t)$ (obtained by using the Laplace transform, for example) is

$$T_0(x, t) = \frac{1}{2}\mathrm{e}^{-(x-t)}\,\mathrm{erfc}\left(\frac{x - 2t}{2\sqrt{t}}\right) + \frac{1}{2}\mathrm{erfc}\left(\frac{x}{2\sqrt{t}}\right), \tag{5.30}$$

where erfc is the *complementary error function*:

$$\mathrm{erfc}(z) = 1 - \frac{2}{\sqrt{\pi}}\int_0^z \exp(-y^2)\,\mathrm{d}y.$$

All this appears to be quite satisfactory, at this stage.

The solution for T_0, (5.30), can now be used to initiate the procedure for finding the next term. In particular, (5.29b) becomes

$$\dot{X}_1 = \frac{1}{2}\mathrm{erfc}\left(\frac{1}{2}\sqrt{t}\right) - \frac{1}{\sqrt{\pi t}}\mathrm{e}^{-t/4} - 1$$

which means that

$$\dot{X} \sim 1 + \varepsilon\left\{\tfrac{1}{2}\mathrm{erfc}\left(\tfrac{1}{2}\sqrt{t}\right) - \mathrm{e}^{-t/4}\big/\sqrt{\pi t} - 1\right\} \tag{5.31}$$

and this asymptotic expansion is not uniformly valid as $t \to 0^+$; indeed, we see that it breaks down when $t = \mathrm{O}(\varepsilon^2)$. But we still have $\dot{X} = \mathrm{O}(1)$, so $X = \mathrm{O}(\varepsilon^2)$; further, for the general ε-term in equation (5.26) to be O(1) then, retaining $T = \mathrm{O}(1)$ which is necessary in order to accommodate (5.25), we see that $x = \mathrm{O}(\varepsilon)$ in this region. Thus we define the new variables

$$\tau = \varepsilon^2 t, \quad X = \varepsilon^2\chi, \quad x = \varepsilon\xi, \quad T = \Theta(\xi, \tau; \varepsilon) \tag{5.32}$$

which produces the problem in this region as

$$\frac{\partial \Theta}{\partial \tau} = \frac{\partial^2 \Theta}{\partial \xi^2} \tag{5.33}$$

with $$\Theta(\xi, 0; \varepsilon) = 0, \quad \Theta(\xi, \tau; \varepsilon) \to 0 \quad \text{as } \xi \to \infty, \quad \Theta(\varepsilon\chi, \tau; \varepsilon) = 1, \tag{5.34a,b,c}$$

where $$\frac{d\chi}{d\tau} = 1 + \frac{\partial\Theta}{\partial\xi}(\varepsilon\chi, \tau; \varepsilon) \quad \text{with} \quad \chi(0; \varepsilon) = 0. \tag{5.35}$$

One essential difficulty that we have overlooked thus far is that the process of heating the material, starting from ambient conditions at $t = 0$, requires the temperature to be raised before *any* vaporisation occurs, and therefore before the hole can begin to form. Because we have a failure of our original asymptotic expansion only when $t = O(\varepsilon^2)$, this should contain the time period over which the initial heating phase occurs. Let the hole be unformed for $0 \le \tau < \tau_0$ (and we assume that $\tau_0 = O(1)$; if it is smaller, we will rescale), then in this time interval we have $\dot{\chi} = 0$ (i.e. $\chi = 0$) and so the boundary condition (5.35) now reads

$$\frac{\partial\Theta}{\partial\xi}(0, \tau; \varepsilon) = -1. \tag{5.36}$$

The resulting problem for Θ (which excludes (5.34c) and uses (5.36) in place of (5.35)) no longer contains ε and it can be solved exactly:

$$\Theta = 2\sqrt{\frac{\tau}{\pi}} \exp\left(-\frac{\xi^2}{4\tau}\right) - \xi \operatorname{erfc}\left(\frac{\xi}{2\sqrt{\tau}}\right). \tag{5.37}$$

(Many of these standard solutions that we use in theories of heat conduction can be found in Carslaw & Jaeger, 1959.) The additional boundary condition (5.35c) now becomes the condition which, when attained, heralds the formation of the hole. From (5.37), we see that $\Theta = 1$ on $\xi = 0$ at the time

$$\tau_0 = \pi/4 \ (=O(1)), \tag{5.38}$$

and after this time the hole develops.

Finally, we complete the formulation of the problem for $t = O(\varepsilon^2)$, but for times larger than $t = \varepsilon^2\tau_0 = \varepsilon^2\pi/4$; we introduce $\tau = (\pi/4) + \hat{\tau}$, and write $\Theta = \hat{\Theta}(\xi, \hat{\tau}; \varepsilon)$ with $\chi = \hat{\chi}(\hat{\tau}; \varepsilon)$, to give

$$\frac{\partial\hat{\Theta}}{\partial\hat{\tau}} = \frac{\partial^2\hat{\Theta}}{\partial\xi^2} \tag{5.39}$$

with $$\hat{\Theta}(\xi, 0; \varepsilon) = \exp(-\xi^2/\pi) - \xi \operatorname{erfc}\left(\xi/\sqrt{\pi}\right) \quad \text{(from(5.37))}; \tag{5.40}$$

$$\hat{\Theta}(\xi, \hat{\tau}; \varepsilon) \to 0 \quad \text{as } \xi \to \infty \tag{5.41}$$

and $$\hat{\Theta}(\varepsilon\hat{\chi}, \hat{\tau}; \varepsilon) = 1 \quad \text{with} \quad \frac{d\hat{\chi}}{d\hat{\tau}} = 1 + \frac{\partial\hat{\Theta}}{\partial\xi}(\varepsilon\hat{\chi}, \hat{\tau}; \varepsilon). \tag{5.42a,b}$$

The leading-order problem, as $\varepsilon \to 0^+$, is therefore (with $\hat{\Theta} \sim \hat{\Theta}_0$; $\hat{\chi} \sim \hat{\chi}_0$) equations (5.39), (5.40) and (5.41), and (5.42a,b) replaced by

$$\hat{\Theta}_0(0, \hat{\tau}) = 1 \ (\hat{\tau} \geq 0); \quad \frac{d\hat{\chi}_0}{d\hat{\tau}} = 1 + \frac{\partial \hat{\Theta}_0}{\partial \xi}(0, \hat{\tau}). \tag{5.43a,b}$$

The solution for $\hat{\Theta}_0(\xi, \hat{\tau})$ can be found (by using the Laplace transform again) and then used to determine $\hat{\chi}_0$; some details of this calculation are given in Crank (1984). Although the form of $\hat{\Theta}_0$ is cumbersome, the resulting expression for $\partial\hat{\Theta}_0/\partial\xi$ on $\xi = 0$ is very straightforward, yielding

$$\frac{d\hat{\chi}_0}{d\hat{\tau}} = \frac{2}{\pi} \arcsin\left(\sqrt{\frac{\hat{\tau}}{\hat{\tau} + (\pi/4)}}\right) \sim 1 - \frac{1}{\sqrt{\pi \hat{\tau}}} \quad \text{as } \hat{\tau} \to \infty$$

which matches precisely with (5.31) when this is expanded for $t \to 0$. The difficulties in $X(t; \varepsilon)$, as $t \to 0$, have been overcome.

This example has required us to undertake some quite intricate analysis in terms of singular perturbation theory, coupled with a careful appreciation of the details of the physical processes involved. This problem, perhaps more than the previous three, shows how powerful these techniques can be in illuminating the details. We now turn to a far more routine type of calculation, although the equation and physical background are important, and the resulting solution has far-reaching consequences.

E5.5 The van der Pol/Rayleigh oscillator

This classical example requires a fairly routine application of the method of multiple scales to a nearly linear oscillator (cf. E 4.2), although the solution that we obtain takes a quite dramatic form. The equation first came to prominence following the work of van der Pol (1922) on the self-sustaining oscillations of a triode circuit (for which the anode current-voltage law takes the form of a cubic relation). However, essentially the same equation had already been discussed by Rayleigh (1883), as a model for 'maintained' vibrations in, for example, organ pipes. (A simple transformation takes Rayleigh's equation into the van der Pol equation.) We will write the equation in the Rayleigh form

$$\ddot{x} + x - \varepsilon\left(\dot{x} - \tfrac{1}{3}\dot{x}^3\right) = 0, \quad \varepsilon > 0, t \geq 0, \tag{5.44}$$

for $x(t; \varepsilon)$; in the context of the van der Pol problem, $\dot{x}$ is proportional to the grid voltage, V. [A circuit diagram is given in figure 11, and the governing equations for this triode circuit are

$$L\frac{di}{dt} + iR + V - M\frac{di_a}{dt} = 0; \quad C\frac{dV}{dt} = i; \quad i_a = \lambda V(1 - \mu V^2),$$

where λ and μ are positive constants.]

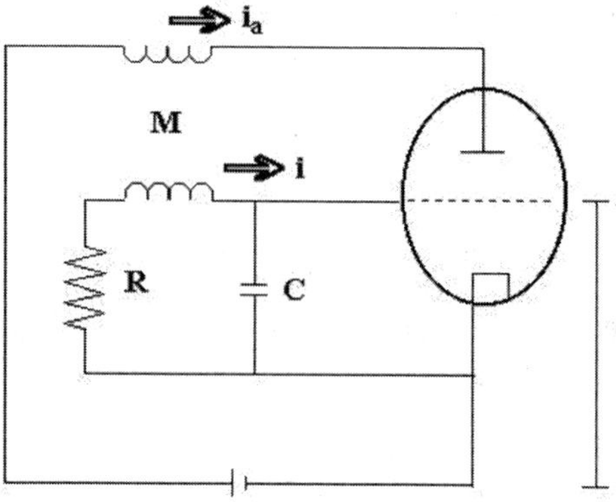

Figure 11. Circuit diagram for the triode oscillator.

The method of multiple scales leads us to introduce

$$T = \omega t, \quad \text{where } \omega \sim 1 + \sum_{n=2}^{\infty} \varepsilon^n \omega_n, \quad \text{and} \quad \tau = \varepsilon t,$$

and then equation (5.44), for $x(t;\varepsilon) \equiv X(T, \tau; \varepsilon)$, becomes

$$\omega^2 X_{TT} + 2\varepsilon\omega X_{T\tau} + \varepsilon^2 X_{\tau\tau} + X - \varepsilon\left[\omega X_T + \varepsilon X_\tau - \tfrac{1}{3}(\omega X_T + \varepsilon X_\tau)^3\right] = 0.$$

We seek a solution in the form

$$X(T, \tau; \varepsilon) \sim \sum_{n=0}^{\infty} \varepsilon^n X_n(T, \tau)$$

which is periodic in T and uniformly valid as $\tau \to \infty$. The equations for the first three terms are

$$X_{0TT} + X_0 = 0; \quad X_{1TT} + X_1 + 2X_{0T\tau} - \left(X_{0T} - \tfrac{1}{3}X_{0T}^3\right) = 0; \tag{5.45a,b}$$

$$X_{2TT} + X_2 + 2\omega_2 X_{0TT} + 2X_{1T\tau} + X_{0\tau\tau} - X_{1T} - X_{0\tau} + (X_{1T} + X_{0\tau})X_{0T}^2 = 0. \tag{5.46}$$

The general solution of equation (5.45a) is

$$X_0(T, \tau) = A_0(\tau)\sin(T + \phi_0(\tau))$$

(for arbitrary functions A_0 and ϕ_0) and then equation (5.45b) can be written

$$X_{1TT} + X_1 = A_0\cos\theta - 2(A_0'\cos\theta - A_0\phi_0'\sin\theta) - \tfrac{1}{12}A_0^3(3\cos\theta + \cos 3\theta),$$

where $\theta = T + \phi_0$. For $X_1(T, \tau)$ to be periodic in T, i.e. in θ, we require

$$A_0 - 2A_0' - \tfrac{1}{4}A_0^3 = 0; \quad A_0\phi_0' = 0$$

which may be integrated to give

$$A_0(\tau) = a\sqrt{\frac{e^{\tau}}{1 + \frac{1}{4}a^2(e^{\tau} - 1)}}, \quad \phi_0(\tau) = \text{constant},$$

where $A_0(0) = a$. Hence, for *any* initial amplitude $a \neq 0$, $|A_0(\tau)| \to 2$ as $\tau \to \infty$: the solution exhibits a *limit cycle* (ultimately an oscillation in T with amplitude 2). This is the *raison d'être* of the triode circuit.

If we proceed with the analysis (the details of which are left as an exercise) we find, first , that

$$X_1(T, \tau) = A_1(\tau)\sin(T + \phi_1(\tau)) + \tfrac{1}{96}A_0^3\cos[3(T + \phi_0)]$$

where A_1 and ϕ_1 are additional arbitrary functions. (The constant ϕ_0 is fixed by the initial data $x(0; \varepsilon)$, $\dot{x}(0; \varepsilon)$.) At the next order, we deduce that the amplitude $A_1(\tau)$ remains bounded as $\tau \to \infty$ only if $\omega_2 = -1/16$.

Another problem with an electrical background, but with a rather different asymptotic structure, will now be described.

E5.6 A diode oscillator with a current pump

In this problem, which contains two small parameters (one of which is used to simplify some of the intermediate results, as expedient), we seek the initial condition which leads to a periodic solution. The circuit (figure 12) is represented by the equations

$$i = C\frac{dV}{dt}, \quad i_1 = \left(e^{\alpha V} - 1\right)I_s, \quad i_2 = \left(1 - e^{-\alpha V}\right)I_s$$

and then Kirchhoff's law gives

$$C\frac{dV}{dt} + \left(e^{\alpha V} - e^{-\alpha V}\right)I_s = I\sin\omega t$$

which leads to the non-dimensional equation ($x \propto e^{\alpha V}$)

$$\delta\dot{x} = x\sin t - x^2 + \varepsilon, \quad x(0; \varepsilon) = a \quad (1 > a > 0). \tag{5.47}$$

Typical values of the parameters are: $\delta \approx 0.03$, $\varepsilon \approx 10^{-10}$. (This equation was brought to the author's attention by a colleague, Dr Armstrong.)

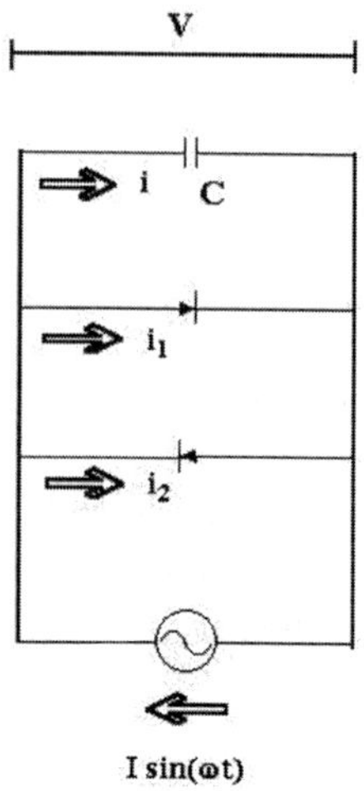

Figure 12. Circuit diagram for the diode oscillator with a current pump.

We seek the first term of an asymptotic expansion for $\varepsilon \to 0^+$:

$$x(t;\varepsilon) = x_0(t) + o(1),$$

for δ fixed (and we may take advantage of small δ to simplify some of the details, but this is not essential). Thus the problem for $x_0(t)$, from (5.47), becomes

$$\delta \dot{x}_0 = x_0 \sin t - x_0^2, \quad x_0(0) = a, \tag{5.48}$$

which can be solved exactly (by writing $x_0 = u(t)\exp(-\cos t/\delta)$); it is convenient to express the initial value as

$$a = \exp\{-(1 - \cos t_0)/\delta\} \quad \text{for } 0 < t_0 \le \pi/2,$$

and then we obtain

$$x_0(t) = \frac{\exp\{(\cos t_0 - \cos t)/\delta\}}{1 + \delta^{-1}\int_0^t \exp[(\cos t_0 - \cos t')/\delta]\,\mathrm{d}t'}. \tag{5.49}$$

The nature of this solution is, perhaps, not immediately apparent, but it becomes more transparent if we invoke $\delta \to 0^+$. We provide approximations to $x_0(t)$, for various t, below (the details of which are left as an exercise):

(a) $t = 0 : x_0(0) = \exp[(\cos t_0 - 1)/\delta]$; (5.50a)
(b) $0 < t < t_0 : x_0 \sim \exp[(\cos t_0 - \cos t)/\delta]$ (exponentially small); (5.50b)

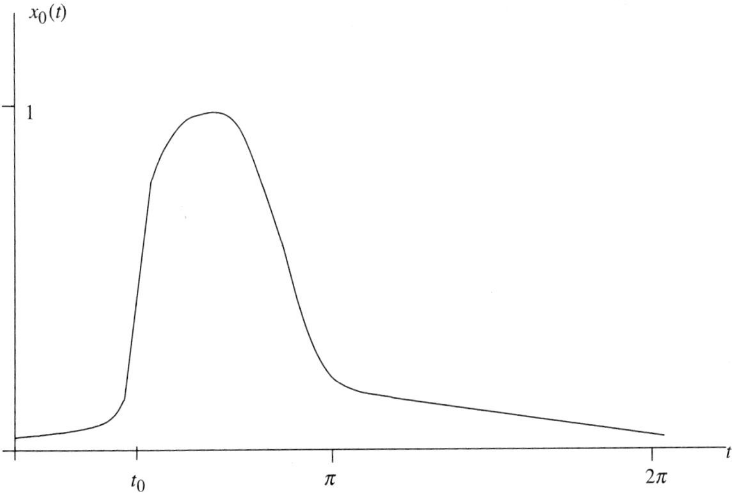

Figure 13. Sketch of the solution (5.49) for small δ.

(c) near $t = t_0$, set $t = t_0 + \delta T : x_0 \sim \dfrac{\exp(T \sin t_0)}{1 + (\text{cosec} t_0) \exp(T \sin t_0)}$; (5.50c)

(d) $t_0 < t < \pi : x_0 \sim \sin t$; (5.50d)

(e) $t = \pi : x_0 \sim \sqrt{2\delta/\pi}$; (5.50e)

(f) $\pi < t < 2\pi : x_0 \sim \sqrt{\delta/2\pi} \exp[-(1 + \cos t)/\delta]$ (exponentially small); (5.50f)

(g) $t = 2\pi : x_0 \sim \sqrt{\delta/2\pi} \exp(-2/\delta)$(exponentially small). (5.50g)

(Most of these results require the use of asymptotic estimates of $\int_0^t \exp(-\cos t'/\delta) dt'$ for various t, as $\delta \to 0^+$; see e.g. Olver, 1974, or Copson, 1967; cf. Q1.16.) This (approximate) solution is sketched in figure 13. Clearly $x_0(0) \neq x_0(2\pi)$, and so this first term is not periodic; however, for $\pi < t \leq 2\pi$, the solution is exponentially small (as $\delta \to 0^+$) and if this becomes as small as $O(\varepsilon)$, then the term ε, omitted in (5.48) cannot be ignored. (This observation also follows when we find the next term in the asymptotic expansion and seek a breakdown; this is a rather tiresome process here, so we treat the problem as one of rescaling the differential equation, as discussed in §2.5.)

For $\pi < t \leq 2\pi$, although the precise domain is not yet known, we write $x(t; \varepsilon) = \varepsilon X(t; \varepsilon)$, and then (5.47) becomes

$$\delta \dot{X} = 1 + X \sin t - \varepsilon X^2. \tag{5.51}$$

We seek a solution

$$X(t; \varepsilon) = X_0(t) + o(1), \quad \varepsilon \to 0^+,$$

and so, from (5.51), we obtain the equation

$$\delta \dot{X} = 1 + X_0 \sin t$$

which can be integrated directly:

$$X_0(t) = \frac{1}{\delta} \exp\{-(\cos t)/\delta\} \int_{t_1}^{t} \exp\left[(\cos t')/\delta\right] dt' \qquad (5.52)$$

where t_1 is an arbitrary constant. This solution is to match to the asymptotic expansion for $x(t;\varepsilon)$ and, in particular, to the first term, (5.49). Further, we are seeking the condition(s) that ensure the existence of a periodic solution, so we also impose

$$x(0;\varepsilon) = \varepsilon X(2\pi;\varepsilon). \qquad (5.53)$$

In principle, we are able to match (5.49) and (5.52) for arbitrary δ (>0), but the result is not particularly useful; further, the periodicity condition implies that, for small ε, then δ must be small. Thus we again invoke $\delta \to 0^+$, and from (5.52) we obtain (for $0 < t_1 < \pi$)

$$X_0(t) \sim (\operatorname{cosec} t_1) \exp\{(\cos t_1 - \cos t)/\delta\},$$

which matches with (5.50f) if we choose

$$\varepsilon(\operatorname{cosec} t_1) \exp(\cos t_1/\delta) = \sqrt{\delta/2\pi} \exp(-1/\delta). \qquad (5.54)$$

Indeed, this match is valid for $\forall t \in (\pi,\ 2\pi - t_1)$. The periodicity requirement, (5.53), now becomes $x_0(0) = \varepsilon X_0(2\pi)$, and thus from (5.52) (with $\delta \to 0^+$) and (5.50a) we obtain

$$\varepsilon\sqrt{\pi/(2\delta)} = \exp[(\cos t_0 - 1)/\delta]. \qquad (5.55)$$

Thus (5.54) and (5.55) imply that we must have

$$\exp[(\cos t_0 + \cos t_1)/\delta] = \tfrac{1}{2} \sin t_1$$

where $\delta \to 0^+$, and so

$$t_1 \sim \pi - t_0 - \delta(\operatorname{cosec} t_0) \ln\left(\tfrac{1}{2} \sin t_0\right) \qquad (5.56)$$

i.e. for the given $0 < t_0 < \pi/2$, we have $\pi/2 < t_1 < \pi$.

Thus for a periodic solution to exist, we require the initial amplitude, a, to be restricted in value, a conclusion that can also be reached on the basis of an examination of the direction field for this equation. For a given ε, the amplitude $a = \exp[-(1 - \cos t_0)/\delta]$ is chosen by selecting t_0 and δ which satisfy (5.54) and (5.55), where t_1 is given by (5.56). The problem of the existence of solutions to this set—solutions do exist!—is left as an additional investigation.

This example has demonstrated how we can extract fairly simple estimates from a solution with a complicated structure, even though the governing differential equation may have persuaded us that no serious difficulties would be encountered.

Finally, we apply the method of multiple scales to a partial differential equation of some importance.

E5.7 Klein-Gordon equation

The general form of the Klein-Gordon equation, written in one spatial dimension, is

$$u_{tt} - u_{xx} + V'(u) = 0,$$

where $V(u)$ (which can be taken as a potential, in quantum-mechanical terms) is, typically, a function with nonlinearity more severe than quadratic. We will consider the problem for which

$$V'(u) = u + \varepsilon u^3, \qquad \varepsilon > 0,$$

and so introduce a parameter which we will allow to satisfy $\varepsilon \to 0^+$. The equation with this choice arises, for example, in the study of wave propagation in a cold plasma. (The choice $V(u) = -\cos u$ gives rise to the so-called *sine-Gordon equation*—a pun on the original name—and this equation has exact 'soliton' solutions; see e.g. Drazin & Johnson, 1992.) We follow the technique described in §4.4, and so we introduce

$$X = \varepsilon x, \qquad T = \varepsilon t, \qquad \theta_x = k(X, T; \varepsilon), \qquad \theta_t = -\omega(X, T; \varepsilon),$$

and then with $u(x, t; \varepsilon) \equiv U(\theta, X, T; \varepsilon)$ we obtain the equation

$$\omega^2 U_{\theta\theta} - 2\varepsilon\omega U_{\theta T} - \varepsilon\omega_T U_\theta + \varepsilon^2 U_{TT} - k^2 U_{\theta\theta} - 2\varepsilon k U_{\theta X} - \varepsilon k_X U_\theta - \varepsilon^2 U_{XX} + U + \varepsilon U^3 = 0. \tag{5.57}$$

We seek a solution which is to be periodic in θ, in the form

$$U(\theta, X, T; \varepsilon) \sim \sum_{n=0}^{\infty} \varepsilon^n U_n(\theta, X, T)$$

and so we obtain the set of equations

$$\omega^2 U_{0\theta\theta} - k^2 U_{0\theta\theta} + U_0 = 0; \tag{5.58}$$

$$\omega^2 U_{1\theta\theta} - k^2 U_{1\theta\theta} + U_1 - 2\omega U_{0\theta T} - 2k U_{0\theta X} + U_0^3 = 0, \text{ etc.}, \tag{5.59}$$

where we have elected to take k (and hence ω) as constants, for the purposes of this discussion. A suitable solution of equation (5.58) is

$$U_0(\theta, X, T) = A_0(X, T)\sin(\theta + \phi_0(X, T))$$

with $\omega^2 = 1 + k^2$; A_0 and ϕ_0 are arbitrary functions. Equation (5.59), for U_1, now becomes

$$(\omega^2 - k^2)U_{1\theta\theta} + U_1 = 2k(A_{0X}\cos\phi - A_0\phi_{0X}\sin\phi) + 2\omega(A_{0T}\cos\phi - A_0\phi_{0T}\sin\phi)$$
$$- \tfrac{1}{4}(3\sin\phi - \sin 3\phi)A_0^3,$$

where $\phi = \theta + \phi_0$. We use $\omega^2 - k^2 = 1$ and then observe that U_1 is periodic in θ, i.e. in ϕ, only if

$$k\phi_{0X} + \omega\phi_{0T} + \tfrac{3}{8}A_0^2 = 0; \quad kA_{0X} + \omega A_{0T} = 0,$$

which have the general solutions

$$A_0 = F(\omega X - kT), \quad \phi_0 = -\tfrac{3}{16\omega k}(\omega X + kT)F^2 + \alpha_0(\omega X - kT)$$

for arbitrary functions F and α_0. Thus the leading term in the asymptotic expansion is

$$U(\theta, X, T; \varepsilon) \sim F(\xi)\sin\left[\theta - \tfrac{3}{16\omega k}\eta F^2(\xi) + \alpha_0(\xi)\right]$$

where $\xi = \omega X - kT$ and $\eta = \omega X + kT$; we observe that we may write the oscillatory term as

$$\sin\left[\left(k - \tfrac{3\varepsilon}{16k}F^2\right)x - \left(\omega + \tfrac{3\varepsilon}{16\omega}F^2\right)t + \alpha_0(\omega X - kT)\right].$$

Both the wave number (k) and the frequency (ω) have small-amplitude corrections—we assume that $F(\xi)$ is a bounded function—which produces an approximate speed of the carrier wave of

$$\frac{\omega}{k}\left\{1 + \frac{3\varepsilon(\omega^2 + k^2)}{16\omega^2 k^2}F^2\right\}$$

which shows that the inclusion of the nonlinearity in the equation is to change the speed of the wave. Indeed, larger waves travel faster—a typical observation in many wave propagation phenomena. Note also that the amplitude function, $F(\omega X - kT)$, represents propagation at the speed k/ω which is precisely the *group speed*, $\mathrm{d}\omega/\mathrm{d}k$; see §4.4. The initial data for this solution must take the form

$$u(x, 0; \varepsilon) \sim F(\varepsilon\omega x)\sin(kx)$$

where k is a given constant (rather than appear in the more general form $\sin[\int k(\varepsilon x)\mathrm{d}x]$) and for a suitable amplitude function.

This concludes the set of examples that have been taken from a rather broad spectrum of simple mechanical and electrical systems. We will now consider the more specialised branch of classical mechanics.

5.2 CELESTIAL MECHANICS

We present three typical problems that arise from planetary, or related, motions: E5.8 The Einstein equation (for Mercury); E5.9 Planetary rings; E5.10 Slow decay of a satellite orbit.

E5.8 The Einstein equation (for Mercury)

Classical Newtonian (Keplerian) mechanics leads to an equation for a single planet around a sun of the form

$$\frac{\mathrm{d}^2 u}{\mathrm{d}\theta^2} + u = h \ (>0, \text{constant})$$

where θ is the polar angle of the orbit, u is inversely proportional to the radial coordinate of the orbit and h measures the angular momentum of the planet. However, when a correction based on Einstein's theory of gravitation is added, the equation becomes

$$\frac{\mathrm{d}^2 u}{\mathrm{d}\theta^2} + u = h + \varepsilon u^2, \tag{5.60}$$

where ε (>0) is a small parameter (about 10^{-7} for Mercury, the planet for which the equation was first introduced). The plan is to find an asymptotic solution of equation (5.60), using the method of multiple scales (cf. E4.2), subject to

$$u = u_0 \quad \text{and} \quad \mathrm{d}u/\mathrm{d}\theta = 0 \quad \text{at } \theta = 0.$$

(It happens that equation (5.60) can be integrated as it stands, in terms of Jacobian elliptic functions, but this tends to obscure the character of the solution and is therefore; hardly worth the effort since ε is so small.)

We introduce

$$T = \omega\theta \quad \text{where } \omega(\varepsilon) \sim 1 + \sum_{n=2}^{\infty} \varepsilon^n \omega_n; \quad \tau = \varepsilon\theta$$

and then the equation for $u(\theta; \varepsilon) \equiv U(T, \tau; \varepsilon)$ becomes

$$\omega^2 U_{TT} + 2\varepsilon\omega U_{T\tau} + \varepsilon^2 U_{\tau\tau} + U = h + \varepsilon U^2.$$

We seek a solution, periodic in T and bounded in τ, in the familiar form

$$U(T, \tau; \varepsilon) \sim \sum_{n=0}^{\infty} \varepsilon^n U_n(T, \tau)$$

and so obtain

$$U_{0TT} + U_0 = h; \quad U_{1TT} + U_1 + 2U_{0T\tau} = U_0^2; \tag{5.61a,b}$$

$$U_{2TT} + U_2 + 2\omega_2 U_{0TT} + 2U_{1T\tau} + U_{0\tau\tau} = 2U_0 U_1, \tag{5.61c}$$

and so on. The initial conditions give

$$U_0(0, 0) = u_0; \quad U_n(0, 0) = 0, \quad n \geq 1; \quad U_{0T}(0, 0) = 0; \tag{5.62a,b,c}$$

$$U_{nT}(0, 0) + U_{(n-1)\tau}(0, 0) = 0, \quad n = 1, 2. \tag{5.62d}$$

The general solution for U_0, from (5.61a), is

$$U_0(T, \tau) = h + A_0(\tau)\cos(T + \phi_0(\tau))$$

for arbitrary functions A_0 and ϕ_0; the initial conditions, (5.62a,c), require that

$$h + A_0(0)\cos[\phi_0(0)] = u_0; \quad A_0(0)\sin[\phi_0(0)] = 0,$$

and so we select

$$\phi_0(0) = 0 \quad \text{and} \quad A_0(0) = u_0 - h. \tag{5.63}$$

Equation (5.61b) then becomes

$$U_{1TT} + U_1 = 2(A_0' \sin\psi + A_0\phi_0' \cos\psi) + h^2 + 2hA_0\cos\psi + \tfrac{1}{2}(1 + \cos 2\psi)A_0^2$$

where $\psi = T + \phi_0(\tau)$; U_1 is periodic in T, i.e. in ψ, only if

$$A_0' = 0; \quad A_0(h + \phi_0') = 0.$$

Then, with conditions (5.63), we see that

$$A_0 = u_0 - h = \text{constant}, \qquad \phi_0(\tau) = -h\tau,$$

which leaves the solution for U_1 as

$$U_1(T, \tau) = A_1(\tau)\cos(T + \phi_1(\tau)) + h^2 + \tfrac{1}{2}(u_0 - h)^2 - \tfrac{1}{6}(u_0 - h)^2 \cos[2(T - h\tau)].$$

It is left as an exercise to show that the solution of equation (5.61c), for $U_2(T, \tau)$, is periodic in T and bounded as $\tau \to \infty$ if

$$\omega_2 = -\tfrac{1}{2}\left[3h^2 + \tfrac{5}{6}(u_0 - h)^2\right]$$

with $A_1 =$ constant and $\phi_1 = -h\tau +$ constant.

This example has proved to be a particularly straightforward application of the method of multiple scales; the next is a rather less routine problem that contains a turning point (see E4.6).

E5.9 Planetary rings

In a study of a model for differentially rotating discs (Papaloizou & Pringle, 1987), the radial structure of the azimuthal velocity component for large azimuthal mode number (essentially ε^{-1} here) satisfies an equation of the form

$$\varepsilon \frac{\mathrm{d}}{\mathrm{d}r}\left[r\frac{\mathrm{d}}{\mathrm{d}r}(rv)\right] = \left[1 - \left(\frac{r}{r_0}\right)^2\right]v, \qquad 0 < r < \infty. \tag{5.64}$$

This equation clearly possesses a turning point at $r = r_0$ (see §2.8); let us examine the solution near this point first. For the neighbourhood of $r = r_0$ we set $r = r_0(1 + \delta R)$, where $\delta(\varepsilon) \to 0^+$ as $\varepsilon \to 0^+$, and ignore the scaling of v (because the equation is linear), so equation (5.64) becomes (with $v(r;\varepsilon) \equiv V(R;\varepsilon)$)

$$\frac{\varepsilon}{\delta^2}\frac{\mathrm{d}}{\mathrm{d}R}\left\{(1 + \delta R)\frac{\mathrm{d}}{\mathrm{d}R}[(1 + \delta R)V]\right\} = -\delta R(2 + \delta R)V.$$

Thus we select $\delta = \varepsilon^{1/3}$ and with $V(R;\varepsilon) \sim V_0(R)$ we obtain the leading-order equation

$$\frac{\mathrm{d}^2 V_0}{\mathrm{d}R^2} + 2RV_0 = 0$$

which is an Airy equation (see equations (4.52), (4.54)) with a bounded solution

$$V_0(R) = C\mathrm{Ai}(2^{1/3}R) \tag{5.65}$$

where C is an arbitrary constant. The solution in $r > r_0$ and $0 < r < r_0$ is now found by using the *WKB* method (see E4.5). (In this example, we will find an approximation to the solution, in each of the three regions, by using only the appropriate local variable; as we have seen in §4.3, all this could be expressed using formal multiple scales.)

For $R > 0$, V_0 is oscillatory and so in $r > r_0$ we seek a solution of equation (5.64) in the form

$$v(r;\varepsilon) = \frac{a(r;\varepsilon)}{r} \sin\{\theta(r)/\Delta + \alpha\} \tag{5.66}$$

where $\Delta(\varepsilon) \to 0^+$ (as $\varepsilon \to 0^+$) is a scaling to be determined, $\alpha = \mathrm{O}(1)$ is a constant and $a(r;\varepsilon)$ is to be written as a suitable asymptotic expansion when we know $\Delta(\varepsilon)$. Thus we obtain

$$\varepsilon\left\{a' \sin\phi + \frac{a\theta'}{\Delta}\cos\phi + r\left[a'' \sin\phi + 2\frac{a'\theta'}{\Delta}\cos\phi + \frac{a\theta''}{\Delta} - \frac{a\theta'^2}{\Delta^2}\sin\phi\right]\right\}$$
$$= \left\{1 - \left(\frac{r}{r_0}\right)^2\right\}\frac{a}{r}\sin\phi$$

where $\phi = (\theta/\Delta) + \alpha$ and the prime denotes the derivative with respect to r. Thus we require $\Delta = \sqrt{\varepsilon}$ and so we write

$$a(r;\varepsilon) \sim \sum_{n=0}^{\infty} \varepsilon^{n/2} a_n(r)$$

and hence we obtain

$$-a_0 r\theta'^2 = \frac{a_0}{r}\left[1 - \left(\frac{r}{r_0}\right)^2\right]; \quad a_0\theta' + r(2a_0'\theta' + a_0\theta'') = 0 \tag{5.67a,b}$$

and so on. These two equations are readily solved, to give

$$v(r;\varepsilon) \sim \frac{A}{r}\left(r^2 - r_0^2\right)^{-1/4} \sin\left\{\frac{1}{\sqrt{\varepsilon}}\frac{1}{r_0}\int_{r_0}^{r} \frac{1}{r'}\sqrt{r'^2 - r_0^2}\,\mathrm{d}r' + \alpha\right\} \tag{5.68}$$

where A is an arbitrary constant.

The corresponding solution for $r < r_0$ (the details of which are left as an exercise) is

$$v(r;\varepsilon) \sim \frac{B}{r}\left(r_0^2 - r^2\right)^{-1/4} \exp\left\{-\frac{1}{\sqrt{\varepsilon}}\frac{1}{r_0}\int_{r}^{r_0} \frac{1}{r'}\sqrt{r_0^2 - r'^2}\,\mathrm{d}r'\right\} \tag{5.69}$$

where B is a second arbitrary constant. The matching of (5.68) and (5.69) with (5.65) (using the results quoted in (4.54)) follows directly. From (5.68) and (5.65) we find that matching is possible if

$$\frac{C}{\sqrt{\pi}}2^{-1/12} = \frac{A}{r_0}\left(2r_0^2\right)^{-1/4}\varepsilon^{-1/12} \quad \text{and} \quad \alpha = \pi/4;$$

from (5.69) and (5.65) we obtain

$$\frac{C}{2\sqrt{\pi}}2^{-1/12} = \frac{B}{r_0}\left(2r_0^2\right)^{-1/4}\varepsilon^{-1/12}$$

which implies the *connection formula* $A = 2B$. (We also see that, if $C = \mathrm{O}(1)$, then A and B are $\mathrm{O}(\varepsilon^{1/12})$ as $\varepsilon \to 0$.)

Our final example under this heading is related to problem E5.1, but now placed in a celestial context. Our presentation is based on that given by Kevorkian & Cole (1981, 1996).

E5.10 Slow decay of a satellite orbit

The equations for a satellite in orbit around a primary (in the absence of all other masses), with a drag proportional to the (speed)^2 are first written down in terms of polar coordinates, (r, θ). These are then transformed to $r^{-1} = u(t)$ and $\theta(t)$ (where t is time), and finally—this is Laplace's important observation—to $u(\theta)$ and $t(\theta)$; we obtain the non-dimensional equations

$$\frac{\mathrm{d}^2u}{\mathrm{d}\theta^2} + u = u^4\left(\frac{\mathrm{d}t}{\mathrm{d}\theta}\right)^2; \quad \frac{\mathrm{d}}{\mathrm{d}\theta}\left(u^2\frac{\mathrm{d}t}{\mathrm{d}\theta}\right) = \varepsilon\frac{\mathrm{d}t}{\mathrm{d}\theta}\sqrt{u^2 + \left(\frac{\mathrm{d}u}{\mathrm{d}\theta}\right)^2} \qquad (5.70\text{a,b})$$

where ε (>0) is a measure of the drag coefficient on the satellite. We seek a solution of this pair of equations, for $\varepsilon \to 0^+$, subject to the initial conditions i.e. conditions prescribed at what we will call $\theta = 0$:

$$u(0;\varepsilon) = 1; \quad t(0;\varepsilon) = 0; \quad \frac{\mathrm{d}u}{\mathrm{d}\theta}(0;\varepsilon) = 0; \quad \frac{\mathrm{d}t}{\mathrm{d}\theta}(0;\varepsilon) = \frac{1}{\nu}\ (<1).$$

Here, ν $(=\mathrm{d}\theta/\mathrm{d}t$ at $\theta = 0$ i.e. $t = 0)$ is the initial component of the velocity vector in the θ-direction; we assume that this is given such that $\nu > 1$ and that it is independent of ε.

The form of equations (5.70), and our experience with problems of this type, suggests that we should introduce new variables (multiple scales) $T = \theta$ and $\tau = \varepsilon\theta$; more general choices for T (e.g. $T = \omega\theta$ or $\mathrm{d}T/\mathrm{d}\theta = \omega(\tau)$) are unnecessary in this problem. Before we proceed, observe that equations (5.70) contain $t(\theta;\varepsilon)$ only through

$\mathrm{d}t/\mathrm{d}\theta$, so it is convenient to solve for u and $\mathrm{d}t/\mathrm{d}\theta$ first (and then t follows after an integration or, at least, by quadrature). Thus, for the purposes of constructing a solution, let us set $\mathrm{d}t/\mathrm{d}\theta = \omega(T, \tau; \varepsilon)$ with $\omega(0; \varepsilon) = \nu^{-1}$ (and the condition on t at $\theta = 0$ is redundant at this stage). Thus our equations (with $u(\theta; \varepsilon) \equiv U(T, \tau; \varepsilon)$) become

$$U_{TT} + 2\varepsilon U_{T\tau} + \varepsilon^2 U_{\tau\tau} + U = \omega^2 U^4; \tag{5.71a}$$

$$\left(\frac{\partial}{\partial T} + \varepsilon\frac{\partial}{\partial \tau}\right)(\omega U^2) = \varepsilon\omega\sqrt{U^2 + (U_T + \varepsilon U_\tau)^2}, \text{ etc.,} \tag{5.71b}$$

and we seek a solution

$$U(T, \tau; \varepsilon) \sim \sum_{n=0}^{\infty} \varepsilon^n U_n(T, \tau); \quad \omega(T, \tau; \varepsilon) \sim \sum_{n=0}^{\infty} \varepsilon^n \omega_n(T, \tau)$$

which is periodic in T.

From equations (5.71) we obtain

$$U_{oTT} + U_0 = \omega_0^2 U_0^4; \quad \left(\omega_0 U_0^2\right)_T = 0; \tag{5.72a,b}$$

$$U_{1TT} + U_1 + 2U_{0T\tau} = 2\omega_0 U_0^3\left(\omega_1 U_0 + 2\omega_0 U_1\right); \tag{5.73a}$$

$$\left(\omega_1 U_0^2 + 2\omega_0 U_0 U_1\right)_T + \left(\omega_0 U_0^2\right)_\tau = \omega_0\sqrt{U_0^2 + U_{0T}^2}, \tag{5.73b}$$

and so on.

The exact solution of equations (5.72), which describe a Keplerian ellipse, is usually written in the form

$$U_0(T, \tau) = p^{-2}\{1 + e\cos(T - T_0)\}; \quad \omega_0(T, \tau) = p^3\{1 + e\cos(T - T_0)\}^{-2} \tag{5.74a,b}$$

where $e = e(\tau)$ is the eccentricity, $T = T_0(\tau)$ denotes the position of the pericentre (i.e. at $T = T_0$) and $p = p(\tau)$ is the angular momentum. (We may write $p(\tau) = \sqrt{a(1 - e^2)}$ where $a = a(\tau)$ is the semi-major axis, if this is useful.) Equation (5.73b) now becomes (with appropriate use of equations (5.74))

$$\left(\omega_1 U_0^2 + 2\omega_0 U_0 U_1\right)_T - \frac{p'}{p^2} = p\frac{\sqrt{1 + e^2 + 2e\cos(T - T_0)}}{\{1 + e\cos(T - T_0)\}^2}$$

which may be integrated, at least formally, to give

$$\omega_1 U_0^2 + 2\omega_0 U_0 U_1 - T\frac{p'}{p^2} = p\int_{T_0}^{T} \frac{\sqrt{1 + e^2 + 2e\cos(T' - T_0)}}{\{1 + e\cos(T' - T_0)\}^2}\,\mathrm{d}T' + A(\tau),$$

where A is an arbitrary function. This expression can be used directly in (5.73a) to give

$$U_{1TT} + U_1 - 2\frac{\partial}{\partial\tau}\left\{\frac{e}{p^2}\sin(T - T_0)\right\} = \frac{2}{p}\left\{p\int_{T_0}^{T}\frac{\sqrt{1 + e^2 + 2e\cos(T' - T_0)}}{[1 + e\cos(T' - T_0)]^2}\,dT' + T\frac{p'}{p^2} + A\right\} \qquad (5.75)$$

and we impose the condition that U_1 be periodic in T. This can be done quite generally by writing $U_1 = U_{0T}F(T, \tau)$, solving for $F(T, \tau)$ and then imposing periodicity. However, by virtue of the integral term in (5.75), this produces a somewhat involved and far-from-transparent result. In order to make some headway, and to produce useful solutions, let us suppose that the satellite orbit is initially almost circular—a fairly common situation—so that the value of $e(0)$ is small. In particular, the initial conditions give

$$p(0) = \nu, \quad e(0) = \nu^2 - 1 \quad \text{and} \quad \text{we choose } T_0(0) = 0,$$

and so now we are assuming that ν is close to 1; it will soon become clear that this approximation holds as $\tau \to \infty$ because we will show that $e(\tau)$ decreases to zero from its initial value.

For small e, we find that

$$\frac{\sqrt{1 + e^2 + 2e\cos(T - T_0)}}{\{1 + e\cos(T - T_0)\}^2} = 1 - e\cos(T - T_0) + \frac{1}{4}e^2[3 + \cos 2(T - T_0)] + O(e^3)$$

and so we expand equation (5.75) as

$$\begin{aligned} &U_{1TT} + U_1 - 2(ep^{-2})'\sin(T - T_0) + ep^{-2}T_0'\cos(T - T_0) \\ &= 2\left\{(1 + \tfrac{3}{4}e^2)T - e\sin(T - T_0) + \tfrac{1}{8}e^2\sin[2(T - T_0)] + O(e^3)\right\} + 2\frac{p'}{p^3}T + \hat{A}(\tau), \end{aligned} \qquad (5.76)$$

where $\hat{A}$ is a new arbitrary function (replacing A). Now the solution, U_1, of equation (5.76) is periodic in T if

$$(ep^{-2})' = e + O(e^3); \quad ep^{-2}T_0' = O(e^3); \quad p'p^{-3} = -\left(1 + \tfrac{3}{4}e^2\right) + O(e^3),$$

which have solutions (correct to this order in e):

$$T_0 = \text{constant} = 0; \quad p(\tau) = \nu\Big/\sqrt{1 + 2\nu^2\tau + \tfrac{3}{4}e_0^2\ln(1 + 2\nu^2\tau)}; \quad e(\tau) = e_0\Big/\sqrt{1 + 2\nu^2\tau},$$

where $e_0 = e(0) = \nu^2 - 1$ (which is small).

We see, therefore, that the small drag in this model leaves the pericentre unaffected ($T_0(\tau) = 0$), the eccentricity decreases towards zero (from its already assumed small value) and the semi-major axis $a(\tau) = p^2/(1 - e^2)$ also approaches zero as $\tau \to \infty$. Thus the orbit gradually spirals in and, as it does so, it becomes more circular.

We have seen how we might tackle the problem of orbits that graze the atmosphere of a planet, although here it was expedient to assume that the orbit was initially nearly circular. If this is not the case, then $e_0 = e(0)$ will not be small and we face a more exacting calculation, although the essential principles are unaltered.

5.3 PHYSICS OF PARTICLES AND OF LIGHT

In this section, we will examine some problems that arise from fairly elementary physics; these will touch on quantum mechanics, light propagation and the movement of particles. In particular, we discuss: E5.11 Perturbation of the bound states of Schrödinger's equation; E5.12 Light propagating through a slowly varying medium; E5.13 Raman scattering: a damped Morse oscillator; E5.14 Quantum jumps: the ion trap; E5.15 Low-pressure gas flow through a long tube.

E5.11 Perturbation of the bound states of Schrödinger's equation

This is a classical problem in elementary quantum mechanics; it involves the time-independent, one-dimensional Schrödinger equation

$$-\psi'' + V(x;\varepsilon)\psi = E\psi, \quad -\infty < x < \infty, \tag{5.77}$$

where $V(x;\varepsilon)$ is the given potential and E is the energy (i.e. the *eigenvalues* of the differential equation). We seek solutions for which $\psi \to 0$ as $|x| \to \infty$ and $\int_{-\infty}^{\infty} \psi^2 \mathrm{d}x$ is finite (and conventionally, we choose $\int_{-\infty}^{\infty} \psi^2 \mathrm{d}x = 1$ to provide a normalisation of the eigenfunction, ψ). In this example, we choose

$$V(x;\varepsilon) = V_0(x) + \varepsilon V_1(x),$$

and this is to be a uniformly valid approximation as $\varepsilon \to 0^+$, for $\forall x \in (-\infty, \infty)$. The problem then becomes

$$-\psi'' + [V_0(x) + \varepsilon V_1(x)]\psi = E\psi, \quad -\infty < x < \infty,$$

with $\psi \to 0$ as $|x| \to \infty$ and $\int_{-\infty}^{\infty} \psi^2 \mathrm{d}x = 1$.

We seek a solution by assuming a straightforward expansion, and we will comment on the conditions that ensure a uniform expansion valid for all x; see §2.3. Thus we write

$$\psi(x;\varepsilon) \sim \sum_{n=0}^{\infty} \varepsilon^n \psi_n(x); \quad E \sim \sum_{n=0}^{\infty} \varepsilon^n E_n$$

and so equation (5.77) gives

$$\psi_0'' - V_0\psi_0 = -E_0\psi_0; \quad \psi_1'' - (V_0\psi_1 + V_1\psi_0) = -(E_0\psi_1 + E_1\psi_0); \tag{5.78a,b}$$

$$\psi_2'' - (V_0\psi_2 + V_1\psi_1) = -(E_0\psi_2 + E_1\psi_1 + E_2\psi_0), \text{ etc.,} \tag{5.78c}$$

with

$$\psi_n(x) \to 0 \quad \text{as} \quad |x| \to \infty \quad \text{for } n = 0,\ 1,\ 2, \ldots. \tag{5.79}$$

and $$\int_{-\infty}^{\infty} \psi_0^2 \mathrm{d}x = 1; \quad \int_{-\infty}^{\infty} \psi_0\psi_1 \mathrm{d}x = 0; \quad \int_{-\infty}^{\infty} \left(\psi_1^2 + 2\psi_0\psi_2\right)\mathrm{d}x = 0, \text{ etc.} \tag{5.80a,b,c}$$

We assume that a solution exists for $\psi_0(x)$—we will give an example shortly—then (5.78b) can be written

$$\psi_0\psi_1'' - \left(V_0\psi_0\psi_1 + V_1\psi_0^2\right) = -\left(E_0\psi_0\psi_1 + E_1\psi_0^2\right)$$

which is integrated over all x. After using integration by parts on the first term, and invoking the decay conditions at infinity, we obtain

$$\int_{-\infty}^{\infty} \psi_1(\psi_0'' - V_0\psi_0)\,\mathrm{d}x - \int_{-\infty}^{\infty} V_1\psi_0^2 \mathrm{d}x = -E_0 \int_{-\infty}^{\infty} \psi_0\psi_1 \mathrm{d}x - E_1 \int_{-\infty}^{\infty} \psi_0^2 \mathrm{d}x$$

which reduces to

$$E_1 = \int_{-\infty}^{\infty} V_1\psi_0^2 \mathrm{d}x$$

when we make use of (5.78a) and (5.80b). Thus the correction to the energy is known (and we assume that V_1 and ψ_0 are such as to ensure that this correction is finite). The same procedure applied to equation (5.78c) yields

$$E_2 = \int_{-\infty}^{\infty} V_1\psi_0\psi_1 \mathrm{d}x.$$

A simple potential is that associated with the harmonic oscillator, namely $V_0(x) = x^2$; then equation (5.78a) becomes

$$\psi_0'' - x^2\psi_0 = -E_0\psi_0$$

which can be rewritten in terms of $\mathrm{H}_m(x)$: $\psi_0(x) = \exp(-\frac{1}{2}x^2)\mathrm{H}_m(x)$ to give

$$\mathrm{H}_m'' - 2x\mathrm{H}_m' + (E_0 - 1)\mathrm{H}_m = 0.$$

This is *Hermite's equation* with solutions that guarantee $\psi_0 \to 0$ at infinity only if $E_0 = 1 + 2m$ ($m = 0, 1, 2, \ldots$). In particular,

$$\mathrm{H}_0(x) = \pi^{-1/4}; \quad \mathrm{H}_1(x) = \pi^{-1/4}\sqrt{2}x; \quad \mathrm{H}_2(x) = \pi^{-1/4}2^{-1/2}(1 - 2x^2),$$

and each of these solutions has been chosen to satisfy the normalisation condition on ψ_0, (5.80a). A typical choice for $V_1(x)$ is

$$V_1(x) = x^p \exp(-\alpha x^2),$$

but in order to ensure that $V_0 + \varepsilon V_1$ is uniformly valid for all x, then $p \geq 2$ (to avoid a breakdown as $x \to 0$) and $\alpha > 0$ (to avoid a breakdown as $|x| \to \infty$, for $p > 2$; the case $p = 2$ and $\alpha = 0$ is equivalent to $\varepsilon = 0$). Let us calculate E_1 for $m = 0$ (so we have $\psi_0(x) = \pi^{-1/4}\exp[-1/(2x^2)]$ and $E_0 = 1$) and for a perturbation with $p = 2$ and $\alpha > 0$; thus

$$E_1 = \frac{1}{\sqrt{\pi}}\int_{-\infty}^{\infty} x^2 \exp[-(1+\alpha)x^2]\,\mathrm{d}x = \frac{1}{2}(1+\alpha)^{-3/2}$$

and so the energy becomes

$$E \sim 1 + \tfrac{1}{2}\varepsilon(1+\alpha)^{-3/2} \quad \text{as } \varepsilon \to 0.$$

Observe that, at this order, we do not need to determine ψ_1 to find E_1.

Calculations of this type, for various $V_0(x)$ and $V_1(x)$, can be found in any good text on quantum mechanics.

E5.12 Light propagating through a slowly varying medium

Fermat's principle states that light travels between any two points on a path which minimises the time of propagation. If the path, in two dimensions, is written as $y = y(x)$, and the speed of light at any point is $c(x, y)$, then $y(x)$ must satisfy

$$c\frac{\mathrm{d}^2y}{\mathrm{d}x^2} + \left(c_y - c_x\frac{\mathrm{d}y}{\mathrm{d}x}\right)\left[1 + \left(\frac{\mathrm{d}y}{\mathrm{d}x}\right)^2\right] = 0. \tag{5.81}$$

This equation can be obtained either from the *eikonal* equation (see Q4.27) for rays or as the relevant *Euler-Lagrange equation* in the calculus of variations. [In the special case where the medium varies only in x, so that $c = c(x)$, we obtain

$$\frac{y'/c(x)}{\sqrt{1+y'^2}} = \text{constant};$$

if, for a light ray, we set $y'(x) = \tan[\alpha(x)]$, then

$$\frac{c(x)}{\sin[\alpha(x)]} = \text{constant},$$

which is *Snell's law*.]

Let us suppose that the properties of the medium slowly change on the scale ε, in the form

$$c(x, y; \varepsilon) = c_0 + \varepsilon c_1(\varepsilon x, \varepsilon y) \quad (c_0 = \text{constant}) \tag{5.82}$$

and so the inherent difficulty of this problem is now evident: we seek $y(x; \varepsilon)$, yet y appears in the function c. In the case that $c = $ constant, the light rays are straight lines e.g.

$$y = x \tan\alpha \quad (\text{for } y(0) = 0 \text{ and } y'(0) = \tan\alpha);$$

let us seek a solution of (5.81) which satisfies precisely these conditions i.e.

$$y(0; \varepsilon) = 0; \quad y'(0; \varepsilon) = \tan\alpha \quad (0 < \alpha < \pi/2).$$

For the given c, (5.82), and with $\varepsilon \to 0^+$, we see that both c_x and c_y are $O(\varepsilon^2)$ and so, from (5.81), we have that $y'' = O(\varepsilon^2)$; a cursory analysis of the problem suggests that we write the solution in the implicit form

$$y = x \tan\alpha + A(X, Y; \varepsilon), \quad A(X, Y; 0) = O(1), \tag{5.83}$$

where $X = \varepsilon x$, $Y = \varepsilon y$. Thus

$$y' = t + \varepsilon(A_X + y' A_Y) \quad (t = \tan\alpha = \text{constant}) \tag{5.84}$$

and we will assume that $c_1(X, Y)$ is such that A, and all its relevant derivatives, lead to uniform asymptotic expansions in the domain where the material exists. This ensures, for example, that we may write

$$y' = \frac{t + \varepsilon A_X}{1 - \varepsilon A_Y} \sim t \quad \text{as } \varepsilon \to 0 \tag{5.85}$$

for all X, Y in the domain. Differentiation of (5.84) yields

$$y'' = \varepsilon^2 (A_{XX} + 2y' A_{XY} + y'^2 A_{YY}) + \varepsilon y'' A_Y, \tag{5.86}$$

which can be solved for y'' (given y' from (5.85)). We seek a solution in the form

$$A(X, Y; \varepsilon) \sim \sum_{n=0}^{\infty} \varepsilon^n A_n(X, Y)$$

and then (5.86) and (5.85) in equation (5.81) produces, at leading order, the equation for $A_0(X, Y)$:

$$c_0(A_{0XX} + 2tA_{0XY} + t^2 A_{0YY}) + (c_{1Y} - tc_{1X})(1 + t^2) = 0.$$

This equation can be solved in general (by introducing $\xi = Y - tX$ and expressing the solution in terms of (ξ, Y), once c_1 is expressed in these same variables); this is left as an exercise. We present a simple example: $c_1(X, Y) = YC(X)$, for which the solution can be written (on introducing ξ)

$$A_0 = -\frac{(1 + t^2)}{c_0} \int_0^X \int_0^{X'} \left[C(\hat{X}) - t(\xi + t\hat{X})C'(\hat{X})\right] \mathrm{d}\hat{X}\mathrm{d}X'$$

which satisfies $A_0 = A_{0X} = 0$ on $\xi = X = 0$ as required by the condition at the origin.

This problem of finding the path of a light ray has used the idea of multiple scales in a less routine way; we now examine an equation for which a more familiar approach (§4.2) is applicable.

E5.13 Raman scattering: a damped Morse oscillator

Under certain circumstances, a small fraction of the incidence light propagating through a medium may be scattered so that the wavelength of this light differs from that of the incident light—usually it is of greater wavelength. This is called *Raman scattering*. An example of this (Lie & Yuan, 1986), which incorporates the Morse (exponential) model for the potential energy of atoms as a function of their separation, is

$$\ddot{x} + \varepsilon\dot{x} + (1 - \mathrm{e}^{-x})\mathrm{e}^{-x} = 0, \quad t \geq 0. \tag{5.87}$$

This equation also includes a (weak) linear damping term, which we will characterise by $\varepsilon \to 0^+$; we wish to find a solution, subject to the initial conditions

$$x(0; \varepsilon) = a, \quad \dot{x}(0; \varepsilon) = 0. \tag{5.88}$$

Because equation (5.87) is nonlinear (although the underlying solution—valid for $\varepsilon =$ 0—can be expressed in terms of elementary functions), we must expect a development along the lines of that described in E4.3.

We introduce fast and slow scales according to

$$\frac{\mathrm{d}T}{\mathrm{d}t} = \omega(\tau) \quad \text{and} \quad \tau = \varepsilon t$$

and then seek a solution which has a *constant* period (in T). Equation (5.87) therefore becomes, with $x(t;\varepsilon) \equiv X(T,\tau;\varepsilon)$,

$$\omega^2 X_{TT} + 2\varepsilon\omega X_{T\tau} + \varepsilon\omega_\tau X_T + \varepsilon^2 X_{\tau\tau} + \varepsilon\,(\omega X_T + \varepsilon X_\tau) + (1 - \mathrm{e}^{-X})\mathrm{e}^{-X} = 0,$$

where the solution is to be

$$X(T,\tau;\varepsilon) \sim \sum_{n=0}^{\infty} \varepsilon^n X_n(T,\tau).$$

The equations for the X_n are therefore

$$\omega^2 X_{0TT} + (1 - \mathrm{e}^{-X_0})\mathrm{e}^{-X_0} = 0; \tag{5.89}$$

$$\omega^2 X_{1TT} + (2\mathrm{e}^{-X_0} - 1)\mathrm{e}^{-X_0} X_1 + 2\omega X_{0T\tau} + \omega_\tau X_{0T} + \omega X_{0T} = 0, \tag{5.90}$$

and so on. The most general solution of equation (5.89), with a constant period (2π), is

$$X_0(T,\tau) = \ln\left\{\frac{1 + \sqrt{1-\omega^2}\,\sin\,(T + \phi_0(\tau))}{\omega^2}\right\}, \quad 0 < \omega \le 1, \tag{5.91}$$

for arbitrary $\omega(\tau)$ and $\phi_0(\tau)$; note that, for a constant period, we will require that $\phi_0(\tau) =$ constant. The initial conditions, (5.88), now on $T = \tau = 0$, are satisfied by the choices

$$\phi_0(0) = \pi/2, \quad \omega(0) = \mathrm{e}^{-a}\sqrt{2\mathrm{e}^a - 1},$$

and the existence of a real, oscillatory solution (of the form (5.91)) requires that $0 > a > -\ln 2$.

The periodicity condition, which will define $\omega(\tau)$, can be obtained from (5.90) by first writing $X_1(T,\tau) = X_{0T}F(T,\tau)$ and then using the T-derivative of (5.89); this yields

$$\omega^2(2X_{0TT}F_T + X_{0T}F_{TT}) + (\omega + \omega_\tau)X_{0T} + 2\omega X_{0T\tau} = 0.$$

This can be integrated once directly, when multiplied by X_{0T}, so that we now have

$$\omega^2 \frac{\partial}{\partial T}\left(X_{0T}^2 F_T\right) + (\omega + \omega_\tau)X_{0T}^2 + \omega\frac{\partial}{\partial\tau}\left(X_{0T}^2\right) = 0$$

i.e.

$$\omega^2 X_{0T}^2 F_T + (\omega + \omega_\tau)\int^T X_{0T}^2 \mathrm{d}T + \omega\int^T \frac{\partial}{\partial\tau}\left(X_{0T}^2\right)\mathrm{d}T = 0.$$

For X_1 to be periodic, then so must be both F and F_T; with the period prescribed as 2π, we therefore obtain the periodicity condition

$$\omega \int_0^{2\pi} X_{0T}^2 \mathrm{d}T + \left(\omega \int_0^{2\pi} X_{0T}^2 \mathrm{d}T\right)' = 0.$$

Finally, the evaluation of the integral (which is left as an exercise) yields

$$\int_0^{2\pi} X_{0T}^2 \mathrm{d}T = 2\pi(\omega^{-1} - 1)$$

and so we have, after solving for $\omega(\tau)$,

$$\omega(\tau) = 1 + (\omega_0 - 1)\mathrm{e}^{-\tau}$$

where $\omega_0 = \omega(0) = \mathrm{e}^{-a}\sqrt{2\mathrm{e}^a - 1}$. We observe that $\omega(\tau) \to 1^-$ as $\tau \to \infty$ and that

$$T = \varepsilon^{-1}[\tau + (\omega_0 - 1)(1 - \mathrm{e}^{-\tau})] = t + \varepsilon^{-1}(\omega_0 - 1)(1 - \mathrm{e}^{-\tau})$$

with $\phi_0(\tau) = \pi/2$. This describes the evolution (shift) of the frequency, $\omega(\tau)$, as the damping progressively affects the solution.

E5.14 Quantum jumps: the ion trap

In the study of the discontinuous emission or absorption of energy (quantum jumps), a single ion is trapped (in an electromagnetic device called a *Paul trap*; see Cook, 1990) and its motion is governed by an equation of the form

$$\varepsilon(\mathrm{i}\phi_t + \phi_{xx}) = \phi v(x)\cos(t/\varepsilon).$$

Here, $\varepsilon > 0$ is a parameter, $v(x)$ is a given function (sufficient for the existence of ϕ) and we seek the complex-valued function $\phi(x, t; \varepsilon)$ for $\varepsilon \to 0^+$. In this case, we see that the oscillatory term on the right oscillates rapidly and so we use the method of multiple scales in the form

$$T = t, \quad \tau = t/\varepsilon \quad \text{with} \quad \phi(x, t; \varepsilon) \equiv \Phi(x, T, \tau; \varepsilon)$$

which gives the equation

$$\mathrm{i}\Phi_\tau + \varepsilon(\mathrm{i}\Phi_T + \Phi_{xx}) = \Phi v(x)\cos\tau. \tag{5.92}$$

We seek a solution

$$\Phi(x, T, \tau; \varepsilon) \sim \sum_{n=0}^{\infty} \varepsilon^n \Phi_n(x, T, \tau)$$

which is uniformly valid as $\tau \to \infty$; from (5.92) we obtain

$$\mathrm{i}\Phi_{0\tau} = \Phi_0 \nu \cos\tau; \quad \mathrm{i}\Phi_{1\tau} + \mathrm{i}\Phi_{0T} + \Phi_{0xx} = \Phi_1 \nu \cos\tau, \tag{5.93a,b}$$

and so on.

The solution of (5.93a) is immediately seen to be

$$\Phi_0(x, T, \tau) = A_0(x, T)\exp[-\mathrm{i}\nu(x)\sin\tau]$$

where A_0 is an arbitrary function; then (5.93b) becomes

$$\mathrm{i}\Phi_{1\tau} + (\mathrm{i}A_{0T} + A_{0xx} - 2\mathrm{i}\nu' A_{0x}\sin\tau - \mathrm{i}\nu'' A_0 \sin\tau - \nu'^2 A_0 \sin^2\tau)\exp[-\mathrm{i}\nu\sin\tau] = \Phi_1\nu\cos\tau.$$

This can be written as

$$\mathrm{i}\left[\Phi_1 \exp(\mathrm{i}\nu\sin\tau)\right]_\tau + \mathrm{i}A_{0T} + A_{0xx} - \mathrm{i}(2\nu' A_{0x} + \nu'' A_0)\sin\tau - \tfrac{1}{2}\nu'^2(1 - \cos 2\tau)A_0 = 0$$

and hence for Φ_1 to remain bounded as $\tau \to \infty$—this also ensures periodicity in τ—we require

$$\mathrm{i}A_{0T} + A_{0xx} - \tfrac{1}{2}\nu'^2 A_0 = 0.$$

A solution of this equation can be expressed as

$$A_0(x, T) = \mathrm{e}^{-\mathrm{i}kT}\alpha(x) \quad \text{where } \alpha'' + \left(k - \tfrac{1}{2}\nu'^2\right)\alpha = 0,$$

but very little headway can be made, at this stage, without some knowledge of $\nu(x)$. We do note, however, that for $k < 0$ the solutions for $\alpha(x)$ are essentially exponential (growing and decaying), but for ν' bounded and $k > 0$ large enough, the solutions are oscillatory.

Our final example is a problem of a gas flow but, because the density is so low, the model is based on an approach that invokes the ideas of statistical mechanics.

E5.15 Low-pressure gas flow through a long tube

The flow of a gas through a long circular tube, i.e. radius/length is small, where molecular collisions are assumed to occur only with the wall of the tube, can be represented by the *Clausing integral equation*

$$N(x) = N_0(x) + \int_0^L K(x - y)N(y)\mathrm{d}y.$$

Here, $N(x)\Delta x$ is the rate of molecular collisions (with the wall) between x and $x + \Delta x$, and $N_0(x)\Delta x$ the rate contributed by those molecules that have their *first* collision between these same stations. The kernel, $K(x - y)$, measures the probability that a molecule which has collided with the wall at $x = y$ will collide again between the stations. (This type of process is called a *free-molecular* or *Knudsen* flow.)

When we introduce the appropriate models for $N_0(x)$ and $K(x - y)$, non-dimensionalise and use the symmetry of $n(x)$ (i.e. $n(x) + n(-x) = 0$ so that $n(0) = 0$), we obtain the equation

$$\begin{aligned} n(x;\varepsilon) = {} & \tfrac{1}{2}\varepsilon^{-1}\left[\sqrt{\left(x+\tfrac{1}{2}\right)^2+\varepsilon^2} - \sqrt{\left(\tfrac{1}{2}-x\right)^2+\varepsilon^2} - 2x\right] \\ & - \tfrac{1}{4}\varepsilon\left[1\Big/\sqrt{\left(x+\tfrac{1}{2}\right)^2+\varepsilon^2} - 1\Big/\sqrt{(\tfrac{1}{2}-x)^2+\varepsilon^2}\right] \\ & + \varepsilon^{-1}\int_{-1/2}^{1/2}\left[1 - \frac{|x-y|}{\sqrt{(x-y)^2+\varepsilon^2}} - \frac{\varepsilon^2\,|x-y|}{2\{(x-y)^2+\varepsilon^2\}^{3/2}}\right] n(y;\varepsilon)\,\mathrm{d}y, \end{aligned} \tag{5.94}$$

with the normalised boundary condition $n(1/2;\varepsilon) = -1/2$. (See Pao & Tchao, 1970, and DeMarcus, 1956 & 1957, and for more general background information, Patterson, 1971.) At first sight, equation (5.94) looks quite daunting and very different from anything we have examined so far in this text. However, the first terms on the right do indicate the presence of boundary layers near $x = \pm 1/2$, so perhaps our familiar techniques can be employed.

For x away from the ends of the domain, the expansion of the first terms in (5.94) leads to the asymptotic form of the equation:

$$\begin{aligned} n(x;\varepsilon) = {} & \tfrac{1}{16}\varepsilon^3\left\{\left(x+\tfrac{1}{2}\right)^{-3} - \left(\tfrac{1}{2}-x\right)^{-3}\right\} + \mathrm{O}(\varepsilon^4) \\ & + \varepsilon^{-1}\int_{-1/2}^{1/2}\left[1 - \frac{|x-y|}{\sqrt{(x-y)^2+\varepsilon^2}} - \frac{\varepsilon^2\,|x-y|}{2\{(x-y)^2+\varepsilon^2\}^{3/2}}\right] n(y;\varepsilon)\,\mathrm{d}y. \end{aligned} \tag{5.95}$$

Now we must estimate the integral, for which we use ideas discussed in §2.2 and exercise Q2.8. This is accomplished by expressing the domain of integration as $(-1/2, x - \delta)$, $(x - \delta, x + \delta)$ and $(x + \delta,\ 1/2)$, where $\delta \to 0^+$ but such that $\varepsilon/\delta \to 0$. It is left as an exercise (which involves considerable effort) to show that (5.95) eventually can be written as

$$n(x;\varepsilon) = \left\{\tfrac{1}{16}\varepsilon^3\left[\left(x+\tfrac{1}{2}\right)^{-3} - \left(\tfrac{1}{2}-x\right)^{-3}\right] + \mathrm{O}(\varepsilon^4)\right\} + \left\{n + \tfrac{1}{3}\varepsilon^2 n'' + \mathrm{O}(\varepsilon^3)\right\}$$

and so $n(x;\varepsilon) \sim n_0(x)$ as $\varepsilon \to 0^+$ must satisfy

$$n_0''(x) = 0 \quad \text{i.e.} \quad n_0 = A_0 x + B_0.$$

But $n_0(0) = 0$ so we have simply

$$n_0(x) = A_0 x;$$

this is the first term in the asymptotic expansion, where A_0 is an arbitrary constant, valid away from the boundary layers (see §2.6). (If we were to apply the boundary condition on $x = 1/2$, then we would deduce that $A_0 = -1$, which turns out to be correct, as we shall see below.)

For the boundary layer near $x = 1/2$, we write $x = \frac{1}{2} - \varepsilon X$ $(X \geq 0)$ and $n(x;\varepsilon) \equiv N(X;\varepsilon)$, which gives (from (5.94))

$$N(X;\varepsilon) = \left\{\tfrac{1}{2}\left[X - \sqrt{1+X^2} + 1/\left(2\sqrt{1+X^2}\right)\right] + \mathrm{O}(\varepsilon^3)\right\} \\ + \int_0^{\varepsilon^{-1}} \left[1 - \frac{|X-Y|}{\sqrt{1+(X-Y)^2}} - \frac{|X-Y|}{2\{1+(X-Y)^2\}^{3/2}}\right] N(Y;\varepsilon)\,\mathrm{d}Y. \quad (5.96)$$

But the dominant contribution to the integral will come from the behaviour of N *outside* the boundary layers i.e. we use $N(Y;\varepsilon) = A_0(\frac{1}{2} - \varepsilon Y)$. When we do this, the integral term yields the result

$$A_0\left\{\tfrac{1}{2} + \tfrac{1}{2}X - \tfrac{1}{2}\sqrt{1+X^2} + 1/\left(4\sqrt{1+X^2}\right) - \varepsilon\left(\tfrac{1}{4} + X + \tfrac{1}{2}X^2 - \tfrac{1}{2}X\sqrt{1+X^2}\right)\right\} + \mathrm{O}(\varepsilon^2),$$

and this is used in (5.96), together with the boundary condition ($N(0;\varepsilon) = -1/2$), to give

$$N(X;\varepsilon) = -\tfrac{1}{2} + \varepsilon\left\{\tfrac{1}{4} + X + \tfrac{1}{2}X^2 - \tfrac{1}{2}X\sqrt{1+X^2}\right\} + \mathrm{O}(\varepsilon^2).$$

The corresponding solution in the boundary layer at the other end is obtained from this result by forming $-N(Y;\varepsilon)$, where $x = -\frac{1}{2} + \varepsilon Y$. Note that, because the boundary condition has been used here, A_0 is now determined (in a way analogous to matching) and so $n(x;\varepsilon) \sim -x$ away from the boundary layers. This concludes all that we will write about this very different type of boundary-layer problem; see Pao & Tchao (1970) for more details.

5.4 SEMI- AND SUPERCONDUCTORS

The study of semiconductors and of superconductors, as it has unfolded over the last 50 years or so, has thrown up any number of interesting and important equations that describe their properties and design characteristics. We will look at three fairly typical examples: E5.16 Josephson junction; E5.17 A p-n junction; E5.18 Impurities in a semiconductor.

E5.16 Josephson junction

The Josephson junction between two superconductors, which are separated by a thin insulator, can produce an AC current when a DC voltage is applied across the junction (this by virtue of the *tunnelling effect*). An equation that models an aspect of this phenomenon (Sanders, 1983) is

$$\ddot{u} + \varepsilon(1 + a\cos u)\dot{u} + \sin u = \varepsilon b, \quad t \geq 0, \tag{5.97}$$

where a and b are given constants, and $u(0;\varepsilon) = \dot{u}(0;\varepsilon) = 0$. We will construct the asymptotic solution, using the method of multiple scales, for $\varepsilon \to 0^+$. Note that, in the absence of the term εb, then $u \equiv 0$ is a solution of the complete problem. We anticipate that the presence of εb will force a non-zero solution which, if it remains bounded, should be $\mathrm{O}(\varepsilon)$ for all time (t); thus we write $u = \varepsilon U$. Further, we introduce

$$T = \omega t, \quad \text{where } \omega \sim 1 + \sum_{n=2}^{\infty} \varepsilon^n \omega_n, \quad \text{and} \quad \tau = \varepsilon t,$$

and so $U(T, \tau; \varepsilon)$ satisfies the equation

$$\omega^2 U_{TT} + 2\varepsilon\omega U_{T\tau} + \varepsilon^2 U_{\tau\tau} + \varepsilon\,[1 + a\cos(\varepsilon U)]\,(\omega U_T + \varepsilon U_\tau) + \varepsilon^{-1}\sin(\varepsilon U) = b.$$

We assume a bounded, periodic solution can be written as

$$U(T, \tau; \varepsilon) \sim \sum_{n=0}^{\infty} \varepsilon^n U_n(T, \tau)$$

and so we may expand

$$\cos(\varepsilon U) \sim 1 - \tfrac{1}{2}\varepsilon^2 U_0^2; \quad \sin(\varepsilon U) \sim \varepsilon U_0 + \varepsilon^2 U_1.$$

Thus we obtain the set of equations

$$U_{0TT} + U_0 = b; \quad U_{1TT} + U_1 + 2U_{0T\tau} + (1 + a)U_{0T} = 0, \tag{5.98a,b}$$

and so on. These equations follow the pattern for a nearly linear oscillator; see §4.1.

The general solution of (5.98a) is

$$U_0(T, \tau) = b + A_0(\tau)\cos\left(T + \phi_0(\tau)\right)$$

with initial conditions

$$A_0(0) = -b, \quad \phi_0(0) = 0;$$

then equation (5.98b) becomes

$$U_{1TT} + U_1 - 2(A_0' \sin\theta + A_0\phi_0' \cos\theta) - (1+a)A_0 \sin\theta = 0,$$

where $\theta = T + \phi_0(\tau)$. Immediately we see that U_1 is periodic in T, i.e. in θ, only if

$$\phi_0' = 0 \quad \text{and} \quad 2A_0' + (1+a)A_0 = 0,$$

and so we have

$$\phi_0(\tau) = \text{constant} = 0; \qquad A_0(\tau) = -b \exp\left[-\tfrac{1}{2}(1+a)\tau\right]$$

which shows that a bounded solution (as $\tau \to \infty$) requires $a \geq -1$. Thus the solution, to this order, is

$$u \sim \varepsilon b\left[1 - (\cos T)\exp\left(-\tfrac{1}{2}(1+a)\tau\right)\right];$$

the construction of higher-order terms is left as a fairly routine exercise.

E5.17 A p-n junction

A p-n junction is where two semiconducting materials meet; such junctions may perform different functions. The one that we describe is a diode. We analyse the device for $x \in [0, 1]$, where the junction sits at $x = 0$ (and, by symmetry, it extends into $-1 \leq x < 0$) and an ohmic contact is placed at $x = 1$. In suitable non-dimensional, scaled variables we have

$$\varepsilon\frac{\mathrm{d}e}{\mathrm{d}x} = p - n + 1; \quad \varepsilon\frac{\mathrm{d}p}{\mathrm{d}x} = ep - \varepsilon I(x); \quad \varepsilon\frac{\mathrm{d}n}{\mathrm{d}x} = -en + \varepsilon I(x) \qquad (5.99\text{a,b,c})$$

where e is the electrostatic field, p the hole density and n the electron density. The term '+1' in (5.99a) is a constant 'doping' density and we will assume that the current density, $I(x)$ (appearing in (5.99b,c)), is given; indeed, in this simple model, we take $I(x) = $ constant. The boundary conditions are

$$p = n \quad \text{at } x = 0; \quad n = p + 1 \quad \text{and} \quad np = k\ (>0) \quad \text{both at } x = 1,$$

and ε is our small parameter (typically about 0.001). (See Shockley, 1949; Roosbroeck, 1950; Vasil'eva & Stelmakh, 1977; Schmeisser & Weiss, 1986.) It is evident that the set (5.99) exhibits the characteristics of a boundary-layer problem (§§2.6, 2.7) because the small parameter multiplies the derivatives in each equation. However, a neat manoeuvre allows one equation to be independent of ε.

Let us introduce $u = np$, then equations (5.99) can be rewritten as

$$\varepsilon\frac{de}{dx} = p - \frac{u}{p} + 1; \quad \varepsilon\frac{dp}{dx} = ep - \varepsilon I; \quad \frac{du}{dx} = \left(p - \frac{u}{p}\right) I \qquad (5.100a,b,c)$$

with

$$p^2 = u \quad \text{at } x = 0; \quad u = k \quad \text{and} \quad p = \tfrac{1}{2}\left(-1 + \sqrt{1+4k}\right) \quad \text{both at } x = 1. \quad (5.101a,b,c)$$

Now away from the boundary layer (whose position is yet to be determined) we write each of $e(x;\varepsilon)$, $p(x;\varepsilon)$ and $u(x;\varepsilon)$ as asymptotic expansions

$$q(x;\varepsilon) \sim \sum_{n=0}^{\infty} \varepsilon^n q_n(x) \quad (q \equiv e,\, p,\, u)$$

and then the leading order (from (5.100)) yields

$$p_0 - \frac{u_0}{p_0} + 1 = 0; \quad e_0 p_0 = 0; \quad \frac{du_0}{dx} = \left(p_0 - \frac{u_0}{p_0}\right) I.$$

Thus we must select the solution

$$e_0 = 0, \quad \frac{du_0}{dx} = -I \quad \text{and} \quad p_0^2 + p_0 - u_0 = 0;$$

it is now evident that the condition $p_0^2 = u_0$ (see (5.101a)) cannot be attained by this solution, so the boundary layer must be at $x = 0$—the position of the junction—and thus we are permitted (in this soluiton) to use the boundary conditions at $x = 1$, to give

$$e_0 = 0; \quad u_0(x) = k + (1-x)I; \quad p_0(x) = \tfrac{1}{2}\left(-1 + \sqrt{1 + 4\{k + (1-x)I\}}\right). \quad (5.102a,b,c)$$

Now from equations (5.100a,b), it is clear that the boundary-layer thickness is $O(\varepsilon)$, and so we introduce $x = \varepsilon X$ ($X \geq 0$) and write $q(x;\varepsilon) \equiv Q(X;\varepsilon)$ for each of e, p and u, to obtain the set

$$\frac{dE}{dX} = P - \frac{U}{P} + 1; \quad \frac{dP}{dX} = EP - \varepsilon I; \quad \frac{dU}{dX} = \varepsilon\left(P - \frac{U}{P}\right) I \qquad (5.103a,b,c)$$

with $P^2 = U$ at $X = 0$ and matching conditions for $X \to \infty$. Thus, from equations (5.103), the leading-order terms (zero subscripts) in the straightforward asymptotic expansions satisfy the equations

$$\frac{dE_0}{dX} = P_0 - \frac{U_0}{P_0} + 1; \quad \frac{dP_0}{dX} = E_0 P_0; \quad \frac{dU_0}{dX} = 0 \qquad (5.104a,b,c)$$

and the last equation simply requires that

$$U_0 = k + I \ (= c = \text{constant})$$

(which is equivalent to the observation, from (5.100c), that there is no boundary-layer structure in the solution for $u(x;\varepsilon)$). The first two equations, (5.104a,b), give an equation for $P_0(X)$:

$$\frac{\mathrm{d}}{\mathrm{d}X}\left(\frac{\mathrm{d}P_0/\mathrm{d}X}{P_0}\right) = P_0 - \frac{c}{P_0} + 1$$

which can be integrated once (by setting $\mathrm{d}P_0/\mathrm{d}X = f(P_0)$) to give

$$\frac{\mathrm{d}P_0}{\mathrm{d}X} = P_0\sqrt{2}\sqrt{P_0 + \frac{c}{P_0} + \ln P_0 + A}$$

where A is an arbitrary constant. Sadly, we cannot integrate once again (so a numerical approach might be considered), but we can make a few observations.

The boundary condition on $X = 0$ becomes $P_0 = \sqrt{c}$ and, in addition, the matching condition is satisfied if

$$P_0 \to \tfrac{1}{2}\left(-1 + \sqrt{1+4c}\right) \ (= \lambda) \quad \text{as } X \to \infty$$

(see (5.102c) which requires the choice

$$A = -\lambda - \frac{c}{\lambda} - \ln\lambda = -\sqrt{1+4c} - \ln\lambda.$$

(It is left as an exercise to show that there is a solution for which

$$P_0(X) \sim \lambda + \alpha\exp[-(1+4c)^{1/4}X] \quad \text{as } X \to \infty.)$$

However, more success in the development of useful analytical detail is possible if we use (5.104a,b) to produce an equation for $E_0(P_0)$:

$$\frac{\mathrm{d}E_0}{\mathrm{d}P_0}\frac{\mathrm{d}P_0}{\mathrm{d}X} = E_0 P_0 \frac{\mathrm{d}E_0}{\mathrm{d}P_0} = P_0 - \frac{c}{P_0} + 1 \quad \text{so} \quad E_0\frac{\mathrm{d}E_0}{\mathrm{d}P_0} = 1 - \frac{c}{P_0^2} + \frac{1}{P_0}.$$

One integration then produces the result

$$\frac{1}{2}E_0^2 = P_0 + \frac{c}{P_0} + \ln P_0 + B$$

where the arbitrary constant must, in order to satisfy the matching condition at infinity, take the same value as A above:

$$B = -\sqrt{1+4c} - \ln \lambda .$$

Thus, although we are unable to write down an expression for $P_0(X)$, in terms of elementary functions, we do have a simple relation between E_0 and P_0; in particular, we see that

$$E_0(0) = \sqrt{4\sqrt{c} - 2\sqrt{1+4c} + \ln(c/\lambda^2)},$$

the electrostatic field at the junction, $x = 0$. More details can be found in the references cited above; also a discussion of similar problems is given in Smith (1985) and O'Malley (1991).

E5.18 Impurities in a semiconductor

A significant issue in the design and operation of semiconductors is the presence, and movement, of impurities. In particular, the level of impurities that diffuse from the outer surface of the material and move to occupy vacant locations within the structure can be modelled (King, Meere & Rogers, 1992) by the equations

$$\frac{\partial c}{\partial t} = \frac{\partial}{\partial x}\left(v\frac{\partial c}{\partial x} - c\frac{\partial v}{\partial x}\right); \quad \frac{\partial v}{\partial t} + \alpha\frac{\partial c}{\partial t} = \varepsilon\frac{\partial^2 v}{\partial x^2}, \quad x > 0, \ t > 0. \tag{5.105}$$

Here, $c(x, t; \varepsilon)$ is the concentration of the impurities, $v(x, t; \varepsilon)$ the concentration of vacancies (holes), and α (<1) and ε are positive constants; the boundary and initial conditions are

$$c(x, 0; \varepsilon) = 0; \quad v(x, 0; \varepsilon) = 1; \quad c(0, t; \varepsilon) = 1; \quad v(0, t; \varepsilon) = \beta; \tag{5.106a,b,c,d}$$

$$c \to 0 \quad \text{and} \quad v \to 1 \quad \text{both as } x \to \infty \ (t \geq 0). \tag{5.106e,f}$$

All these boundary values are constants—and β $(\neq 1-\alpha)$ is positive—and, furthermore, the appearance of the same values at $t = 0$ and as $x \to \infty$ suggests that we could use a relevant similarity solution. The small parameter, ε, is associated only with the $v(x, t; \varepsilon)$ and so we may anticipate the existence of a boundary-layer structure in v, but not in c; cf. E3.3. Indeed, it should be clear that this necessarily must be near $x = 0$ and used to accommodate the boundary value given by (5.106d).

Away from $x = 0$, we seek a solution with

$$c(x, t; \varepsilon) \sim c_0(x, t) \quad \text{and} \quad v(x, t; \varepsilon) \sim v_0(x, t)$$

which, from (5.105), must therefore satisfy the equations

$$c_{0t} = (v_0 c_{0x} - c_0 v_{0x})_x; \quad (v_0 + \alpha c_0)_t = 0 \tag{5.107a,b}$$

and then (5.107b) gives

$$v_0 + \alpha c_0 = A_0(x),$$

where A_0 is an arbitrary function. We may impose the initial conditions, (5.106a,b), and so $A_0(x) = 1$; then (5.107a) becomes simply

$$c_{0t} = c_{0xx},$$

and the (similarity) solution which satisfies (5.106a,c,e) is

$$c_0(x, t) = 1 - \frac{2}{\sqrt{\pi}} \int_0^{x^2/4t} \mathrm{e}^{-y^2}\, \mathrm{d}y \ (=\mathrm{erfc}(x^2/4t))$$

(provided that $t = 0$ is interpreted as $t \to 0^+$). Thus

$$v_0(x, t) = 1 - \alpha c_0 = 1 - \alpha + \frac{2\alpha}{\sqrt{\pi}} \int_0^{x^2/4t} \mathrm{e}^{-y^2}\, \mathrm{d}y \to 1 - \alpha \quad \text{as } x \to 0^+ \ (t > 0),$$

which does not satisfy the boundary value on $x = 0$ $(1 - \alpha \neq \beta)$, and so we require the boundary layer near here.

Let us introduce $x = \sqrt{\varepsilon} X$, and write

$$c(\sqrt{\varepsilon}X, t; \varepsilon) \equiv C(X, t; \varepsilon) \quad \text{and} \quad v(\sqrt{\varepsilon}X, t; \varepsilon) \equiv V(X, t; \varepsilon),$$

then equations (5.105) become

$$\varepsilon C_t = (VC_X - CV_X)_X; \quad V_t + \alpha C_t = V_{XX};$$

the leading-order problem (zero subscript) therefore satisfies

$$(V_0 C_{0X} - C_0 V_{0X})_X = 0 \quad \text{and} \quad V_{0t} + \alpha C_{0t} = V_{0XX}. \tag{5.108a,b}$$

The solution of this pair is to satisfy the matching conditions

$$C_0 \to 1 \quad \text{and} \quad V_0 \to 1 - \alpha \quad \text{both as} \quad X \to \infty$$

and thus (5.108a) gives directly

$$C_0(X, t) = \left(\frac{1}{1-\alpha}\right) V_0(X, t)$$

and then (5.108b) becomes

$$\left(\tfrac{1}{1-\alpha}\right) V_{0t} = V_{0XX}.$$

The appropriate solution of this equation, which satisfies both the matching condition and (5.106d), is

$$V_0(X, t) = \beta + (1 - \alpha - \beta)\frac{2}{\sqrt{\pi}} \int_0^{X^2/4(1-\alpha)t} e^{-y^2} \, dy.$$

Higher-order terms, in both the outer and boundary-layer solutions, can be found altogether routinely (although the calculations are rather tedious).

This completes our few examples in this group; we now turn to one of the areas where singular perturbation theory has played a very significant rôle.

5.5 FLUID MECHANICS

The study of fluid mechanics is broad and deep and it often has far-reaching consequences. Many of the classical techniques of singular perturbation theory were first developed in order to tackle particular difficulties that were encountered in this field. Examples that are available are numerous, and any number could have been selected for discussion here (and some have already appeared as examples in earlier chapters). We will content ourselves with just four more very different problems that give a flavour of what is possible, but these are all fairly classical examples of their type. Many others can be found in most of the texts already cited earlier. We will discuss: E5.19 Viscous boundary layer on a flat plate; E5.20 Very viscous flow past a sphere; E5.21 A piston problem; E5.22 A variable-depth Korteweg-de Vries equation for water waves.

E5.19 Viscous boundary layer on a flat plate

The solution of this problem (about 1905), with Poincaré's work on celestial mechanics, together laid the foundations for singular perturbation theory. In this example, we consider an incompressible, viscous fluid (in $y > 0$) flowing over a flat plate, $x \geq 0$, the flow direction at infinity being parallel to the plate. The governing equations are the *Navier-Stokes equation* (in the absence of gravity) and the equation of mass conservation:

$$u_t + uu_x + vu_y = -p_x + \mathrm{R}_e^{-1}(u_{xx} + u_{yy}); \qquad (5.109a)$$

$$v_t + uv_x + vv_y = -p_y + \mathrm{R}_e^{-1}(v_{xx} + v_{yy}); \quad u_x + v_y = 0 \qquad (5.109b,c)$$

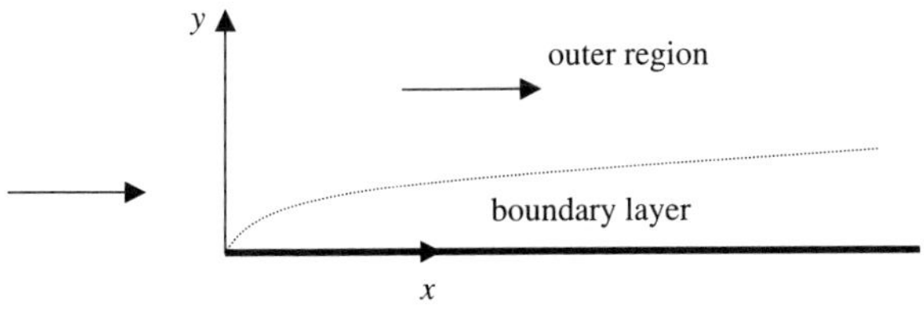

Figure 14. Sketch of the viscous boundary layer on a flat plate.

where R_e is the *Reynolds number* (and we have used subscripts throughout to denote partial derivatives). We will consider the problem of steady flow $(\partial/\partial t \equiv 0)$ with $R_e \to \infty$; the boundary conditions for uniform flow at infinity are

$$u \to 1 \quad \text{and} \quad v \to 0 \quad \text{as } x \to -\infty \quad \text{and} \quad \text{as } y \to \infty; \tag{5.110a,b}$$

$$u = 0 \quad \text{and} \quad v = 0 \quad \text{on} \quad y = 0 \quad \text{in } x > 0, \tag{5.110c,d}$$

which imply that the pressure $p \to$ constant away from the plate (and we will not analyse the nature of the flow near $x = 0$); the plate will extend to infinity $(x \geq 0)$. The presence of the small parameter R_e^{-1}, multiplying the highest derivatives, is the hallmark of a boundary-layer problem. In particular, the (inviscid) problem can satisfy $v = 0$ on $y = 0$, but not $u \to 0$ as $y \to 0$ (in $x > 0$), so we expect a boundary-layer scaling in y; see figure 14.

Outside the boundary layer, the solution is written

$$u(x, y; R_e) \sim u_0(x, y), \quad v(x, y; R_e) \sim v_0(x, y), \quad p(x, y; R_e) \sim p_0(x, y)$$

and so equations (5.109) give

$$u_0 u_{0x} + v_0 u_{0y} = -p_{0x}; \quad u_0 v_{0x} + v_0 v_{0y} = -p_{0y}; \quad u_{0x} + v_{0y} = 0$$

subject to the boundary conditions (5.110a,b,d), for zero-subscripted variables; this has the solution

$$u_0(x, y) = 1, \quad v_0(x, y) = 0, \quad p_0(x, y) = \text{constant} = 0 \quad \text{(say)} \ (y > 0).$$

(It is clear that this solution has additional problems for $x \to 0^-$ on $y = 0$, where the *stagnation point* exists at the leading edge of the plate.) Note that this solution can be expressed in terms of the *stream function*: $\psi_0(x, y) = y$ (where, in general, $u = \psi_y$, $v = -\psi_x$).

The region of the boundary layer is described by the scaled variable $y = \delta Y$, where $\delta(R_e) \to 0^+$ as $R_e \to \infty$, and x is unscaled. (We would need to scale x near exceptional points such as the leading edge, a point of separation and the trailing edge of a finite plate.) By virtue of the existence of a stream function, we see that we must also scale $v = \delta V$; we write

$$u \equiv U(x, Y; \mathrm{R_e}), \quad v \equiv \delta V(x, Y; \mathrm{R_e}), \quad p \equiv P(x, Y; \mathrm{R_e})$$

and then we must choose $\delta = \mathrm{R_e}^{-1/2}$ to give

$$UU_x + VU_Y = -P_x + U_{YY} + \mathrm{R_e}^{-1} U_{xx};$$
$$\mathrm{R_e}^{-1}(UV_x + VV_Y) = -P_Y + \mathrm{R_e}^{-1}\left(V_{YY} + \mathrm{R_e}^{-1} V_{xx}\right); \quad U_x + V_Y = 0.$$

This set is to be solved, subject to the boundary conditions (5.110c,d), written in boundary-layer variables, and matching conditions for $Y \to \infty$. The leading-order problem (zero subscripts) satisfies

$$U_0 U_{0x} + V_0 U_{0Y} = -P_{0x} + U_{0YY}; \quad P_{0Y} = 0; \quad U_{0x} + V_{0Y} = 0 \qquad (5.111\text{a,b,c})$$

with
$$U_0 = V_0 = 0 \quad \text{on} \quad Y = 0 \quad (x > 0) \qquad (5.112)$$

and the matching conditions

$$U_0 \to 1, \quad V_0 \to 0, \quad P_0 \to 0 \quad \text{as} \quad Y \to \infty \quad \text{and} \quad \text{as} \quad x \to -\infty \qquad (5.113)$$

To solve equations (5.111), we note, first, that $P_0 = P_0(x)$ and then the matching condition requires $P_0 = 0$ throughout this region. Next we use (5.111c) to allow the introduction of a stream function ($U_0 = \Psi_{0Y}$, $V_0 = -\Psi_{0x}$) and then (5.111a) can be written

$$\Psi_{0Y}\Psi_{0xY} - \Psi_{0x}\Psi_{0YY} = \Psi_{0YYY}$$

with
$$\Psi_0/Y \to 1 \quad \text{as} \quad Y \to \infty \quad \text{and} \quad \text{as} \quad x \to -\infty$$

and
$$\Psi_0 = \Psi_{0Y} = 0 \quad \text{on} \quad Y = 0 \quad (x > 0).$$

The relevant solution (Blasius, 1908) takes a *similarity* form: $\Psi_0(x, Y) = \sqrt{x} f(\eta)$, $\eta = Y/2\sqrt{x}$ $(x > 0)$; direct substitution then yields the ordinary differential equation for $f(\eta)$:

$$ff'' + f''' = 0$$

with
$$f(0) = f'(0) = 0 \quad \text{and} \quad f'(\eta) \to 2 \quad \text{as } \eta \to \infty.$$

This equation must be solved numerically; the properties of the solution agree well with experimental data for laminar flows. It is left as an exercise to show that there are solutions of the form

$$f(\eta) = \alpha\eta^2 + O(\eta^5) \quad \text{as } \eta \to 0$$

and $$f(\eta) = 2\eta - \beta + \text{exponentially small terms} \quad \text{as } \eta \to \infty;$$

the values of the constants, α and β, are obtained from the numerical solution as $\alpha \approx 0.664$, $\beta \approx 1.721$. From the behaviour as $\eta \to \infty$, we see that the solution outside the boundary layer must now match to

$$\psi \sim y - \frac{1}{\sqrt{R_e}}\beta\sqrt{x} \quad \text{as } y \to 0$$

which shows that we require a term $O(R_e^{-1/2})$ in the asymptotic expansion valid in the outer region.

Thus we seek a solution of the set (5.109) in the form

$$q(x, y; R_e) = q_0(x, y) + R_e^{-1/2} q_1(x, y) + o\left(R_e^{-1/2}\right)$$

where q represents each of u, v and p. The problem for the second terms in this region therefore becomes the set

$$u_0 u_{1x} + u_1 u_{0x} + v_0 u_{1y} + v_1 u_{0y} = -p_{1x};$$

$$u_0 v_{1x} + u_1 v_{0x} + v_0 v_{1y} + v_1 v_{0y} = -p_{1y}; \quad u_{1x} + v_{1y} = 0$$

with $$(u_1, v_1, p_1) \to (0, 0, 0) \quad \text{as} \quad y \to \infty \quad \text{and} \quad \text{as} \quad x \to -\infty$$

and, in terms of the stream function ($u_1 = \psi_{1y}$, $v_1 = -\psi_{1x}$),

$$\psi_1 \to \begin{cases} -\beta\sqrt{x}, & x > 0 \\ 0, & x < 0 \end{cases} \quad \text{as } y \to 0.$$

This is a classical problem in inviscid flow theory, where the exterior flow is distorted by the presence of a parabolic surface—the effect of the boundary layer which grows on the plate. The exact solution can be expressed in terms of the complex variable $z = x + iy$ (and $\Re$ denotes the real part):

$$\psi_1(x, y) = -\beta\Re(\sqrt{z}) \quad (\text{and } p_1 = 0)$$

which gives $$u_1 = -\tfrac{1}{2}\beta\Re(i/\sqrt{z}), \quad v_1 = \tfrac{1}{2}\beta\Re(1/\sqrt{z}).$$

(It can be shown that, in order to match, the boundary-layer solution must now contain a term R_e^{-1}—and not $R_e^{-1/2}$ as might have been expected. For more general surfaces than a flat plate, the next term in the boundary-layer expansion is indeed $O(R_e^{-1/2})$).

We will not proceed further with this analysis here (but far more detail is available in many other texts e.g. van Dyke, 1975), although we should add one word of warning. The details that we have presented suggest that we may continue, fairly routinely, to find the next term in the boundary-layer expansion, and then the next in the outer, and so on, and that these will develop according to the asymptotic sequence $\{R_e^{-n/2}\}$. However, this is not the case: a term $R_e^{-1}\ln R_e$ appears and this considerably complicates the procedure (again, see e.g. van Dyke, 1975).

The two essential types of problem that are usually of most interest in fluid mechanics are associated with (a) $R_e \to \infty$ (the previous example) and (b) $R_e \to 0$ (the next example). Problems for small Reynolds number (sometimes referred to as *Stokes flow* or *slow flow*) have become of increasing interest because this limit relates to important problems in, for example, a biological context. Thus the movement of platelets in the blood, and the propulsion of bacteria using ciliary hairs, are examples of these small-Reynolds number flows. We will describe a simple, classical problem of this type.

E5.20 Very viscous flow past a sphere

In this example, we take $R_e \to 0^+$ (and since $R_e = Ud/\nu$, where U is a typical speed of the flow, d a typical dimension of the object in the flow and ν the kinematic viscosity, this limit can be interpreted as 'highly viscous' or 'slow flow' or flow past a 'small object'). We consider the axisymmetric flow, produced by a uniform flow at infinity parallel to the chosen axis, past a solid sphere; see figure 15. (This could be used as a simple model for flow past a raindrop.) It is convenient to introduce a stream function ψ (usually called a *Stokes stream function*, in this context), eliminate pressure from the Navier-Stokes equation and hence work with the (non-dimensional) equation

$$\frac{1}{r^2\sin\theta}\left\{\psi_\theta\frac{\partial}{\partial r}-\psi_r\frac{\partial}{\partial\theta}+2(\cot\theta)\psi_r-\frac{2}{r}\psi_\theta\right\}D^2\psi=\frac{1}{R_e}D^4\psi, \qquad (5.114)$$

for $1 \le r < \infty$ and $0 \le \theta \le \pi$, where

$$D^2 \equiv \frac{\partial^2}{\partial r^2}+\frac{1}{r^2}\frac{\partial^2}{\partial\theta^2}-\frac{\cot\theta}{r^2}\frac{\partial}{\partial\theta}$$

with

$$\psi = \psi_r = 0 \quad \text{on} \quad r = 1 \qquad (5.115)$$

and

$$\psi/r^2 \to \frac{1}{2}\sin^2\theta \quad \text{as} \quad r \to \infty, \qquad (5.116)$$

this latter condition ensuring that there is an axisymmetric flow of speed one at infinity. (The subscripts here denote partial derivatives; we have mixed the notation because, we submit, this is the neatest way to express this equation.) The velocity components

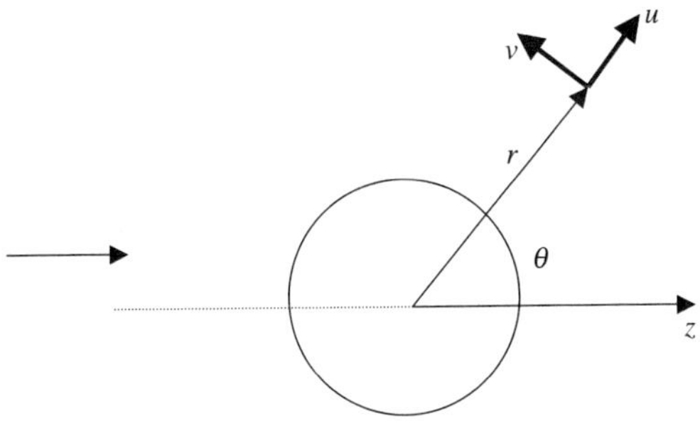

Figure 15. Coordinates and velocity components for the uniform flow past a sphere.

(see figure 15) are given by

$$u = \frac{1}{r^2 \sin\theta}\psi_\theta \quad \text{and} \quad v = -\frac{1}{r \sin\theta}\psi_r.$$

Note that equation (5.114) does not exhibit the conditions for a boundary-layer structure, as $R_e \to 0$, because the highest derivatives *are* retained in this limit—indeed, this term dominates. It is therefore unclear what difficulties we may encounter.

Let us seek a solution

$$\psi(r, \theta; R_e) = \psi_0(r, \theta) + o(1) \quad \text{as } R_e \to 0^+,$$

then from (5.114) we simply have that

$$D^4\psi_0 = 0$$

and all the boundary conditions appear to be available. Indeed, there is an exact solution (Stokes, 1851) which satisfies all the given conditions:

$$\psi_0(r, \theta) = \left(\frac{1}{2}r^2 - \frac{3}{4}r + \frac{1}{4}r^{-1}\right)\sin^2\theta, \tag{5.117}$$

and for a number of years this was thought to be acceptable, and that higher-order terms would simply provide small corrections in the case $R_e \to 0$. However, difficulties were encountered when a more careful analysis was undertaken, and a little thought suggests why this should be so. At infinity the motion (*convective* terms) dominate, i.e. the left-hand side of the equation, but near the sphere the viscous terms dominate (the right-hand side); thus an approximation which uses only the right-hand side (as ψ_0 above does) cannot be uniformly valid—it must break down as $r \to \infty$.

We introduce $r = R/\delta$, where $\delta(R_e) \to 0^+$ as $R_e \to 0^+$, and then from (5.117) we see that we must also scale ψ: $\psi = \Psi/\delta^2$. Equation (5.114) yields immediately that the appropriate choice is $\delta = R_e$ (Oseen, 1910), but unfortunately this scaling recovers the *full* equation—the small parameter is removed identically! However, the good news is that this scaling (obviously) is associated with the region far away from the sphere (the 'far field'), where the uniform flow exists and, presumably, this should be the first term in an asymptotic solution valid here; we expect, therefore, that

$$\Psi(R, \theta; R_e) \sim \Psi_0(R, \theta) = \frac{1}{2} R^2 \sin^2 \theta. \tag{5.118}$$

Of course, (5.118) and (5.117) match directly, and we should now regard (5.117) as valid only for $r = O(1)$ (the 'near field') and then (5.118) is valid for $r = O(R_e^{-1})$. When we express (5.117) in far-field variables, we obtain

$$\Psi \sim \frac{1}{2} R^2 \sin^2 \theta - \frac{3}{4} R_e R \sin^2 \theta \tag{5.119}$$

and so we require a term $O(R_e)$ in the far-field expansion; let us write

$$\Psi(R, \theta; R_e) = \tfrac{1}{2} R^2 \sin^2 \theta + R_e \Psi_1(R, \theta) + o(R_e) \quad \text{as } R_e \to 0^+.$$

The equation for Ψ_1, from (5.114), is

$$\left(\cos\theta \frac{\partial}{\partial R} - \frac{1}{R} \sin\theta \frac{\partial}{\partial \theta} \right) D^2 \Psi_1 = D^4 \Psi_1, \tag{5.120a}$$

where

$$D^2 \equiv \frac{\partial^2}{\partial R^2} + \frac{1}{R^2} \frac{\partial^2}{\partial \theta^2} - \frac{\cot\theta}{R^2} \frac{\partial}{\partial \theta}$$

with

$$\Psi_1 / R \to -\frac{3}{4} \sin^2 \theta \quad \text{as } R \to 0 \tag{5.120b}$$

and

$$\Psi_1 / R^2 \to 0 \quad \text{as } R \to \infty, \tag{5.120c}$$

the former condition being given by the matching, and the latter ensuring that the flow at infinity is unchanged.

The relevant solution of equation (5.120a) is

$$\Psi_1(R, \theta) = A \cos\theta + B + (C \cos\theta + D) R + 2E(1 + \cos\theta) \exp\left[-\frac{1}{2} R(1 - \cos\theta) \right],$$

where $A - E$ are arbitrary constants, and then $\Psi \sim \Psi_0 + R_e \Psi_1$, written in near-field variables, gives

$$\psi \sim \frac{1}{2} r^2 \sin^2 \theta + (C \cos\theta + D) r - E r \sin^2 \theta + R_e^{-1} [A \cos\theta + B + 2E(1 + \cos\theta)].$$

The term in R_e^{-1} is unmatchable, and so it must be removed, and otherwise this expression is to match to (5.119); this requires that

$$A = B = -\frac{3}{2}, \quad C = D = 0, \quad E = \frac{3}{4}$$

i.e.
$$\Psi_1(R, \theta) = \frac{3}{2}(1 + \cos\theta)\left\{-1 + \exp\left[-\frac{1}{2}R(1 - \cos\theta)\right]\right\}.$$

Finally, to be consistent with the development of this asymptotic expansion, we set

$$\psi(r, \theta; \mathrm{R_e}) = \psi_0(r, \theta) + \mathrm{R_e}\psi_1(r, \theta) + \mathrm{o}(\mathrm{R_e})$$

and then ψ_1 satisfies

$$D^4\psi_1 = \frac{1}{r^2 \sin\theta}\left[\psi_{0\theta}\frac{\partial}{\partial r} - \psi_{0r}\frac{\partial}{\partial\theta} + 2(\cot\theta)\psi_{0r} - \frac{2}{r}\psi_{0\theta}\right]D^2\psi_0$$

where $\psi_0(r, \theta)$ is given by (5.117). It is left as an exercise to show that the solution of this equation, which satisfies the boundary and matching conditions, is

$$\psi_1(r, \theta) = \frac{3}{8}\left(\frac{1}{2}r^2 - \frac{3}{4}r + \frac{1}{4}r^{-1}\right)\sin^2\theta - \frac{3}{32}\left(2r^2 - 3r + 1 - r^{-1} + r^{-2}\right)\sin^2\theta\cos\theta.$$

And, to complete our presentation, we comment that $\psi(r, \theta; \mathrm{R_e})$, expanded for $r \to \infty$, produces

$$\psi \sim \frac{1}{2}r^2\left[1 + \frac{3}{8}\mathrm{R_e}(1 - \cos\theta)\right]\sin^2\theta$$

and the term here in $\mathrm{R_e}$ contributes, apparently, to a change in the uniform flow at infinity—which is impossible—and hence the need for a matched solution in the far field. It was this observation that first alerted the earlier researchers to the difficulties inherent in this problem; this complication is typical of flows in the limit $\mathrm{R_e} \to 0$.

Another general area of study in fluid mechanics is *gas dynamics*, where the compressibility of the fluid cannot be ignored. We have already seen some of these problems (E3.2, E3.5 and Q3.9–3.11); we now look at another classical example.

E5.21 A piston problem

We consider the one-dimensional flow of a gas in a long, open-ended tube. The gas is brought into motion by the action of a piston at one end, which moves forward at a speed which is much less than the sound speed in the gas. (This is usually called the *acoustic problem*.) The gas is modelled by the isentropic law for a perfect gas (pressure $\propto$ (density)$^\gamma$, $1 < \gamma < 2$), and is described by the equations

$$a_t + ua_x + \frac{1}{2}(\gamma - 1)au_x = 0; \quad u_t + uu_x + \frac{2}{\gamma - 1}aa_x = 0,$$

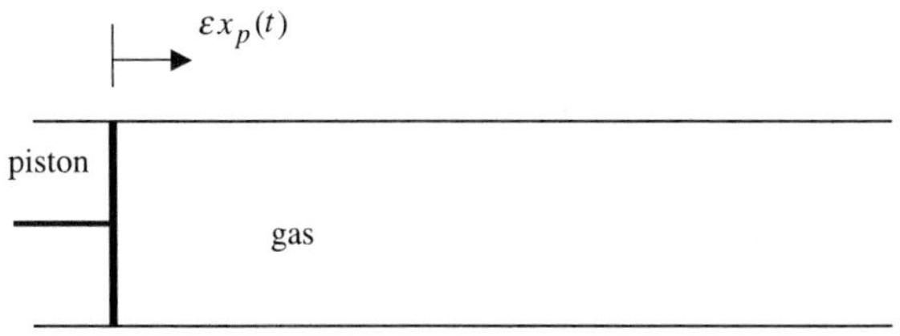

Figure 16. Sketch of a piston moving a gas (according to $x = \varepsilon x_p(t)$) in an open tube.

where a is the (local) sound speed in the gas and u its speed along the tube. The initial and boundary conditions are

$$u = 0 \quad \text{and} \quad a = 1 \quad \text{at} \quad t = 0, \quad x > \varepsilon x_p(t)$$

and

$$u = \varepsilon V(t) \quad \text{on} \quad x = \varepsilon x_p(t), \quad t \geq 0,$$

with $x_p(t) = \int_0^t V(t')\,\mathrm{d}t'$ and $V(0) = 0$; see figure 16. (The problem of a closed tube, described in a Lagrangian framework and using the method of multiple scales, is discussed by Wang & Kassoy, 1990.)

We are already familiar with the result, in small-disturbance theories—which this is for $\varepsilon \to 0$—that the simple, near-field wave-propagation problem is not uniformly valid as t (or x) $\to \infty$; see E3.2. In particular, for disturbances propagating down the tube (into $x > x_p$), which are described by e.g. $u(x,t) \sim F(x-t)$, there will be a breakdown where $x = \mathrm{O}(\varepsilon^{-1})$ (or, equivalently, when $t = \mathrm{O}(\varepsilon^{-1})$) with $x - t = \mathrm{O}(1)$. Note that, in this problem, because the tube is open, there can be no disturbances propagating back towards the piston. We introduce $\xi = t - x$ and $\tau = \varepsilon t$ (to make the evaluation on $x = \varepsilon x_p(t)$ as simple as possible) and then write $u \equiv \varepsilon U(\xi, \tau; \varepsilon)$ and $a = 1 + \varepsilon A(\xi, \tau; \varepsilon)$ to give

$$A_\xi + \varepsilon A_\tau - \varepsilon U A_\xi - \frac{1}{2}(\gamma - 1)(1 + \varepsilon A)U_\xi = 0; \tag{5.121a}$$

$$U_\xi + \varepsilon U_\tau - \varepsilon U U_\xi - \left(\frac{2}{\gamma - 1}\right)(1 + \varepsilon A)A_\xi = 0 \tag{5.121b}$$

with

$$U(-x, 0; \varepsilon) = A(-x, 0; \varepsilon) = 0, \quad x > \varepsilon x_p \tag{5.122a}$$

and

$$U(t - \varepsilon x_p(t), \tau; \varepsilon) = V(t). \tag{5.122b}$$

We seek an asymptotic solution, which is to be uniformly valid as $\tau \to \infty$, in the form

$$U(\xi, \tau; \varepsilon) \sim \sum_{n=0}^{\infty} \varepsilon^n U_n(\xi, \tau); \quad A(\xi, \tau; \varepsilon) \sim \sum_{n=0}^{\infty} \varepsilon^n A_n(\xi, \tau)$$

which gives (from (5.121)

$$A_{0\xi} - \frac{1}{2}(\gamma - 1)U_{0\xi} = 0; \quad U_{0\xi} - \left(\frac{2}{\gamma - 1}\right) A_{0\xi} = 0; \tag{5.123a,b}$$

$$A_{1\xi} - \frac{1}{2}(\gamma - 1)U_{1\xi} + A_{0\tau} - U_0 A_{0\xi} - \frac{1}{2}(\gamma - 1) A_0 U_{0\xi} = 0; \tag{5.124a)}$$

$$U_{1\xi} - \left(\frac{2}{\gamma - 1}\right) A_{1\xi} + U_{0\tau} - U_0 U_{0\xi} - \left(\frac{2}{\gamma - 1}\right) A_0 A_{0\xi} = 0, \tag{5.124b)}$$

and so on. Further, we assume that the boundary condition at the piston can be expressed as a Taylor expansion about t:

$$U_0(t, \tau) + \varepsilon U_1(t, \tau) - \varepsilon x_p(t) U_{0\xi}(t, \tau) + \mathrm{O}(\varepsilon^2) = V(t) \tag{5.125}$$

the validity of which certainly requires that $x_p(t)$ remains finite as $t \to \infty$ (and so $\int_0^\infty V(t)\mathrm{d}t$ must be finite).

From (5.123) we find that

$$U_0 - \left(\frac{2}{\gamma - 1}\right) A_0 = C_0(\tau), \tag{5.126}$$

where C_0 is an arbitrary function, but U_0 and A_0 are otherwise undetermined at this stage. From (5.124), we multiply the first by $2/(\gamma - 1)$ and then add to it (5.124b), which eliminates U_1 and A_1 to produce

$$U_{0\tau} + \left(\frac{2}{\gamma - 1}\right) A_{0\tau} - (U_0 + A_0) \left\{ U_{0\xi} + \left(\frac{2}{\gamma - 1}\right) A_{0\xi} \right\} = 0.$$

The terms in A_0 are now replaced by using (5.126) to give the equation for $U_0(\xi, \tau)$:

$$U_{0\tau} - \frac{1}{2}(\gamma + 1) U_0 U_{0\xi} = \frac{1}{2} C_0' - \frac{1}{2}(\gamma - 1) C_0 U_{0\xi}. \tag{5.127}$$

This is a nonlinear equation which, for given $C_0(\tau)$, is readily solved. However, this solution is incomplete without the weak acoustic *shock wave* that propagates ahead of this solution; we must therefore write down the conditions for the insertion of a shock (discontinuity).

First, from (5.122), this initial condition requires that $C_0(0) = 0$. To further determine $C_0(\tau)$ we impose the *Rankine-Hugoniot* conditions that define the jump conditions across the shock. The conditions ahead are undisturbed; let the conditions behind the shock be denoted by the subscript 's' and write the speed of the shock as $\dot{S}$. In this problem, these conditions (see e.g. Courant & Friedrichs, 1967) can be written as

$$\dot{S} = \frac{(1+\varepsilon R_s)U_s}{R_s} \quad (\text{where } (1+\varepsilon A)^2 = (1+\varepsilon R)^{\gamma-1}) \quad \text{with} \quad \dot{S}^2 = 1 + \frac{1}{2}\varepsilon(\gamma+1)U_s\dot{S}$$

and so

$$\dot{S} \sim 1 + \frac{1}{4}\varepsilon(\gamma+1)U_s; \quad U_s \sim R_s \sim \left(\frac{2}{\gamma-1}\right)A_s \quad (\text{as } \varepsilon \to 0). \tag{5.128}$$

(Here, εR is the perturbation of the density.) This latter result confirms, from (5.126), that $C_0(\tau) = 0$ behind the shock and, since $U_0 = A_0 = 0$ ahead of the shock, we have $C_0(\tau) = 0$ for $\tau \geq 0$. Thus from (5.127) we obtain the implicit result

$$U_0(\xi, \tau) = F\left(\xi - \frac{1}{2}(\gamma+1)\tau U_0\right)$$

where F is an arbitrary function which, from (5.125), can be determined to give

$$U_0(\xi, \tau) = V\left(\xi - \frac{1}{2}(\gamma+1)\tau U_0\right)$$

(since, at the piston, $\xi - \frac{1}{2}(\gamma+1)\tau U_0 = t - \varepsilon x_p - \frac{1}{2}\varepsilon(\gamma+1)tU_0 \sim t$ as $\varepsilon \to 0$). Thus the near-field solution ($\tau = \mathrm{O}(\varepsilon)$ is recovered, although this needs to be written to accommodate the existence of the wave front there i.e.

$$u \sim V(\xi)\mathrm{H}(-\xi)$$

where H is the *Heaviside step function*: $\mathrm{H}(x) = 1, x \geq 0$; $\mathrm{H}(x) = 0, x < 0$. We conclude with the observation that the shock wave travels *faster* than the local sound speed behind the shock; that is, from (5.128), $\dot{S} \sim 1 + \frac{1}{4}\varepsilon(\gamma+1)U_s$ as compared with $a \sim 1 + \frac{1}{2}\varepsilon(\gamma-1)U_s$ (and remember that $1 < \gamma < 2$). Much more detail can be found in any good text on gas dynamics.

As our final example, we use a similar technique to that employed in the previous problem, but now in a quite different context: waves on the surface of water. (See Q3.4 for a much simpler but related exercise.)

E5.22 A variable-depth Korteweg-de Vries equation for water waves

We consider the one-dimensional propagation of waves over water (incompressible), which is modelled by an inviscid fluid without surface tension. The water is stationary in the absence of waves, but the local depth varies on the same scale (ε) that is used to measure the weak nonlinearity and dispersive effects in the governing equations. For right-running waves, the appropriate far-field coordinates are $\xi = \varepsilon^{-1}\chi(X) - t$ and

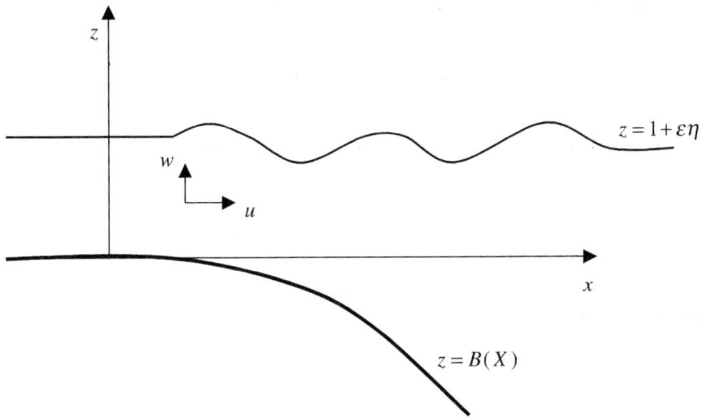

Figure 17. Wave propagation in stationary water over variable depth.

$X = \varepsilon x$, where $\chi(X)$ is to be determined. The non-dimensional equations are

$$-u_\xi + \varepsilon[u(\chi' u_\xi + \varepsilon u_X) + w u_z] = -(\chi' p_\xi + \varepsilon p_X); \tag{5.129a}$$

$$\varepsilon[-w_\xi + \varepsilon\{u(\chi' w_\xi + \varepsilon w_X) + w w_z\}] = -p_z; \quad \chi' u_\xi + \varepsilon u_X + w_z = 0 \tag{5.129b,c}$$

with $$p = \eta \quad \text{and} \quad w = -\eta_\xi + u(\chi' \eta_\xi + \varepsilon \eta_X) \quad \text{on} \quad z = 1 + \varepsilon\eta \tag{5.129d,e}$$

and $$w = \varepsilon u B'(X) \quad \text{on} \quad z = B(X). \tag{5.129f}$$

Here (u, w) are the velocity components of the flow, p the pressure in the fluid relative to the hydrostatic pressure (with pressure constant at the surface) and $z = 1 + \varepsilon\eta$ is the surface of the water. The bottom is represented by the function $z = B(X)$; see figure 17.

We seek a solution in the familiar form:

$$q \sim \sum_{n=0}^{\infty} \varepsilon^n q_n \quad \text{as } \varepsilon \to 0,$$

where q represents each of u, w, p and η. The leading-order problem gives, from (5.129),

$$u_{0\xi} = \chi' p_{0\xi}; \quad p_{0z} = 0; \quad \chi' u_{0\xi} + w_{0z} = 0$$

with $$p_0 = \eta_0 \quad \& \quad w_0 = -\eta_{0\xi} \quad \text{on} \quad z = 1, \quad \text{and} \quad w_0 = 0 \quad \text{on} \quad z = B(X).$$

This set is easily solved; the relevant solution (in which u_0 is not a function of z) is

$$p_0 = \eta_0; \quad u_0 = \eta_0 \chi' \quad \text{and} \quad w_0 = (B - z)\chi'^2 \eta_{0\xi} \quad (\text{all for } 0 \le z \le 1)$$

where the surface boundary condition finally gives

$$\chi'^2 = 1/D(X) \quad \text{where } D(X) = 1 - B(X)\ (>0),$$

and $\eta_0(\xi, X)$ is arbitrary at this stage. For rightward propagation, we select

$$\chi(X) = \int_0^X \frac{\mathrm{d}X'}{\sqrt{D(X')}}$$

so that the characteristic variable becomes

$$\xi = \varepsilon^{-1} \int_0^X \frac{\mathrm{d}X'}{\sqrt{D(X')}} - t$$

which, with $D = 1$ (constant depth), recovers the standard result: $\xi = x - t$. Note that, in this calculation, we have taken the evaluation at the surface to be on $z = 1$; see below.

At the next order, equations (5.129) give

$$-u_{1\xi} + \chi' u_0 u_{0\xi} + w_0 u_{0z} = -(\chi' p_{1\xi} + p_{0X});$$

$$w_{0\xi} = p_{1z}; \quad \chi' u_{1\xi} + u_{0X} + w_{1z} = 0$$

with $$p_1 + \eta_0 p_{0z} = \eta_1 \quad \& \quad w_1 + \eta_0 w_{0z} = -\eta_{1\xi} + \chi' u_0 \eta_{0\xi} \quad \text{on} \quad z = 1$$

and $$w_1 = u_0 B'(X) \quad \text{on} \quad z = B(X).$$

The boundary conditions at the surface, $z = 1 + \varepsilon\eta$, have been written by invoking Taylor expansions, and so become the corresponding boundary conditions evaluated on $z = 1$, valid as $\varepsilon \to 0$ (and for sufficiently smooth surface waves). Again, this set is fairly easily solved; the details are left as an exercise, but some of the intermediate results are

$$p_1 = \frac{1}{D}\left\{(z-1)B + \frac{1}{2}(1-z^2)\right\}\eta_{0\xi\xi} + \eta_1$$

and

$$w_1 = (B - z)\left\{\left(\eta_0\eta_0/\sqrt{D}\right)_X + \frac{1}{D^2}\eta_0\eta_{0\xi} + \frac{1}{\sqrt{D}}\eta_{0X} + \frac{1}{D}\eta_{1\xi}\right\}$$
$$- \frac{1}{D^2}\left\{z\left(\frac{1}{2}z - 1\right)B + \frac{1}{2}z\left(1 - \frac{1}{3}z^2\right) - \frac{1}{3}B^3 + B^2 - \frac{1}{2}B\right\}\eta_{0\xi\xi\xi} - \frac{D'}{\sqrt{D}}\eta_0.$$

Finally, the surface boundary condition for w_1 gives the equation for η_0 — η_1 identically cancels—in the form of a variable-coefficient Korteweg-de Vries equation (see E3.1 and Q3.4):

$$2\sqrt{D}\eta_{0X} + \frac{D'}{2\sqrt{D}}\eta_0 + \frac{3}{D}\eta_0\eta_{0\xi} + \frac{D}{3}\eta_{0\xi\xi\xi} = 0.$$

This is usually expressed in terms of $H_0(\xi, X) = D^{1/4}\eta_0(\xi, X)$, to give

$$2H_{0X} + 3D^{-7/4}H_0H_{0\xi} + \frac{1}{3}D^{1/2}H_{0\xi\xi\xi} = 0$$

which recovers the classical Korteweg-de Vries equation for water waves when we set $D = 1$. This equation is the basis for many of the modern studies in water-wave theory; more background to this, and related problems in water waves, can be found in Johnson (1997).

5.6 EXTREME THERMAL PROCESSES

This next group of problems concerns phenomena that involve explosions, combustion and the like. The two examples that we will describe are: E5.23 A model for combustion; E5.24 Thermal runaway.

E5.23 A model for combustion

A model that aims to describe ignition, followed by a rapid combustion, requires a slow development over a reasonable time scale that precedes a massive change on a very short time scale, initiated by the attainment of some critical condition. A simple (non-dimensional) model for such a process (Reiss, 1980) is the equation

$$\dot{c} = c^2(1-c) \quad \text{with} \quad c = \varepsilon \quad \text{at} \quad t = 0, \tag{5.130}$$

where $c(t;\varepsilon)$ is the concentration of an appropriate chemical that takes part in the combustive reaction. The whole process is initiated by the small disturbance $\varepsilon(\to 0^+)$ at time $t = 0$. (It should be fairly apparent that this equation can be integrated completely to give the solution for c, but in implicit form and so the detailed structure as $\varepsilon \to 0^+$ is far from transparent; this integration is left as an exercise.)

By virtue of the initial value (ε), we first seek a solution in the form

$$c(t;\varepsilon) \sim \sum_{n=1}^{\infty} \varepsilon^n c_n(t),$$

and then from (5.130) we obtain

$$\dot{c}_1 = 0; \quad \dot{c}_2 = c_1^2; \quad \dot{c}_3 = 2c_1c_2 - c_1^2, \text{ etc.,}$$

with
$$c_1(0) = 1 \quad \text{and} \quad c_n(0) = 0 \quad \text{for } n = 2, 3, 4 \ldots.$$

This set is very easily solved to give the asymptotic solution

$$c(t;\varepsilon) \sim \varepsilon + \varepsilon^2 t + \varepsilon^3(t^2 - t) \tag{5.131}$$

and it is immediately evident that this expansion breaks down when $\varepsilon t = O(1)$; let us write $T = \varepsilon t$ and $c \equiv \varepsilon C(T;\varepsilon)$ to produce the new equation

$$\dot{C} = C^2(1 - \varepsilon C). \tag{5.132}$$

Again, we seek a straightforward solution

$$C(T;\varepsilon) \sim \sum_{n=0}^{\infty} \varepsilon^n C_n(T)$$

so that (5.132) gives

$$\dot{C}_0 = C_0^2; \quad \dot{C}_1 = 2C_0C_1 - C_0^2, \tag{5.133a,b}$$

and so on, together with the requirement to match to (5.131).

The general solution to equation (5.133a) is

$$C_0(T) = \frac{1}{A - T}$$

where A is an arbitrary constant; with $T = \varepsilon t$, we obtain (for $A \neq 0$)

$$C_0 \sim 1/A \quad \text{as} \quad \varepsilon \to 0^+ (t = O(1))$$

and hence matching to the first term in (5.131) requires the choice $A = 1$. Thus we have

$$C_0(T) = 1/(1 - T);$$

the next term can be found similarly and leads to

$$C(T;\varepsilon) \sim \frac{1}{1 - T} + \varepsilon \frac{\ln(1 - T)}{(1 - T)^2}, \quad T = O(1), \tag{5.134}$$

which clearly exhibits a catastrophic breakdown as $T \to 1^-$. Thus we have a gradual acceleration of the process, until the time $t \sim \varepsilon^{-1}$ is reached and then—presumably—combustion occurs. This breakdown is not simple: it is at a time given by $1 - T = \mathrm{O}(\varepsilon \ln(1 - T))$. When logarithms (or exponentials) arise, we have learnt (§2.5) to return to the original equation and seek a relevant scaling (although the presence of a logarithm here indicates that $\ln \varepsilon$ terms are likely to appear in the asymptotic expansion). Let us set $1 - T = -\delta\tau$, then $C = \mathrm{O}(\delta^{-1})$ from (5.134), where $\delta(\varepsilon) \to 0^+$ as $\varepsilon \to 0^+$; thus we write $C = \delta^{-1}\kappa(\tau; \varepsilon)$ and it is immediately clear from (5.132) that we must choose $\delta(\varepsilon) = \varepsilon$. The equation for κ is therefore

$$\dot{\kappa} = \kappa^2(1 - \kappa), \tag{5.135}$$

the original equation! In order to proceed, we need an appropriate solution—but this is no longer required to satisfy the initial condition. The general solution to equation (5.135) can be written as

$$-\frac{1}{\kappa} + \ln\left(\frac{\kappa}{1 - \kappa}\right) = \tau + B \tag{5.136}$$

and for small κ this gives $\kappa \sim -1/(\tau + B)$ and so to match we require the arbitrary constant, B, to be zero (but we will return to this origin shift below). In passing, we note that for κ close to unity, we obtain $\kappa \sim 1 - \mathrm{e}^{-\tau}$ and so the state of full combustion ($\kappa = 1$) is attained as $\tau \to \infty$.

Finally, we reconsider the matching of (5.136), with $B = 0$, to the expansion (5.134). From (5.136) we obtain

$$\kappa \sim -\frac{1}{\tau} - \frac{1}{\tau^2}\ln(-\tau) \quad \text{as } \tau \to -\infty \tag{5.137}$$

which with $\tau = (T - 1)/\varepsilon$ gives

$$C \sim \frac{1}{1 - T} + \frac{\varepsilon}{(1 - T)^2}\left[\ln(1 - T) - \ln \varepsilon\right]$$

and the matching is not possible, as it stands, because of the presence of the $\ln \varepsilon$ term. However, this suggests that the variable used in this region of rapid combustion ($\tau = \mathrm{O}(1)$) should include an origin shift. If we write, now, $\tau = (T - 1 - \Delta)/\varepsilon$. Then (5.137) produces

$$C \sim \frac{1}{1 - T} + \frac{\varepsilon}{(1 - T)^2}\left[\ln(1 - T) - \ln \varepsilon - \frac{\Delta}{\varepsilon}\right]$$

and matching with (5.134) requires that $\Delta(\varepsilon) = -\varepsilon \ln \varepsilon$ ($\to 0$ as $\varepsilon \to 0^+$). Thus the combustion occurs in an $\mathrm{O}(1)$ neighbourhood of the time $t = \varepsilon^{-1} + \ln \varepsilon$; the appearance of shifts expressed in terms of $\ln \varepsilon$ are quite typical of these problems.

E5.24 Thermal runaway

A phenomenon that can be encountered in certain chemical reactions involves the release of heat (*exothermic*) which increases the temperature, and the temperature normally controls the reaction rate. It is possible, therefore, to initiate a reaction, then heat is released which raises the temperature and so increases the rate of reaction which releases even more heat, and so on; this is called *thermal runaway*. In the most extreme cases, there is no theoretical limit to the temperature, although physical reality intervenes e.g. the containing vessel might melt or the products explode. A standard model used to describe this (Szekely, Sohn & Evans, 1976; see also Fowler, 1997, which provides the basis for the discussion presented here) is the equation

$$T_t = \nabla^2 T + \lambda \exp\left[T/(1+\varepsilon T)\right]$$

where T is the temperature and λ is a parameter. In this example, we will examine the nature of the *steady-state* temperature in one dimension i.e. the solution of

$$T_{xx} + \lambda \exp\left[T/(1+\varepsilon T)\right] = 0 \tag{5.138}$$

as $\varepsilon \to 0^+$ for various λ. The boundary condition is that the external temperature is maintained; we will represent this by $T = 0$ on $x = \pm 1$, and so we seek a solution for $x \in [-1, 1]$ (and we will assume symmetry of the temperature distribution about $x = 0$).

An important property of the solution of (5.138) can be derived by first writing

$$T(x;\varepsilon) \sim \sum_{n=0}^{\infty} \varepsilon^n T_n(x), \quad \varepsilon \to 0^+,$$

where $T_0(x)$ satisfies the equation

$$T_0'' + \lambda \exp(T_0) = 0 \tag{5.139}$$

with the boundary conditions $T_0(\pm 1) = 0$. This problem has the exact solution

$$T_0(x) = T_m - 2\ln\left\{\cosh\left[x\sqrt{\frac{1}{2}\lambda e^{T_m}}\right]\right\}$$

where T_m is the maximum temperature (attained at $x = 0$) defined by

$$e^{T_m/2} = \cosh\left(\sqrt{\frac{1}{2}\lambda e^{T_m}}\right), \tag{5.140}$$

which follows immediately when evaluation on $x = \pm 1$ is imposed. It is left as an exercise (which may require a graphical approach) to confirm that (5.140) has zero, one or two solutions for T_m, given λ, depending on whether $\lambda > \lambda_c$, $\lambda = \lambda_c$ or $\lambda < \lambda_c$, respectively, where the critical value (λ_c) is the solution of

$$1 = \sqrt{\frac{1}{2}\lambda}\,\sinh\left\{\sqrt{\frac{1}{2}(\lambda+2)}\right\};$$

it can be shown that there does exist just one λ_c (≈ 0.878). It turns out that the consequences of this are fundamental: for any λ, if the initial temperature is high enough, or for any temperature if $\lambda > \lambda_c$, then the time-dependent problem produces a temperature that increases without bound—indeed, $T \to \infty$ in a finite time. What we will do here is to examine the temperature attained according to the steady-state equation, (5.138), for various λ, although we will approach this by considering different sizes of temperature (as measured by ε).

It is immediately apparent that the approximation that led to (5.139) cannot be valid if the temperature is as large as $O(\varepsilon^{-1})$; see equation (5.138). Let us therefore write $T \equiv \varepsilon^{-1}\theta(x;\varepsilon)$, and then (5.138) becomes

$$\theta'' + \varepsilon\lambda \exp[\varepsilon^{-1}\{\theta/(1+\theta)\}] = 0 \qquad (5.141)$$

and so if we seek a solution $\theta(x;\varepsilon) \sim \theta_0(x)$ (with $\varepsilon\lambda \to 0$ as $\varepsilon \to 0$) we obtain simply

$$\theta_0'' = 0 \quad \text{or} \quad \theta_0(x) = Ax + B \qquad (5.142)$$

where A and B are arbitrary constants. Such a solution is unable to accommodate a maximum temperature at $x = 0$ (if the solution is to be differentiable, and $\theta =$ constant $\neq 0$ for all x does not satisfy (5.141)). Thus (5.142) can describe the solution only away from $x = 0$, but then we may impose the boundary conditions on $x = \pm 1$, so

$$\theta_0(x) = A(1 \mp x), \quad 1 \geq x > 0 \text{ (upper sign)}, \; -1 \leq x < 0 \text{ (lower sign)}, \qquad (5.143)$$

and the single arbitary constant, A, may be used in both solutions by virtue of the symmetry. Near $x = 0$, let $x = \delta X$ and seek a solution $\theta \equiv \theta_m + \Delta\Theta(X;\varepsilon)$, where θ_m ($= O(1)$) is the (scaled) maximum temperature attained in the limit $\varepsilon \to 0^+$; note that, at this stage, we do not know the scalings $\delta(\varepsilon)$ and $\Delta(\varepsilon)$. Equation (5.141) becomes

$$\frac{\Delta}{\delta^2}\Theta'' + \varepsilon\lambda \exp\left[\varepsilon^{-1}\left(\frac{\theta_m + \Delta\Theta}{1 + \theta_m + \Delta\Theta}\right)\right] = 0 \qquad (5.144)$$

and for an appropriate solution to exist, to leading order for $X = \mathrm{O}(1)$, we must have

$$\varepsilon\lambda \exp\left[\varepsilon^{-1}\left\{\theta_m/(1+\theta_m)\right\}\right] = \frac{\Delta}{\delta^2}, \tag{5.145}$$

which implies that λ is exponentially small as $\varepsilon \to 0^+$. Let us write $\Theta(X;\varepsilon) \sim \Theta_0(X)$, then from (5.144) with (5.145) included, we obtain the equation for Θ_0 as

$$\Theta_0'' + \exp\left[\frac{\Delta}{\varepsilon}\frac{\Theta_0}{(1+\theta_m)^2}\right] = 0$$

and this gives a meaningful first approximation, independent of ε, only if we choose $\Delta = \mathrm{O}(\varepsilon)$ e.g. $\Delta = \varepsilon$. This equation then has the general solution

$$\Theta_0(X) = (1+\theta_m)^2\left\{\ln(C\beta^2) + \beta X - 2\ln\left[1 + \frac{C}{2(1+\theta_m)^2}\mathrm{e}^{\beta X}\right]\right\}$$

for arbitrary constants β and C. This solution is to be symmetric about $X = 0$, so $\Theta_0'(0) = 0$ (which is satisfied with $C = 2(1+\theta_m)^2$), and $\theta_m + \varepsilon\Theta_0$ is to match to (5.143). Thus we must have $A = \theta_m$ and then we obtain

$$\theta \sim \theta_m + \varepsilon(1+\theta_m)^2\beta\frac{x}{\delta}$$

which matches only if, first, we choose the scaling $\delta(\varepsilon) = \varepsilon$ and then $\beta = \mp\theta_m/(1+\theta_m)^2$ (valid in $X > 0$, $X < 0$, respectively). Thus, in particular, we find that

$$\Theta_0(0) = (1+\theta_m)^2\ln\left[\frac{\theta_m^2}{2(1+\theta_m)^2}\right]$$

and so the maximum temperature becomes

$$\varepsilon^{-1}\theta_m + (1+\theta_m)^2\ln\left[\frac{\theta_m^2}{2(1+\theta_m)^2}\right] + \mathrm{O}(\varepsilon) \tag{5.146}$$

where, from (5.145), we have

$$\theta_m = -\frac{\varepsilon\ln(\varepsilon^2\lambda)}{1+\varepsilon\ln(\varepsilon^2\lambda)} \quad (= \mathrm{O}(1)) \tag{5.147}$$

for given λ; we may write this equivalently as $\lambda = \varepsilon^{-2}\ln(-\mu/\varepsilon)$ where $\mu = \theta_m/(1+\theta_m)$, and so λ determines θ_m is the interpretation that we employ.

The calculation thus far indicates that, for suitable λ, the resulting steady-state temperature (if it can be attained through a time-dependent evolution) is already very large, namely $\mathrm{O}(\varepsilon^{-1})$. But it is also clear from (5.147) that even smaller λs exist that give

$\theta_m \to \infty$, and then the temperature expansion, (5.146), is not uniformly valid; in particular this expansion breaks down when $\theta_m = \mathrm{O}(\varepsilon^{-1})$, that is, for $\lambda = \mathrm{O}(\varepsilon^{-2}\mathrm{e}^{-1/\varepsilon})$. Let us therefore rescale $\theta = \varepsilon^{-1}\phi$, write $\lambda = \varepsilon^{-2}\mathrm{e}^{-1/\varepsilon}\Lambda$, and then equation (5.141) becomes

$$\phi'' + \Lambda \exp\left[-1/(\phi + \varepsilon)\right] = 0 \tag{5.148}$$

with $\phi = 0$ at $x = \pm 1$. This branch of the solution, interpreted as a function of λ, is usually called the *hot branch*. We seek a solution $\phi(x;\varepsilon) \sim \phi_0(x)$, to give

$$\phi_0'' + \Lambda \exp(-1/\phi_0) = 0,$$

but we cannot use the boundary conditions here because of the evident non-uniformity as $\phi \to 0$ in equation (5.148). However, if we set $\phi_0 = \phi_m$ at $x = 0$ (where $\phi_0' = 0$, by symmetry) we obtain

$$\phi_0'^2 = 2\Lambda \int_{\phi_0}^{\phi_m} \mathrm{e}^{-1/y}\mathrm{d}y = 2\Lambda \left\{\phi_m \mathrm{e}^{-1/\phi_m} - \phi_0 \mathrm{e}^{-1/\phi_0} - \int_{1/\phi_m}^{1/\phi_0} \frac{\mathrm{e}^{-u}}{u}\,\mathrm{d}u\right\} \tag{5.149}$$

(where we have set $u = 1/y$) and this latter integral can be expressed in terms of $\mathrm{E}_1(u)$, an *exponential integral*, if that is useful. A solution with the property that $\phi_0 \to 0$ as $x \to \pm 1$ (which is necessary if matching is to be possible to the solution valid near $x = \pm 1$, where ϕ'' is exponentially small i.e. $\phi \sim B(1 \pm x)$) must satisfy

$$A^2 = 2\Lambda \left\{\phi_m \mathrm{e}^{-1/\phi_m} - \mathrm{E}_1(1/\phi_m)\right\},$$

where $\phi_0' \sim \pm A$ i.e. $\phi_0 \sim A(1 \pm x)$ as $x \to 1$ or -1. (Of course, matching is then trivial, for we simply choose $A = B$.) Thus we may integrate (5.149) from $x = 0$ to, say, $x = 1$:

$$\int_0^{\phi_m} \frac{\mathrm{d}\psi}{\sqrt{\int_\psi^{\phi_m} \mathrm{e}^{-1/y}\mathrm{d}y}} = \sqrt{2\Lambda}$$

and now we find that ϕ_m increases as Λ increases. So once we have reached this 'hot branch', which is accessed by using $\lambda = \mathrm{O}(\varepsilon^{-2}e^{-1/\varepsilon})$, the temperature will increase without bound (or, rather, until some other physics intervenes). Even more details of this problem, and related thermal processes, can be found in the excellent text on modelling by Fowler (1997).

The final group of problems bear some relation to those just considered, for they also involve chemical processes, but we include in this section some mention of biochemical processes as well.

5.7 CHEMICAL AND BIOCHEMICAL REACTIONS

The examples that are presented here are intended to show that it is possible to model, and describe using perturbation theory, some very complex processes that, perhaps fifty years ago, were thought to be mathematically unresolvable. Certainly, some extensive simplification is necessary in the development of the model—and this requires considerable skill and knowledge—but the resulting differential equations remain, generally, quite daunting. We will describe: E5.26 Kinetics of a catalysed reaction; E5.27 Enzyme kinetics; E5.28 The Belousov-Zhabotinskii reaction.

E5.26 Kinetics of a catalysed reaction

In a model (the *Langmuir-Hinshelwood model*; see Kapila, 1983) for the kinetics of a particular type of catalysed reaction, the concentration of the reactant varies according to the equation

$$\varepsilon \frac{dc}{dx} = 1 - \frac{2}{xc}(1-c)(\gamma + c), \quad 0 < x \leq 1,$$

with

$$c(1; \varepsilon) = 0;$$

γ (where $0 < \gamma < 1$) is a given constant. The parameter ε (>0) is a rate constant and, for $\varepsilon \to 0^+$, we observe that the equation reduces to a purely algebraic problem which is readily solved:

$$c = -\tfrac{1}{2}\left(\tfrac{1}{2}x + \gamma - 1\right) + \tfrac{1}{2}\sqrt{4\gamma + \left(\tfrac{1}{2}x + \gamma - 1\right)^2}. \tag{5.150}$$

We have selected the positive sign so that $c \geq 0$. On $x = 1$, this gives the value

$$\tfrac{1}{2}\left[\tfrac{1}{2} - \gamma + \sqrt{\gamma^2 + 3\gamma + \tfrac{1}{4}}\right] \neq 0, \tag{5.151}$$

and so we will require a boundary layer near $x = 1$; see §§2.6, 2.7. (Note that, from this solution, $c = 1$ on $x = 0$, so we can expect that $c \to 1$ as $x \to 0$ in the solution of the full equation.) Let us introduce $x = 1 - \varepsilon X$ and write $c(1 - \varepsilon X; \varepsilon) \equiv C(X; \varepsilon)$; the equation for C is then

$$-C' = 1 - \frac{2(1-C)(\gamma + C)}{(1-\varepsilon X)C}.$$

We seek a solution

$$C(X; \varepsilon) \sim \sum_{n=0}^{\infty} \varepsilon^n C_n(X)$$

and so $C_0(X)$ satisfies

$$C_0' = \frac{2}{C_0}(1 - C_0)(\gamma + C_0) - 1$$

with $C_0(0) = 0$. The general solution of this equation can be found, albeit in implicit form, as

$$(C_0 + \lambda + \mu)^{\lambda+\mu}(C_0 + \mu - \lambda)^{\lambda-\mu} = Ae^{-4\lambda X}$$

where $\mu = \frac{1}{2}(\gamma - \frac{1}{2})$ and $\lambda = \frac{1}{2}\sqrt{\gamma^2 + 3\gamma + \frac{1}{4}}$; A is an arbitrary constant which is evaluated as

$$A = (\lambda + \mu)^{\lambda+\mu}(\mu - \lambda)^{\lambda-\mu}$$

when the boundary condition on $X = 0$ is imposed. As $X \to \infty$ (which will give the behaviour outside the boundary layer), we see that

$$C_0(X) \to -\mu + \lambda = \tfrac{1}{2}\left(\tfrac{1}{2} - \gamma\right) + \tfrac{1}{2}\sqrt{\gamma^2 + 3\gamma + \tfrac{1}{4}}$$

which therefore automatically matches with the solution already found in the region away from the boundary layer, (5.151). It is left as a straightforward exercise to find higher-order terms, or to write down a composite expansion valid for $\forall x \in (0,\ 1]$; see §1.10.

This first example has presented us with a very routine exercise in elementary boundary-layer theory; the next has a similar structure, but in only one of the two components.

E5.27 Enzyme kinetics

A standard process in enzyme kinetics concerns the conversion of a substrate (x) into a product, by the action of an enzyme, *via* a substrate-enzyme complex (y); this is the *Michaelis-Menton reaction*. A model for this process is the pair of equations

$$\dot{x} = -x + (x + \alpha - \beta)y; \quad \varepsilon\dot{y} = x - (x + \alpha)y, \tag{5.152}$$

where α and β are positive constants and the initial conditions are

$$x = 1 \quad \text{and} \quad y = 0 \quad \text{at} \quad t = 0.$$

The parameter ε measures the rate of the production of y; we consider the problem posed above, with $\varepsilon \to 0^+$. It is clear that we should expect a boundary-layer structure in the solution for y, but not for x. A suitable boundary-layer variable is $\tau = \varepsilon^{-1} f(t)$ (rather than simply t/ε), and we will take the opportunity to use the method of multiple scales here, using $T = t$ and τ (see §4.6). We introduce $x(t; \varepsilon) \equiv X(T, \tau; \varepsilon)$ and $y(t; \varepsilon) \equiv Y(T, \tau; \varepsilon)$, and then equations (5.152) become

$$\varepsilon^{-1} f' X_\tau + X_T = -X + (X + \alpha - \beta)Y; \quad f' Y_\tau + \varepsilon Y_T = X - (X + \alpha)Y \tag{5.153}$$

with
$$X(0, 0; \varepsilon) = 1 \quad \text{and} \quad Y(0, 0; \varepsilon) = 0. \tag{5.154}$$

The asymptotic solutions are expressed in the usual way:

$$X(T, \tau; \varepsilon) \sim \sum_{n=0}^{\infty} \varepsilon^n X_n(T, \tau); \quad Y(T, \tau; \varepsilon) \sim \sum_{n=0}^{\infty} \varepsilon^n Y_n(T, \tau)$$

which gives immediately that $X_{0\tau} = 0$ i.e. $X_0 = X_0(T)$ only; then we obtain

$$f' Y_{0\tau} = X_0 - (X_0 + \alpha) Y_0; \quad f' Y_{1\tau} + Y_{0T} = X_1 - (X_0 + \alpha) Y_1 - X_1 Y_0; \tag{5.155a,b}$$

$$f' X_{1\tau} + X_{0T} = -X_0 + (X_0 + \alpha - \beta) Y_0; \tag{5.155c}$$

$$f' X_{2\tau} + X_{1T} = -X_1 + (X_0 + \alpha - \beta) Y_1 + X_1 Y_0, \tag{5.155d}$$

and so on.

In (5.155a) we choose to make the τ-dependence as simple as possible—the real purpose behind the introduction of $f(T) = f(t)$—and so we write

$$f'(T) = X_0 + \alpha \; : \; f(T) = \int_0^T [X_0(T') + \alpha] \, dT'$$

and then we have simply

$$Y_{0\tau} + Y_0 = \frac{X_0}{X_0 + \alpha}.$$

This is solved with ease, to produce

$$Y_0(T, \tau) = \frac{X_0(T)}{\alpha + X_0(T)} + A_0(T) e^{-\tau} \tag{5.156}$$

where A_0 is an arbitrary function. The initial conditions, (5.154), require that

$$A_0(0) = -\frac{X_0(0)}{\alpha + X_0(0)} = -\frac{1}{1 + \alpha}. \tag{5.157}$$

From (5.155c) we now have

$$X_{1\tau} = -\frac{X_0 + X_0'}{X_0 + \alpha} + \left(\frac{X_0 + \alpha - \beta}{X_0 + \alpha}\right)\left(\frac{X_0}{X_0 + \alpha} + A_0 e^{-\tau}\right)$$

which integrates to give

$$X_1(T, \tau) = \left\{\frac{X_0(X_0 + \alpha - \beta)}{(X_0 + \alpha)^2} - \frac{X_0 + X_0'}{X_0 + \alpha}\right\}\tau - \left(\frac{X_0 + \alpha - \beta}{X_0 + \alpha}\right) A_0 e^{-\tau} + B_1(T)$$

and then uniformity as $\tau \to \infty$ requires that

$$X_0' + X_0 = \frac{X_0(X_0 + \alpha - \beta)}{X_0 + \alpha}$$

which defines $X_0(T)$. Thus

$$X_0' = -\beta \frac{X_0}{X_0 + \alpha} \quad \text{or} \quad X_0 + \alpha \ln X_0 = C - \beta T$$

where C is an arbitrary constant. The initial condition, (5.154), then yields the result $C = 1$; the implicit solution for $X_0(T)$ is therefore described by

$$X_0 + \alpha \ln X_0 = 1 - \beta T.$$

(Although this equation appears unsatisfactory and rather involved, its solution has a simple interpretation: $X_0(T)$ starts at $X_0 = 1$ and decreases to zero, eventually exponentially like $X_0(T) \sim \exp[(1 - \beta T)/\alpha]$ as $T \to \infty$.)

Finally, we look at equation (5.155b), which can be written as

$$Y_{1\tau} + Y_1 = \frac{(1 - Y_0)X_1 - Y_{0T}}{X_0 + \alpha};$$

the important information we require from this equation is the condition that defines $A_0(T)$ (in (5.156)). This requires that the term in $e^{-\tau}$, on the right-hand side, is removed (because otherwise Y_1 will contain a term $\tau e^{-\tau}$, which leads to a non-uniformity as $\tau \to \infty$). The relevant term is

$$\left\{A_0' + \left(\frac{\alpha}{X_0 + \alpha}\right)\left(\frac{X_0 + \alpha - \beta)}{X_0 + \alpha}\right) A_0 + A_0 B_1\right\} e^{-\tau} = 0,$$

but this contains $B_1(T)$, so now we must call on equation (5.155d) to find this term. In this equation, the avoidance of non-uniformities requires the removal of all terms that depend on only T (for otherwise we will have $X_2 \propto \tau$). It is left as an exercise to show that this condition produces the equation for $B_1(T)$:

$$B_1' + \frac{\alpha\beta}{(X_0+\alpha)^2} B_1 = \alpha\beta \frac{X_0(X_0+\alpha-\beta)}{(X_0+\alpha)^4}$$

or

$$\left(\frac{X_0+\alpha}{X_0} B_1\right)' = \alpha\beta \frac{X_0+\alpha-\beta}{(X_0+\alpha)^2};$$

this, and then the equation for A_0, can be integrated, and solutions expressed in terms of $X_0(T)$.

In our final example, we are able to use ideas from singular perturbation theory in only a rather superficial way, but this is such an important problem that we could not ignore it.

E5.28 The Belousov-Zhabotinskii reaction

This is a famous and much-studied phenomenon, first demonstrated by Belousov in 1951. He discovered that a steady oscillation of the concentration of a catalyst, between its oxidised and reduced states, was possible. In a suitable medium, this can be exhibited as a dramatic change in colour—a colour being associated with a state—with a period of a minute or so. A set of model equations for this chemical reaction is

$$\varepsilon\dot{x} = x + y - xy - \varepsilon\lambda x^2; \quad \varepsilon\mu\dot{y} = \nu z - y - xy; \quad \dot{z} = x - z, \qquad (5.158\text{a,b,c})$$

where the concentration of the catalyst is represented by z (Tyson, 1985). The constants λ, μ and ν are positive and independent of ε; ε measures the rate constant for the production of x, and gives the size of the corresponding constant for y and of the nonlinearity in (5.158a). We will consider the problem with $\varepsilon \to 0^+$. The initial conditions are relatively unimportant here, but we will assume that they are sufficient to start the oscillatory process. It would be impossible in a text such as ours, with its emphasis on singular perturbation theory, to give a comprehensive description of the relevant solution of the set (5.158). This would involve, for example, a detailed (but local) stability analysis. We will content ourselves with a brief overview that emphasises the various scales that are important.

The first point to note is that equations (5.158a,b) have a boundary-layer structure, so that there is a short time ($\mathrm{O}(\varepsilon)$) during which the initial values are lost and the solution settles to a (local) steady state (sometimes referred to as *quasi-equilibrium*). Note, however, that (5.158c) allows a relatively long evolution time, so for times of size O(1) we have (with $\varepsilon \to 0^+$)

$$\text{I:}\quad \begin{cases} x \sim \dfrac{y}{y-1} \quad (\text{for } y > 1) \quad \text{or} \quad y \sim \dfrac{x}{x-1} \quad (\text{from (5.158a)}) \\ y \sim \dfrac{\nu z}{1+x} \quad (\text{from (5.158b)}) \quad \text{and} \quad \dot{z} = x - z. \end{cases}$$

These we will label state I.

From I it is clear that, if $y \to 1^+$, then both x and z must increase, but then eventually the nonlinearity in equations (5.158) will become important (and most particularly the term $\varepsilon\lambda x^2$ in (5.158a)). This observation provides the basis for the scaling in this situation: we write $y \equiv Y = \mathrm{O}(1)$, $x \equiv X/\varepsilon$ and $z \equiv Z/\varepsilon$, and so equations (5.158) become

$$\varepsilon\dot{X} = X - XY - \lambda X^2 + \varepsilon Y; \quad \varepsilon^2\mu\dot{Y} = \nu Z - XY - \varepsilon Y; \quad \dot{Z} = X - Z. \qquad (5.159\text{a,b,c})$$

These three equations each have a *different* time scale, so another story unfolds here. On a very short time interval ($\mathrm{O}(\varepsilon^2)$), Y evolves from its initial value (close to 1) to a (local) steady state governed by $XY \sim \nu Z$. Then on a longer—but still short!—time interval ($\mathrm{O}(\varepsilon)$), X now evolves so that $Y \sim (1 - \lambda X)$ $(X \neq 0)$, and then on an O(1) time scale we have $\dot{Z} = X - Z$; these together constitute state II:

$$\text{II:} \quad XY \sim \nu Z; \quad Y \sim 1 - \lambda X; \quad \dot{Z} = X - Z.$$

In terms of the relevant scales, this is sufficient for the oscillatory process. Of course, how the solution actually evolves, and jumps between states I and II, needs more discussion than we are able to present here. A careful analysis demonstrates that, provided $1/2 < \nu < 1 + \sqrt{2}$, then the solution slowly (O(1) scales) evolves in state I, becomes unstable and jumps to state II; this also slowly evolves until *it* becomes unstable and reverts to state I. This is the essence of the oscillatory process (sometimes called a *relaxation oscillation*, because the solution 'relaxes' back to a former branch of the solution).

We have chosen to ignore many important elements in our presentation of the Belousov-Zhabotinskii reaction, mainly because they require much that goes well beyond the methods of singular perturbation theory. An illuminating discussion of this process can be found in Fowler (1997) and this text, in conjunction with Murray (1993), provide an excellent introduction to many chemical, biochemical and biological models and their solutions.

In this final chapter, we have presented and described a number of examples taken from the physical and chemical sciences, and in each the ideas of singular perturbation theory play a significant rôle. As we implied earlier, such a collection could not be exhaustive—indeed, we can hope to give only an indication of what is possible. Even if the particular applications offered here are of no specific interest to some readers, they do provide a set of additional worked examples that should help to reinforce the ideas that contribute to singular perturbation theory. Other examples, some described in detail and some set as exercises, are available in many of the texts previously cited. In addition, interested readers are encouraged to investigate the references to related material that have been provided throughout this chapter.

APPENDIX: THE JACOBIAN ELLIPTIC FUNCTIONS

Given the integral

$$u = \int_0^{\phi} \frac{\mathrm{d}\theta}{\sqrt{1 - m\sin^2\theta}}, \qquad 0 \leq m \leq 1,$$

where m is usually called the *modulus*, we then define the *Jacobian elliptic functions*

$$\mathrm{cn}(u; m) = \cos\phi; \quad \mathrm{sn}(u; m) = \sin\phi; \quad \mathrm{dn}(u; m) = \sqrt{1 - m\sin^2\phi}.$$

We see immediately that we have the identities

$$\mathrm{sn}^2(u; m) + \mathrm{cn}^2(u; m) = 1; \quad \mathrm{dn}^2(u; m) + m\mathrm{sn}^2(u; m) = 1.$$

Further, cn $\rightarrow$ sech as $m \rightarrow 1^-$ and cn $\rightarrow$ cos as $m \rightarrow 0^+$.

The relevant differential relations are

$$\begin{aligned} \frac{\mathrm{d}}{\mathrm{d}u}[\mathrm{sn}(u; m)] = \frac{\mathrm{d}\phi}{\mathrm{d}u}\frac{\mathrm{d}}{\mathrm{d}\phi}(\sin\phi) &= \sqrt{1 - m\sin^2\phi}\cos\phi \\ &= \mathrm{cn}(u; m)\mathrm{dn}(u; m); \end{aligned}$$

similarly we have

$$\frac{\mathrm{d}}{\mathrm{d}u}[\mathrm{cn}(u; m)] = -\mathrm{sn}(u; m)\mathrm{dn}(u; m)$$

and
$$\frac{\mathrm{d}}{\mathrm{d}u}[\mathrm{dn}(u;m)] = -m\,\mathrm{sn}(u;m)\mathrm{cn}(u;m).$$

The (real) period of the Jacobian elliptic functions is $4K(m)$ where

$$K(m) = \int_0^{\pi/2} \frac{\mathrm{d}\theta}{\sqrt{1 - m\sin^2\theta}}\,; \quad E(m) = \int_0^{\pi/2} \sqrt{1 - m\sin^2\theta}\,\mathrm{d}\theta$$

are the *complete elliptic integrals of the first* and *second kinds*, respectively. (The Jacobian elliptic functions are *doubly-periodic* in the complex plane; for example, the other period of $\mathrm{cn}(u;m)$ is $2\mathrm{i}K(1-m)$.)

The interested reader can find more information in texts that specialise in the functions, such as Lawden (1989) or Byrd & Friedman (1971).

ANSWERS AND HINTS

The answer, where one is given, is designated by the prefix A; for example, the answer to Q1.1 is A1.1. In some cases a hint to the method of solution is included; in a few of the more involved calculations, some intermediate steps are given.

CHAPTER 1

A1.1 (a) $\frac{1}{2} - \frac{x}{8} + \frac{x^2}{16} + \cdots$ for $0 \le |x| < 1$; **(b)** $\frac{1}{2} - \frac{x}{4} + \frac{x^3}{12} + \cdots$ for $-\infty < x < \ln 3$; **(c)** $\cos 1 - x \sin 1 - \frac{x^2}{2}(\sin 1 + \cos 1) + \cdots$ for $\forall x$;
(d) $x + \frac{x^2}{2} - \frac{2x^3}{3} + \cdots$ for $-1/2 \le x \le (\sqrt{5} - 1)/2$; **(e)** $\ln 2 + \frac{x}{2} + \frac{x^2}{8} + \cdots$ for $-\infty < x \le \ln 3$;
(f) $-\ln x + x + \frac{x^2}{2} + \cdots$ (cf. (d)) for $0 < x \le (\sqrt{5} - 1)/2$;
(g) $x \ln x + \frac{1}{2}(x \ln x)^2 + \frac{1}{6}(x \ln x)^3 + \cdots$ for $x > 0$. [N.B. $|x \ln x| \to 0$ as $x \to 0^+$.]

A1.2 (a) $\frac{1}{\sqrt{x}}\left(1 + \frac{x}{2} - \frac{x^2}{8} + \cdots (-1)^{n+1}\left(\frac{x}{2}\right)^n \frac{1.3. \ldots (2n-3)}{n!} + \cdots\right)$ for $0<x<1$;

(b) $\sin 1 \sum_{n=0}^{\infty} (-1)^n \frac{x^{2n}}{(2n)!} + \cos 1 \sum_{n=0}^{\infty} (-1)^n \frac{x^{2n+1}}{(2n+1)!}$ for $\forall x$;

(c) $-\ln x + \sum_{n=1}^{\infty} (-1)^{n-1} \frac{x^n}{n}$ for $0 < x \le 1$.

A1.3 (a) $x(t) = \mathrm{e}^{-\varepsilon t}(\cos t + \varepsilon \sin t)$; **(b)** $x(t) = \frac{1}{2\varepsilon - \varepsilon^2}[\sin(1-\varepsilon)t - (1-\varepsilon)\sin t]$ which is periodic but of large amplitude as $\varepsilon \to 0$; **(c)** $y(x) = \mathrm{e}^{-x/\varepsilon} + \cos x + \frac{1}{\varepsilon}\sin x$;
(d) $y(x) = 1 + \mathrm{e}^{-\sqrt{x}/\varepsilon}$; **(e)** $y(x) = \exp[(x - x^2)/\varepsilon]$ and note the behaviour near $x = 0$ and near $x = 1$; **(f)** $y(x) = \exp[-(x + x^2)/\varepsilon]$ and only near $x = 0$ is interesting (cf. (e)); **(g)** $y(x) = x^2 + (1 - 2\varepsilon)x - (\varepsilon - 2\varepsilon^2)[\ln(x + \varepsilon) - \ln \varepsilon]$ and the P.I. dominates for $x = \mathrm{O}(1)$ as $\varepsilon \to 0^+$; **(h)** first $\varepsilon f'(y) + 1 + f = 0$ and then $y(x) = \varepsilon \ln(1 + \mathrm{e}^{-x/\varepsilon}) - \varepsilon \ln 2$; **(i)** $y(x) = A\left(\frac{1-\mathrm{e}^{2Ax/\varepsilon}}{1+\mathrm{e}^{2Ax/\varepsilon}}\right)$ where $A = 1 + 2\mathrm{e}^{-2/\varepsilon} + \cdots = 1 +$ exp. small terms; **(j)** $y(x) = \frac{\varepsilon\pi}{8}\tan\left(\frac{\pi}{8}(1 + x)\right)$;
(k) $y(x) = \frac{\mathrm{e}^{x/\varepsilon}-1}{\mathrm{e}^{1/\varepsilon}-1}$; **(l)** $y(x) = 1 + \frac{x^3}{12\varepsilon^2}$; **(m)** $y(x) = 1 - \varepsilon \ln\left(1 + \frac{x}{\varepsilon}\right)$;
(n) $y(x) = 1 - \sqrt{1 - 2\varepsilon x}$.

A1.4 (a) both limiting processes give 0—uniform; **(b)** for $x \to 0$ first, the limit is 0; for $y \to 0$ first, the limit is 1—non-uniform; **(c)** for $x \to 0^+$first, the function tends to $+\infty$; for $y \to 0^+$ first, the function tends to $-\infty$—non-uniform; **(d)** for $x \to 0^+$ first, the limit is 0; for $y \to 0^+$ first, the limit is 1—non-uniform.

A1.5 (a) simply perform the integrations; **(b)** multiply the result in (a) by $x^{-\beta}$; **(d)** raise the result in (b) to the power γ.

A1.6 (a) yes; **(b)** yes; **(c)** yes; **(d)** $n^x - 1 = \mathrm{e}^{x \ln n} - 1 \sim x \ln n$;
(e) $\ln\{(x + x^2)/2\} = \ln x + \ln(1 + x) - \ln 2 \sim \ln x$; **(f)** yes—see Q1.5(e);
(g) yes; **(h)** yes; **(i)** $\ln(\frac{1}{2} + x + 2x^{-1}) = \ln 2 - \ln x + \ln[1 + (x/4) + (x^2/2)] \sim -\ln x$; **(j)** yes; **(k)** $x^x(\ln x)^{-1} = \mathrm{e}^{x \ln x}/\ln x$ and so $\mathrm{e}^{x \ln x}/x \ln x \to -\infty$.

A1.7 (a) $x^3\mathrm{e}^{-1/x}$; **(b)** $-1/\ln x$; **(c)** $\mathrm{e}^{1/x}$; **(d)** $2\mathrm{e}^{-1-2/x}$; **(e)** $x \ln x$; **(f)** $-\ln x$;
(g) $F[G(x)] = \ln[1 + \exp\{-1/\ln(1 + \mathrm{e}^{-1/x})\}] \sim \ln[1 + \exp(-1/\mathrm{e}^{-1/x})] \sim \exp(-\mathrm{e}^{1/x})$.

A1.8 (a) suppose that $f/g \to m$, $F/G \to n$, then $fF/gG \to mn$; **(b)** as in (a), but with $F/G \to 0$.

A1.9 (a) form $\frac{\ln(1+x^n)}{\ln(1+x^{n-1})} = \frac{x^n + \cdots}{x^{n-1} + \cdots} \to 0$; **(b)** form $\frac{x^n\mathrm{e}^{-nx}}{x^{n-1}\mathrm{e}^{-(n-1)x}} = x\mathrm{e}^{-x} \to 0$;
(c) form $\frac{x^n(2+\sin x^{-n})}{x^{n-1}(2+\sin x^{-n+1})} = x\frac{2+\sin x^{-n}}{2+\sin x^{-n+1}} \to 0$.

A1.10 $(1 + x)^{-1} = x^{-1}(1 + x^{-1})^{-1} = x^{-1}(1 - x^{-1} + x^{-2} - x^{-3} + \cdots)$ as required for (a); **(b)** write as $(x^{-1} - x^{-2}) + (x^{-3} - x^{-4}) + \cdots = x^{-2}(x - 1) + x^{-4}(x - 1) + \cdots$, so $p(x) = x - 1$.

A1.11 (a) $\frac{2}{\sqrt{\pi}}\left\{x - \frac{x^3}{3} + \frac{x^5}{5.2!} + \cdots (-1)^n\frac{x^{2n+1}}{(2n+1)n!} + \cdots\right\}$ and ratio test gives $\left(\frac{2n-1}{2n+1}\right)\frac{x^2}{n} \to 0$ as $n \to \infty$ for all finite x, so convergent for these xs;
(b) $\exp(-x^2)\left\{\frac{1}{2x} - \frac{1}{4x^3} + \cdots (-1)^{n+1}\frac{(2n-3).(2n-5)\cdots 1}{2^n x^{2n-1}} + \cdots\right\}$ and the ratio test fails for all finite x—divergent. Consider $\frac{1}{2x} - \frac{1}{4x^3} + \frac{3}{8x^5} - \frac{5.3}{16x^7}\ \left(+\frac{7.5.3}{32x^9} + \cdots\right)$ with $x = 2$, then $0.9952 < \mathrm{erf}(2) < 0.9954$. (Accurate value is ≈ 0.99532.)

A1.12 $\frac{\cos x}{x} + \frac{\sin x}{x^2} - \frac{2\cos x}{x^3} + \cdots$; divergent; $R_n = n!\int_x^\infty \frac{\sin t \text{ or } \cos t}{t^{n+1}}\,\mathrm{d}t$, then $|R_n| < n!\int_x^\infty \frac{\mathrm{d}t}{t^{n+1}} = \frac{(n-1)!}{x^n}$.

A1.13 Write as $\int_x^1 \frac{e^{-t}}{t}\,dt + \int_1^\infty \frac{e^{-t}}{t}\,dt = \int_0^1 \left(\frac{e^{-t}-1}{t}\right)dt + \int_1^\infty \frac{e^{-t}}{t}\,dt + \int_x^1 \frac{dt}{t} - \int_0^x \left(\frac{e^{-t}-1}{t}\right)dt = -\ln x - \gamma - \left(-x + \frac{x^2}{2.2!} - \frac{x^3}{3.3!}\right)$ with the general term $(-1)^{n-1}\frac{x^n}{n.n!}$.

A1.14 (a) $\frac{1}{2}\exp(-x^2)\left\{\frac{1}{x^2} + \frac{1}{x^3} - \frac{1}{x^4} - \frac{3/2}{x^5}\right\}$; **(b)** $x^{-1} - \ln x + x + \frac{1}{2}x^2$.

A1.15 (a) $x\ln x - x^2\ln x + \frac{7}{6}x^3\ln x + \frac{5}{18}x^3$; **(b)** $x + (\alpha - \beta) - x^2 e^{-x} + xe^{-x}\ln x$.

A1.16 $\left\{\frac{\varepsilon}{\cos x} - \frac{\varepsilon^2 \sin x}{\cos^3 x}\right\}\exp\left[-(2 - \sin x)/\varepsilon\right]$; fails because $\cos(\pi/2) = 0$.

A1.17 $\frac{1}{x} - \frac{1!}{x^2} + \frac{2!}{x^3} + \cdots + (-1)^{n-1}\frac{(n-1)!}{x^n} + (-1)^n \frac{n!}{x^n}\int_0^\infty \frac{e^{-xt}}{(1+t)^{n+1}}\,dt$; ratio test gives n/x, so diverges; $|R_n| < \frac{n!}{x^n}\int_0^\infty e^{-xt}\,dt = \frac{n!}{x^{n+1}}$ which is minimised by $n = [x]$.

A1.18 $z'' + [1 + (\frac{1}{4} - \nu^2)/x^2]\,z = 0$; $\omega = \pm i$, $\lambda = 0$; $2\omega(n+1)A_{n+1} = [n(n+1) + \frac{1}{4} - \nu^2]A_n$; $z(x) \sim Ae^{ix}\left[1 - i\frac{\mu}{2x} - \frac{\mu(2+\mu)}{4x^2}\right] + Be^{-ix}\left[1 + i\frac{\mu}{2x} - \frac{\mu(2+\mu)}{4x^2}\right]$, where $\mu = \frac{1}{4} - \nu^2$. N.B. More complete descriptions of $J_\nu(x)$ produce $C = 2/\sqrt{\pi}$, $\alpha = -\frac{1}{4}\pi - \frac{1}{2}\nu\pi$.

A1.19 (a) $f \sim 1 - \frac{1}{2}\varepsilon(x^{-1} + x)$; $F \sim \sqrt{\frac{X+e^{-X}}{1+X}}\left(1 - \frac{1}{2}\frac{\varepsilon^2 X^2}{1+X}\right)$ where $x = \varepsilon X$; $\mathfrak{F} \sim \frac{1}{\sqrt{1+\chi}}\left(1 - \frac{1}{2}\frac{\varepsilon^2}{\chi+\chi^2}\right)$ where $x = \chi/\varepsilon$; match 1st & 2nd: $1 - \frac{1}{2}X^{-1} - \frac{1}{2}\varepsilon^2 X$; match 1st & 3rd: $1 - \frac{1}{2}\chi - \frac{1}{2}\varepsilon^2\chi^{-1}$;

(b) $f \sim 1 + \varepsilon\frac{(x + 1/(2\sqrt{x}))}{1+\sqrt{x}}$; $F \sim 1 + \sqrt{\varepsilon}\left(\sqrt{1+X} - \sqrt{X + e^{-X}}\right)$ where $x = \varepsilon X$; $\mathfrak{F} \sim 1 + \sqrt{\varepsilon}\sqrt{\chi}$ where $x = \chi/\varepsilon$; match 1st & 2nd: $1 + \frac{1}{2}(\sqrt{\varepsilon}/\sqrt{X})$; match 1st & 3rd: $1 + \sqrt{\varepsilon}\sqrt{\chi}$;

(c) $f \sim \varepsilon\left(\frac{1+x}{x^2}\right) + \frac{1}{2}\varepsilon^2 x^2$; $F \sim \frac{1}{\varepsilon}\left(\frac{1}{1+X^2}\right) + \frac{X}{1+X^2}$ where $x = \varepsilon X$; $\mathfrak{F} \sim \cosh\chi - 1 + \varepsilon^2\chi^{-1}$ where $x = \chi/\varepsilon$; match 1st & 2nd : $1/(\varepsilon X^2) + X^{-1}$; match 1st & 3rd: $\varepsilon x^{-1} + \frac{1}{2}\varepsilon^2 x^2$;

(d) $f \sim \frac{1}{1+x}\left(1 - \frac{\varepsilon^2 x^3}{1+x}\right)$; $F \sim \frac{1}{1+e^{-X}}\left(1 - \frac{\varepsilon X}{1+e^{-X}}\right)$ where $x = \varepsilon X$; $\mathfrak{F} \sim \frac{\varepsilon}{\chi + \chi^2\sin\chi}\left(1 - \frac{\varepsilon}{\chi + \chi^2 \sin\chi}\right)$ where $x = \chi/\varepsilon$; match 1st & 2nd: $1 - \varepsilon X$; match 1st & 3rd: $\frac{\varepsilon}{\chi}(1 - \chi^2) + \varepsilon^2(2\chi - 1/\chi^2)$;

(e) $f \sim \sqrt{\varepsilon}\sqrt{x^{-1} + 1 + x}\left\{1 - \frac{\varepsilon}{2x}\left(\frac{1+x}{1+x+x^2}\right)\right\}$; $F \sim \frac{1}{\sqrt{1+X}}(1 + \frac{1}{2}\varepsilon X)$ where $x = \varepsilon X$; $\mathfrak{F} \sim \sqrt{\sinh\chi}\left\{1 + \frac{\varepsilon}{2\sinh\chi}\right\}$ where $x = \chi/\varepsilon$; match 1st & 2nd: $\frac{1}{\sqrt{X}}\left(1 - \frac{1}{2}X^{-1}\right)\left(1 + \frac{1}{2}\varepsilon X\right)$; match 1st & 3rd: $\sqrt{\chi}\left(1 + \frac{1}{2}\varepsilon\chi^{-1}\right)$;

(f) $f \sim \sqrt{\varepsilon}\sqrt{x} - \varepsilon\sqrt{\varepsilon}\left(\frac{1}{2}\sqrt{x} + \frac{1}{6}x\sqrt{x}\right)$; $F \sim \varepsilon(\sqrt{X} + e^{-X}) - \frac{1}{2}\varepsilon^2(\sqrt{X} + e^{-X})$ where $x = \varepsilon X$; $\mathfrak{F} \sim \frac{\sin\sqrt{\chi}}{\sqrt{1+\chi^2}}\left(1 - \frac{1}{2}\frac{\varepsilon}{1+\chi^2}\right)$ where $x = \chi/\varepsilon$; $f \sim \frac{\varepsilon}{\sqrt{1+e^{-\xi}}} + \frac{\varepsilon\sqrt{\varepsilon}\sqrt{\xi}}{\sqrt{1+e^{-\xi}}}$ where $x = \varepsilon^2\xi$; match 1st & 2nd: $\sqrt{\varepsilon x} - \frac{1}{2}\varepsilon\sqrt{\varepsilon x}$; match 1st & 3rd: $\sqrt{\chi} - \frac{1}{6}\chi\sqrt{\chi} - \frac{1}{2}\varepsilon\sqrt{\chi}$; match 2nd & 4th: $\varepsilon + \varepsilon\sqrt{X}$.

A1.20 (a) $f \sim 1 - \frac{\varepsilon}{2(1+x)} + \frac{e^{-x/\varepsilon}}{2(1+x)}$; $F \sim \sqrt{1 + e^{-X}}\left(1 + \frac{\varepsilon X}{2(1+e^{-X})} - \frac{1}{2}\varepsilon(1+X)\right)$ where $x = \varepsilon X$; match: $1 + \frac{1}{2}(1 - \varepsilon X)e^{-X} - \frac{1}{2}\varepsilon$;

(b) $f \sim \left(\frac{1+x}{1+2x}\right)\left\{1 + \frac{\varepsilon x}{(1+x)(1+2x)} - 2\left(\frac{1+x}{1+2x}\right)e^{-x/\varepsilon}\right\}$; $F \sim \left(\frac{1}{1+\operatorname{sech}X}\right)\left(1 - \frac{\varepsilon X}{1+\operatorname{sech}X}\right)$ where $x = \varepsilon X$; match: $1 - 2e^{-X} - \varepsilon X(1 - 2e^{-X}) + 2\varepsilon Xe^{-X}$.

A1.21 **(a)** $f \sim 1 + x - \frac{\ln x}{\ln \varepsilon} + \frac{\varepsilon}{\ln \varepsilon}\left(1 - \frac{\ln x}{\ln \varepsilon} - \frac{1}{x}\right)$; $F \sim -\frac{\ln(1+X)}{\ln \varepsilon} + \varepsilon\left\{X - \frac{\ln(1+X)}{(\ln \varepsilon)^2}\right\}$ where $x = \varepsilon X$; match: $-\frac{\ln X}{\ln \varepsilon} - \frac{1}{X \ln \varepsilon} + \varepsilon X - \frac{\varepsilon}{(\ln \varepsilon)^2} \ln X$;
(b) $f \sim \sqrt{\ln(1+x) + x(\ln x - \ln \varepsilon)}\left(1 + \frac{\varepsilon}{\ln(1+x) + x(\ln x - \ln \varepsilon)}\right)$;
$F \sim \sqrt{\varepsilon}\sqrt{1 + X + X\ln(1+X)}\left\{1 - \frac{\frac{1}{4}\varepsilon X^2}{1 + X + X\ln(1+X)}\right\}$ where $x = \varepsilon X$;
match: $\sqrt{\varepsilon}\frac{(1 + X + X\ln X)}{\sqrt{X + X\ln X}}\left\{1 - \frac{\frac{1}{4}\varepsilon X}{1 + \ln X} + \frac{\frac{1}{2}\varepsilon X}{(1 + \ln X)(1 + X + X\ln X)}\right\}$;
(c) $f \sim (1 + x^2)(\ln x - \ln \varepsilon) + \varepsilon\left[x(\ln x - \ln \varepsilon) + x^{-1} + x - \ln x + \ln \varepsilon\right]$; $F \sim \ln(1+X) - \varepsilon \ln X$ where $x = \varepsilon X$; match: $\ln X - \varepsilon \ln X + X^{-1}$.

A1.22 **(a)** $f \sim \frac{x}{1+x} + \frac{\varepsilon}{(1+x)^2}$; $F \sim \varepsilon(1+X)[1 - \varepsilon(1+X)]$ where $x = \varepsilon X$; $\phi_{22} = \frac{x}{1+x} + \frac{\varepsilon}{(1+x)^2} - \varepsilon^2$; **(b)** $f \sim 1 + \frac{\varepsilon}{2}\left(\frac{1+x}{x}\right)$; $F \sim \sqrt{\frac{2+X}{1+X}}\left\{1 + \frac{\varepsilon X}{2(2+X)}\right\}$ where $x = \varepsilon X$; $\phi_{22} = \sqrt{\frac{x+2\varepsilon}{x+\varepsilon}}\left(1 + \frac{\varepsilon x}{2(x+2\varepsilon)}\right)$ which is identical to the expansion for F with $x = \varepsilon X$.

A1.23 **(a)** Φ_{22} does not exist at $x = \frac{1}{2}\left(1 + \sqrt{1 + 4\varepsilon^2}\right)$; **(b)** $\Phi_{22} \equiv \phi_{22}$.

A1.24 Form $|f - \Phi_{21}| = \left|\frac{1}{1 + x + \varepsilon + \mathrm{e}^{-x/\varepsilon}} - \frac{1}{1+x}\left(1 - \frac{\varepsilon}{1+x}\right)\left(\frac{1}{1 + \mathrm{e}^{-x/\varepsilon}}\right)\right|$

$$< \left|\frac{x\mathrm{e}^{-x/\varepsilon} + \varepsilon^2/(1+x)}{(1 + x + \varepsilon + \mathrm{e}^{-x/\varepsilon})(1+x)(1 + \mathrm{e}^{-x/\varepsilon})}\right|$$

$$\leq \left|\frac{x e^{-x/\varepsilon} + \varepsilon^2}{1.1.1}\right| \leq \varepsilon(\mathrm{e}^{-1} + \varepsilon).$$

CHAPTER 2

A2.1 **(a)** $x \sim \pm 1 - \varepsilon \pm \frac{1}{2}\varepsilon^2$; **(b)** $x \sim 1 - \frac{1}{4}\varepsilon + \frac{1}{8}\varepsilon^2$, $x \sim -4\varepsilon^{-1} - 1 + \frac{1}{4}\varepsilon$; **(c)** $x \sim 1 \pm \sqrt{\varepsilon} + \frac{1}{2}\varepsilon$; and for $\varepsilon \to \infty$: **(a)** $x \sim \frac{1}{2}\varepsilon^{-1} - \frac{1}{8}\varepsilon^{-3} + \frac{1}{16}\varepsilon^{-5}$, $x \sim -2\varepsilon - \frac{1}{2}\varepsilon^{-1} + \frac{1}{8}\varepsilon^{-3}$; **(b)** $x \sim \pm 2\varepsilon^{-1/2} - 2\varepsilon^{-1} \pm \varepsilon^{-3/2}$; **(c)** $x \sim \varepsilon + 2 - \varepsilon^{-1}$, $x \sim \varepsilon^{-1} - 2\varepsilon^{-2} + 5\varepsilon^{-3}$.

A2.2 **(a)** $x \sim \frac{3}{2} + \frac{27}{16}\varepsilon$, $x \sim -\frac{1}{2} + \frac{1}{16}\varepsilon$, $x \sim \varepsilon^{-1} - 1$; **(b)** $x \sim 1 \pm \sqrt{\varepsilon}$, $x \sim \varepsilon^{-1} - 2$; **(c)** $x \sim 1 - 2\varepsilon$; **(d)** $x \sim 1 \pm 2\sqrt{\varepsilon}$, $x \sim \varepsilon^{-1/2} + \varepsilon^{-1/4}$; **(e)** $x \sim \varepsilon^2 + \varepsilon^3$, $x \sim \pm\varepsilon^{-1} - \varepsilon$; **(f)** $x \sim 1 + \frac{3}{2}\varepsilon$, $x \sim -2\varepsilon^2 + 2\varepsilon^3$, $x \sim \varepsilon^{-1} - \varepsilon$; **(g)** $x \sim \varepsilon + \frac{3}{2}\varepsilon^2$, $x \sim \varepsilon^{-1} - 1$, $x \sim \varepsilon^{-2} - \frac{1}{2}\varepsilon^{-1}$; **(h)** $x \sim 1 \pm \frac{2}{\sqrt{5}}\sqrt{\varepsilon}$, $x \sim \frac{1}{3}\varepsilon^{-1} - 18\sqrt{3}\varepsilon^{3/2}$; **(i)** $x \sim -\varepsilon^2 + \frac{1}{2}\varepsilon^3$; **(j)** $x \sim \pm 1 - \frac{1}{2}\varepsilon$, $x \sim -\varepsilon^{-1} + \varepsilon$.

A2.3 **(a)** $x \sim 1 + \sqrt{\varepsilon}$, $x \sim \frac{1}{2}\varepsilon^{3/2} + \frac{1}{8}\varepsilon^2$, $\varepsilon^{-1/2} + 2$, $2\varepsilon^{-1/2} + 1$; **(b)** $x \sim \pm(\varepsilon^2 + \frac{1}{2}\varepsilon^3)$, $x \sim -\varepsilon^{-1} \pm \varepsilon^{-1/2}$; **(c)** $x \sim \pm\varepsilon^{3/2}\ln \varepsilon + \frac{1}{2}\varepsilon^2$, $\varepsilon \ln \varepsilon - \varepsilon/\ln \varepsilon$, $x \sim \varepsilon^{-1} + 1$; **(d)** $x \sim \ln 2 - \varepsilon/\{2[1 + (\ln 2)^2]\}$; **(e)** $x \sim -\varepsilon - \frac{1}{3}\varepsilon^3$, $x \sim \pm\varepsilon^{-1} + \frac{1}{3}$.

A2.4 $u \sim nt + e \sin nt + \frac{1}{2}e^2 \sin 2nt$.

A2.5 **(a)** $z \sim -\frac{1}{2} \pm \frac{\sqrt{3}}{2}\mathrm{i} \pm \frac{1}{\sqrt{3}}\varepsilon\,\mathrm{i}$, $z \sim -\varepsilon^{-1} + 1$; **(b)** $z \sim 1 \pm \mathrm{i} \pm 2\varepsilon\,\mathrm{i}$, $z \sim \pm\varepsilon^{-1/2} - 1$; **(c)** $z \sim 1 \pm \mathrm{i} \mp 2\varepsilon\,\mathrm{i}$, $z \sim \pm\mathrm{i}\varepsilon^{-1/2} - 1$; **(d)** $z \sim -\frac{1}{2} \pm \frac{\sqrt{3}}{2}\mathrm{i} - \frac{1}{6}\varepsilon(1 \mp \sqrt{3}\,\mathrm{i})$, $z \sim \pm\mathrm{i}\sqrt{\varepsilon} + \frac{1}{2}\varepsilon$; **(e)** $z \sim 2m\pi\,\mathrm{i} + 4\varepsilon\, m^2\pi^2$; **(f)** $z \sim 2\varepsilon(1 + \mathrm{i}m\pi) + \varepsilon^2\mathrm{e}^{-2}$.

A2.6 **(a)** $I(\varepsilon) \sim \frac{1}{3} + 3\varepsilon$; **(b)** $I(\varepsilon) \sim -\ln\varepsilon - \varepsilon\ln\varepsilon$; **(c)** $I(\varepsilon) \sim 1 + 4\varepsilon$; **(d)** $I(\varepsilon) \sim \ln\varepsilon - 3\varepsilon$.

A2.7 **(a)** $I(\varepsilon) \sim \pi/4 + \varepsilon(3 + \ln 4)$; **(b)** $I(\varepsilon) \sim 2 - \varepsilon$; **(c)** $I(\varepsilon) \sim \pi - 2\sqrt{\varepsilon}$; **(d)** $I(\varepsilon) \sim \varepsilon^{-1} + \ln\varepsilon$; **(e)** $I(\varepsilon) \sim -2\ln\varepsilon - \varepsilon$; **(f)** $I(\varepsilon) \sim \frac{2}{3}(2\sqrt{2} - 1) + 3\varepsilon(\sqrt{2} - 1)$.

A2.8 $I(x;\varepsilon) \sim x\pi$; $I(1 - \varepsilon X;\varepsilon) \sim \pi/2 + \arctan X$; $I(-1 + \varepsilon X;\varepsilon) \sim -\pi/2 - \arctan X$.

A2.9 **(a)** $y(x;\varepsilon) \sim x^2 + \frac{1}{5}\varepsilon x^5$;
(b) $y(x;\varepsilon) \sim \mathrm{e}^{-x} + \frac{1}{6}\varepsilon\left\{(1 + \mathrm{e}^{-1} + \mathrm{e}^{-2})\mathrm{e}^{-x} - \mathrm{e}^{-3x} - \mathrm{e}^{-1} - \mathrm{e}^{-2}\right\}$;
(c) $y(x;\varepsilon) \sim \cos x + \varepsilon\left\{\frac{3\pi}{16}\sin x - \frac{1}{32}\cos x - \frac{3}{8}x\sin x + \frac{1}{32}\cos 3x\right\}$;
(d) $y(x;\varepsilon) \sim x^2 - x^3 + \varepsilon\left\{\frac{1}{2}(11\mathrm{e}^{-1} - 5\mathrm{e})x^3 - \frac{1}{2}(5\mathrm{e} + 11\mathrm{e}^{-1})x^2 + (8 - 3x)x^2\mathrm{e}^x\right\}$;
(e) $x(t;\varepsilon) \sim \sin t + \frac{1}{25}\varepsilon\left\{2\sin t - 14\cos t + [(2 + 5t)\sin t + 2(7 + 5t)\cos t]\mathrm{e}^{-t}\right\}$.

A2.10 **(a)** $\lambda \sim n^2\pi^2(1 - \frac{1}{2}\varepsilon)$,
$y(x;\varepsilon) \sim A\sin n\pi x + \varepsilon A\left\{-\frac{1}{4}x\sin n\pi x + \frac{1}{4}n\pi x(x - 1)\cos n\pi x\right\}$;
(b) $\lambda \sim n^2\pi^2\left(1 - \frac{4\varepsilon n^2\pi^2(1 - \cos 1)}{4n^2\pi^2 - 1}\right)$ $(\equiv \lambda_0 + \varepsilon\lambda_1)$, $y(x;\varepsilon) \sim A\sin n\pi x + \varepsilon A\left\{\left(\frac{2n^3\pi^3}{4n^2\pi^2 - 1} + \frac{\lambda_1 x}{2n\pi}\right)\cos n\pi x - \frac{1}{2}n^2\pi^2\left(\frac{\cos(n\pi - 1)x}{2n\pi - 1} + \frac{\cos(2n\pi + 1)x}{2n\pi + 1}\right)\right\}$;
(c) $\lambda \sim n^2\pi^2 + \frac{3}{4}\varepsilon A^2$, $y(x;\varepsilon) \sim A\sin n\pi x + \varepsilon A^3\frac{\sin 3n\pi x}{32n^2\pi^2}$, $n \neq 0$.

A2.11 **(a)** $y(x;\varepsilon) \sim x^{-2} + \varepsilon\left\{\frac{1}{2}x + x^{-2} - \frac{1}{2}x^{-6}\right\}$, $x = \varepsilon^{1/4}X$, $y = \varepsilon^{-1/2}Y$, $Y(X;\varepsilon) \sim \sqrt{2 + X^4} - X^2$, $y(0;\varepsilon) \sim \sqrt{2/\varepsilon}$;
(b) $y(x;\varepsilon) \sim x^{-1} + \varepsilon\left\{\frac{3}{4}x^{-1} - \frac{1}{2}x^{-3} - \frac{1}{4}x^{-5}\right\}$, $x = \varepsilon^{1/4}X$, $y = \varepsilon^{-1/4}Y$, $Y(X;\varepsilon) \sim \sqrt{\sqrt{2 + X^4} - X^2}$, $y(0;\varepsilon) \sim (2/\varepsilon)^{1/4}$;
(c) $y(x;\varepsilon) \sim x^2 + \varepsilon(1 + x^3)$, $x = \sqrt{\varepsilon}X$, $y = \varepsilon Y$, $Y(X;\varepsilon) \sim 1 + X^2$, $y(0;\varepsilon) \sim \varepsilon$,
N.B. Solution for $x = O(1)$ is correct on $x = 0$; **(d)** $y(x;\varepsilon) \sim (1 + x)^2[(x + x^{-2}) + \varepsilon x^{-2}(\frac{31}{4} - 2\ln x - 3x - \frac{1}{2}x^2 - \frac{4}{3}x^3 - \frac{3}{4}x^4 - x^{-1} - \frac{2}{3}x^{-3} - \frac{1}{2}x^{-4})]$,
$x = \varepsilon^{1/4}X$, $y = \varepsilon^{-1/2}Y$, $Y(X;\varepsilon) \sim \sqrt{2 + X^4} - X^2$, $y(0;\varepsilon) \sim \sqrt{2/\varepsilon}$;
(e) $y(x;\varepsilon) \sim x^{-1} + \varepsilon(x^{-1} - x^{-3} - x)$, $x = \sqrt{\varepsilon}X$, $y = Y/\sqrt{\varepsilon}$, $Y(X;\varepsilon) \sim \frac{1}{2}\left(\sqrt{4 + X^2} - X\right)$, $y(0;\varepsilon) \sim 1/\sqrt{\varepsilon}$.

A2.12 $y(x;\varepsilon) \sim x + \varepsilon(x^2 - x^{-1}\ln x + x^{-1} - x^{-2})$, $x = \varepsilon^{1/3}X$, $y = \varepsilon^{1/3}Y$, $Y(X;\varepsilon) \sim \sqrt{X^2 - 2X^{-1}}$ so $X \geq 2^{1/3}$ i.e. $x_0(\varepsilon) \sim (2\varepsilon)^{1/3}$, $y(2\varepsilon^{1/3};\varepsilon) \sim \sqrt{3}\,\varepsilon^{1/3}$.

A2.13 $y(x;\varepsilon) \sim 1 + \mathrm{e}^{-x} + \varepsilon(3\mathrm{e}^{-x} + 2x - 3 - \frac{1}{2}x^2 + x\mathrm{e}^{-x})$, $x = X/\sqrt{\varepsilon}$, $y = Y$, $Y(X;\varepsilon) \sim \cos X + \frac{1}{2}\sqrt{\varepsilon}(X\cos X - 3\sin X)$, $X = \chi/\sqrt{\varepsilon}$, $Y = z(\chi;\varepsilon)$.

A2.14 **(a)** $x = \varepsilon^{-1/3}X$, $y = \varepsilon^{2/3}Y$, $Y(X;\varepsilon) \sim \sqrt{X^{-1} + X^{-4}}$;
(c) $x = X/\varepsilon$, $y = Y/\varepsilon^2$, $Y(X;\varepsilon) \sim (X + \frac{1}{2}X^2)^2$.

A2.15 $y(x;\varepsilon) \sim A_0 + B_0\mathrm{e}^{-x} + \varepsilon\left\{A_1 + B_1\mathrm{e}^{-x} + (1-x)A_0^2 + 2A_0B_0x\mathrm{e}^{-x} - \frac{1}{2}B_0^2\mathrm{e}^{-2x}\right\}$, where $A_0 = 1 - (1 + B_0)\mathrm{e}^{-1}$, $A_1 = \frac{1}{2}(B_0^2 - 1)\mathrm{e}^{-2} - (B_1 + 2A_0B_0)\mathrm{e}^{-1}$; $x = \varepsilon X$, $y = \varepsilon Y$, $Y(X;\varepsilon) \sim 1 + C_0X + \varepsilon\left\{C_1X - \frac{1}{2}C_0X^2\right\}$; matching gives $A_0 = 1$, $B_0 = -1$, $A_1 = \frac{3\mathrm{e}^{-1}}{2(1-\mathrm{e}^{-1})}$, $B_1 = \frac{1-4\mathrm{e}^{-1}}{2(1-\mathrm{e}^{-1})}$, $C_0 = 1$, $C_1 = \frac{8\mathrm{e}^{-1}-5}{2(1-\mathrm{e}^{-1})}$; the scaling does not produce a *balance* of terms.

A2.16 **(a)** $x = \delta X$, $y = Y/\delta^2$, $\delta = \varepsilon^{1/4}$ $(\delta \to 0)$, $\delta = \varepsilon^{-1/3}$ $(\delta \to \infty)$;
(b) $x = \delta X$, $y = Y/\delta$, $\delta = \varepsilon^{1/4}$ $(\delta \to 0)$, no balance for large δ;
(c) $x = \delta X$, $y = \delta^2 Y$, $\delta = \sqrt{\varepsilon}$ $(\delta \to 0)$, $\delta = \varepsilon^{-1}$ $(\delta \to \infty)$;
(d) $x = \delta X$, $y = Y/\delta^2$, $\delta = \varepsilon^{1/4}$ $(\delta \to 0)$, for $\delta \to \infty$: $x = \delta X$, $y = \delta^3 Y$, $\delta = \varepsilon^{-1}$;(e) $x = \delta X$, $y = Y/\delta$, $\delta = \sqrt{\varepsilon}$ $(\delta \to 0)$, $\delta = \varepsilon^{-1/2}$ $(\delta \to \infty)$.

A2.17 **(a)** $y(x;\varepsilon) \sim 1 + x + \varepsilon(1 + x)^2$, $y(\varepsilon X;\varepsilon) \sim 1 + \mathrm{e}^{-X}$;
(b) $y(x;\varepsilon) \sim (1 + x)^2 + \varepsilon\left\{4(1 + x) - (1 + x)^3\right\}$, $y(\varepsilon X;\varepsilon) \sim 1 + \mathrm{e}^{-X}$;
(c) $y(x;\varepsilon) \sim 2\ln\left[\frac{2}{1+2x}\right] + \varepsilon\left\{\frac{\frac{10}{3} + 2x^2 + \frac{8}{3}x^3 - 8x}{(1+2x)^2}\right\}$, $y(\varepsilon X;\varepsilon) \sim 2\ln 2 + (1 - 2\ln 2)\mathrm{e}^{-X}$; **(d)** $y(x;\varepsilon) \sim \frac{1}{1+x} + \varepsilon$, $y(\varepsilon X;\varepsilon) \sim \frac{3-\mathrm{e}^{-X}}{3+\mathrm{e}^{-X}}$; **(e)** $y(x;\varepsilon) \sim \mathrm{e}^{x-1} + 2\varepsilon(1 - \sqrt{x})\mathrm{e}^{x-1}$, $y(\varepsilon^{2/3}X;\varepsilon) \sim \frac{(3/2)^{1/3}}{\mathrm{e}\Gamma(2/3)}\int_0^X \exp\left(-\frac{2}{3}\hat{X}^{3/2}\right)\mathrm{d}\hat{X}$.

A2.18 **(a)** $y(x;\varepsilon) \sim \frac{1}{1+x} + \varepsilon\left\{\frac{x^2+x-2-2\ln[(1+x)/2]}{(1+x)^2}\right\}$, $y(\varepsilon X;\varepsilon) \sim 1 + \ln(1 + \mathrm{e}^{-X})$;
(b) $y(x;\varepsilon) \sim \sqrt{2 + 2x} - \varepsilon\left\{\exp[-x^2 - 2x]\int_x^1\left(\frac{1+2z+2z^2}{2(1+z)}\right)\exp[z^2 + 2z]\mathrm{d}z\right\}$,
$y(\varepsilon X;\varepsilon) \sim \sqrt{2}\left(\frac{3+\mathrm{e}^{-\sqrt{2}X}}{3-\mathrm{e}^{-\sqrt{2}X}}\right)$;
(c) $y(x;\varepsilon) \sim \mathrm{e}^{-x} - \varepsilon x\mathrm{e}^{-x}$, $y(1 - \varepsilon X;\varepsilon) \sim \mathrm{e}^{-1}\left(\frac{3-\exp[-\mathrm{e}^{-1}X]}{3+\exp[-\mathrm{e}^{-1}X]}\right)$;
(d) $y(x;\varepsilon) \sim 2 + \frac{1}{2}x^2 + \varepsilon\left\{2x + \frac{1}{2}x^2 + \frac{1}{6}x^3 + A_1\right\}$, $y(\varepsilon X;\varepsilon) \sim 2\tanh X$ (and we cannot determine A_1 at this order);
(e) $y(x;\varepsilon) \sim (1 + x)^2 + \varepsilon(1 + x)^3$, $y(\varepsilon X;\varepsilon) \sim \left(\frac{2-\mathrm{e}^{-X}}{2+\mathrm{e}^{-X}}\right)$;
(f) $y(x;\varepsilon) \sim 1 + \cos x - \frac{1}{2}\varepsilon\left(1 + x\cos x + \frac{1}{3}\cos 2x\right)$, $y(\varepsilon X;\varepsilon) \sim hfill$ $2(1 - \mathrm{e}^{-X})$.

A2.19 $y(x;\varepsilon) \sim \frac{1}{5}(2 - x^2) - \varepsilon\frac{8x^3}{25}$; $y(\varepsilon X;\varepsilon) \sim \frac{1}{5}\left(2 + 3\mathrm{e}^{-\sqrt{5}X}\right)$; $y(1 - \varepsilon\chi;\varepsilon) \sim \frac{1}{5}\left(1 + 9\mathrm{e}^{-5\chi}\right)$.

A2.20 $y(x;\varepsilon) \sim 1 + x^2$; $y(\varepsilon X;\varepsilon) \sim 2/(1 + \pi - 2\arctan X)$; $y(\varepsilon^2\chi;\varepsilon) \sim \frac{2}{1+\pi} + \left(\frac{\pi-1}{\pi+1}\right)\mathrm{e}^{-\chi}$.

A2.21 **(a)** boundary layer near $x = 1$; **(b)** boundary layer near $x = 0$; **(c)** transition layer near $x = 1/2$; **(d)** transition layer near $x = 0$; **(e)** solution depends on the particular boundary values.

A2.22 **(a)** $y(x;\varepsilon) \sim (1 - x)^4 + 3\varepsilon\left\{1 - (1 - x)^2\right\}$, for $0 \le x < 1$, and $y(x;\varepsilon) \sim 1 + (1 - x)^4 + 3\varepsilon\{1 - (1 - x)^2\}$, for $1 < x \le 2$, both away from $x = 1$; $y(1 + \sqrt{\varepsilon}X;\varepsilon) \sim \frac{1}{2} + \frac{1}{\sqrt{\pi}}\int_0^X \exp[-z^2]\mathrm{d}z$; **(b)** $y(x;\varepsilon) \sim 1/(1 + x^{6/5})$, for $-1 \le x < 0$, and $y(x;\varepsilon) \sim 1/(2 + x^{6/5})$, for $0 < x \le 1$, both away from $x = 0$; $y(\varepsilon^{5/6}X;\varepsilon) \sim \frac{3}{4} - \frac{(6/5)^{1/6}}{4\Gamma(5/6)}\int_0^X \exp[-\frac{5}{6}z^{6/5}]\mathrm{d}z$.

A2.23 $y(x;\varepsilon) \sim \alpha - x$, for $0 \le x < x_0$, and $y(x;\varepsilon) \sim 1 + \beta - x$, for $x_0 < x \le 1$, both away from $x = x_0$; $y(x_0 + \varepsilon X;\varepsilon) \sim A\left(\frac{Be^X - 1}{Be^X + 1}\right)$ so that the jump is between $\pm A$ and the result follows;
(a) $y(x;\varepsilon) \sim 1 - x$, $y(x;\varepsilon) \sim -x$, $x_0 = 1/2$, $y(\frac{1}{2} + \varepsilon X;\varepsilon) \sim \frac{1}{2}\left(\frac{1 - Be^X}{1 + Be^X}\right)$ and B is unknown at this order; **(b)** $y(x;\varepsilon) \sim 3 - x$, $y(\varepsilon X;\varepsilon) \sim 3\left(\frac{2 - e^{-X}}{2 + e^{-X}}\right)$;
(c) $y(x;\varepsilon) \sim 4 - x$, $y(\varepsilon X;\varepsilon) \sim 4\left(\frac{3 - 5e^{-X}}{3 + 5e^{-X}}\right)$.

A2.24 Choose $2\varepsilon u' + au = 0$, then $F = \frac{\varepsilon}{u}f - \frac{a^2 v}{2} - \frac{\varepsilon v (au)'}{2u}$; special case gives $F = \left[b - \frac{1}{2}a^2 - \frac{1}{2}\varepsilon\left\{(au)'/u\right\}\right]v$ and the turning points are the zeros of $b - \frac{1}{2}a^2 - \frac{1}{2}\varepsilon[(au)'/u]$, if the derivative is non-zero at the zero.

A2.25 $y(\frac{1}{2} + \varepsilon X;\varepsilon) \sim A\mathrm{Ai}(X) + B\mathrm{Bi}(X)$ where Ai and Bi are the *Airy* functions.

A2.26 **(a)** $y(x;\varepsilon) \sim \ln(1 + x) - x\ln 2$, $y(\varepsilon X;\varepsilon) \sim -\varepsilon(1 - \ln 2)(X - 1 + e^{-X})$, $y(1 - \varepsilon\chi;\varepsilon) \sim -\varepsilon(\frac{1}{2} + \ln 2)(\chi - 1 + e^{-\chi})$; **(b)** $y(x;\varepsilon) \sim x(2 - x^2)$, $y(1 + \varepsilon X;\varepsilon) \sim 1$, $y(4 - \varepsilon\chi;\varepsilon) \sim -56(1 - e^{-2\chi})$; **(c)** $y(x;\varepsilon) \sim \frac{1+x}{2+x}$, $y(\varepsilon X;\varepsilon) \sim \frac{1}{2}(1 + e^{-X})$, $y(1 - \varepsilon\chi;\varepsilon) \sim \frac{2}{3}$.

A2.27 $t(z) = \sqrt{\frac{R}{2k}}\sqrt{\frac{gR^2}{k} - R} - \sqrt{\frac{R+z}{2k}}\sqrt{\frac{gR^2}{k} - (R+z)}$
$+ \frac{gR^2}{k\sqrt{2k}}\left\{\arcsin\left(\frac{\sqrt{R+z}}{R\sqrt{g/k}}\right) - \arcsin\left(\sqrt{\frac{k}{gR}}\right)\right\}$
where $k = gR - \frac{1}{2}V^2$ (>0); **(a)** $Z(\tau;\varepsilon) \sim \tau - \frac{1}{2}\tau^2 + \frac{1}{3}\varepsilon\left(\tau^3 - \frac{1}{4}\tau^4\right)$ then max. height $\sim \frac{1}{2} + \frac{1}{4}\varepsilon$ at time $\sim 1 + \frac{2}{3}\varepsilon$, returning at time $\sim 2 + \frac{4}{3}\varepsilon$ (twice the previous time); **(b)** $Z(\tau;\varepsilon) \sim \tau - \frac{1}{2}\tau^2 + \varepsilon\left\{\frac{1}{3}\tau^3 - \frac{1}{12}\tau^4 - \delta\left(\frac{1}{2}\tau^2 - \frac{1}{6}\tau^3\right)\right\}$ then max. height $\sim \frac{1}{2} + \varepsilon\left(\frac{1}{4} - \frac{1}{3}\delta\right)$ at time $\sim 1 + \varepsilon\left(\frac{2}{3} - \frac{1}{2}\delta\right)$, returning at time $\sim 2 + \varepsilon\left(\frac{4}{3} - \frac{2}{3}\delta\right)$ (not twice now); **(c)** $Z(\tau;\delta) \sim \frac{1}{2}\left[(1 + 3\tau)^{2/3} - 1\right] + \delta\left\{\frac{1}{2}(1 + 3\tau) + \exp.\text{ terms}\right\}$ then breakdown where $\tau = \delta^{-3}T$, $Z = \delta^{-2}\varsigma$, which gives $\ddot{\varsigma} = -1/(4\varsigma^2) - \dot{\varsigma}/\varsigma$ (for the dominant terms).

A2.28 $\frac{1}{2}\dot{x}^2 - \frac{\varepsilon}{1-x} - \frac{1-\varepsilon}{x} = $ constant; then
(a) $t(x;\varepsilon) \sim \frac{1}{k\sqrt{2k}}\left[\arcsin\left(\sqrt{kx}\right) - \sqrt{kx(1 - kx)}\right]$
$+\varepsilon\left\{\frac{2-k}{k(1-k)}\sqrt{\frac{x}{2(1-kx)}} - \frac{\sqrt{2}}{k\sqrt{k}}\arcsin\left(\sqrt{kx}\right)\right.$
$\left.- \frac{1}{[2(1-k)]^{3/2}}\ln\left[\frac{1 + (1-2k)x + 2\sqrt{(1-k)(1-kx)x}}{1-x}\right]\right\}$,
which breaks down because the first term approaches a constant, but the second is dominated by $\frac{1}{[2(1-k)]^{3/2}}\ln(1 - x)$, as $x \to 1$;
(b) $t(1 - \varepsilon X;\varepsilon) \sim T_0 + \varepsilon\left\{T_1 - \frac{\sqrt{X\{1 + (1-k)X\}}}{(1-k)\sqrt{2}} + \frac{\ln\left\{\sqrt{(1-k)X} + \sqrt{1 + (1-k)X}\right\}}{\sqrt{2}(1-k)^{3/2}}\right\}$,
where T_0, T_1 are constants, with $T_0 = \frac{\arcsin(\sqrt{k}) - \sqrt{k(1-k)}}{k\sqrt{2k}}$ (and T_1 is also determined).

A2.29 $\lambda \sim n\pi, n = \pm 1, \pm 2, \ldots$ with $y_n(x;\varepsilon) \sim A \sin n\pi x$, $y_n(\varepsilon X;\varepsilon) \sim \varepsilon An\pi(X - 1 + e^{-X})$, $y_n(1 - \varepsilon\chi;\varepsilon) \sim -\varepsilon(-1)^n A(\chi - 1 + e^{-\chi})$.

A2.30 $T(x;\varepsilon) \sim T_1 e^{x-1} - \varepsilon T_1 e^{x-1} \ln x$, $T(\sqrt{\varepsilon} X;\varepsilon) \sim T_0 + (T_1 e^{-1} - T_0) \times \sqrt{\frac{2}{\pi}} \int_0^X \exp[-\frac{1}{2}y^2]\mathrm{d}y$.

A2.31 $P(R;\varepsilon) \sim \frac{1}{2}(\beta - 1)(1 - R^2) + \frac{1}{6}\varepsilon(4 - 5\beta)R(1 - R^2) + \varepsilon^2[\gamma + \delta(1 + R^2)] \times (1 - R^2)$, where $\gamma = 9(1 - \frac{3}{2}\beta) + \frac{1}{2}\alpha(\beta - 1)$, $\delta = \frac{1}{6}(5\beta - 4) - \frac{1}{2}\alpha(\beta - 1)$.

A2.32 $y(x;\varepsilon) \sim \frac{\sinh x}{\sinh 1} - x + \varepsilon\left\{\left(\frac{\sinh x}{3\sinh 1}\right)\cosh 1 + \frac{\sinh 2x}{3(\sinh 1)^2} - \frac{x\cosh x}{\sinh 1}\right\}$.

A2.33 $y(x;\varepsilon) \sim k\cosh(Bx)$, $A(0) = -k^2B^2$, $y(1 - \varepsilon X;\varepsilon) \sim 1 + \varepsilon(1 - X - e^{-X}) \times kB\sinh B$, where $B = \operatorname{arccosh}(k^{-1})$.

A2.34 $p(x;\varepsilon) \sim \frac{h(0)}{h(x)} + \varepsilon\frac{h^2(0)}{h(x)}\left[h'(0) - h'(x)\right]$, $p(1 - \varepsilon X;\varepsilon) \sim P_0(X)$ where $P_0 - 1 + \frac{h(0)}{h(1)} \ln\left|\frac{P_0 - h(0)/h(1)}{1 - h(0)/h(1)}\right| = -\frac{X}{h^2(1)}$.

A2.35 **(a)** $\delta = \sqrt{\varepsilon}$; **(b)** solution for $s(\rho) = \frac{\sqrt{k}}{\rho}S(\rho/\sqrt{k})$ then $S'' - S = 0$; the constant A_0 is given by $\ln A_0 = \frac{1}{2k}\int_0^1 \left\{\frac{1}{\sqrt{c - k\ln(1 + k/c)}} - \frac{\sqrt{2k}}{c}\right\} \mathrm{d}c$.

CHAPTER 3

A3.1 $\psi \sim U\left(r - \frac{a^2}{r}\right)\sin\theta - \varepsilon\frac{a^3U}{r^2}\sin 2\theta - \varepsilon^2\frac{3a^2U}{4}\left(\frac{1}{r}\sin\theta + \frac{a^2}{r^3}\sin 3\theta\right)$; $\mathbf{u} \equiv U\left(2\varepsilon(\cos\theta - \cos 2\theta) - \varepsilon^2\left(\frac{3}{2} - \frac{1}{4}\cos\theta + \frac{3}{2}\cos 2\theta + \frac{5}{4}\cos 3\theta\right),\ -2\sin\theta + 3\varepsilon\sin 2\theta - \varepsilon^2(\sin\theta - \frac{1}{2}\sin 3\theta)\right)$.

A3.2 $\psi \sim U\left(r - \frac{a^2}{r}\right)\sin\theta + \varepsilon\frac{1}{4}U\left[r^2(1 - \cos 2\theta) + \frac{a^4}{r^2}\cos 2\theta - a^2\right]$.

A3.3 $\phi \sim 1 + \ln r - \varepsilon\frac{\sin\theta}{r}$.

A3.4 $u_{0x} + \eta_{0t} = 0$; $u_{0t} + \eta_{0x} = 0$ so e.g. $\eta_0 = u_0 = f(x - t)$; $2U_{0\tau} + 3U_0U_{0\xi} + \frac{1}{3}\lambda U_{0\xi\xi\xi} = 0$.

A3.5 $2U_{0\tau} + 4U_0U_{0\xi} = \mu U_{0\xi\xi}$; $U = \alpha\left[1 - \tanh\left\{\frac{1}{2}\alpha(\xi - \alpha\tau\right\}\right]$.

A3.6 $u \sim F(x - t) - \varepsilon\frac{t}{2}\left(F''' + FF' + \int_0^\xi FF''\,\mathrm{d}\xi\right)$; $2V_{0\tau\eta} + 2V_0V_{0\eta\eta} + V_{0\eta}^2 - V_{0\eta\eta\eta\eta} = 0$; $f(\eta,\tau) = -\frac{1}{2}\int_0^\eta V_0(\eta',\tau)\mathrm{d}\eta'$; $g(\xi,\tau) = \frac{1}{2}\int_0^\xi U_0(\xi',\tau)\mathrm{d}\xi'$.

A3.7 $u_0 = e^{-\lambda\tau}f(\xi + u_0/\lambda)$, $\xi = x - c_1t$, $\tau = \varepsilon t$, $\lambda = (c - c_1)/(c_2 - c_1)$; $u_0 = e^{-\mu\tau}g(\eta + u_0/\mu)$, $\eta = x - c_2t$, $\tau = \varepsilon t$, $\mu = (c - c_2)/(c_1 - c_2)$; $u_{0\tau} + u_0u_{0\varsigma} = (c_2 - c)(c - c_1)u_{0\varsigma\varsigma}$, $\varsigma = \varepsilon^{3/2}(x - ct)$, $\tau = \varepsilon^2 t$.

A3.8 $\eta_0 = u_0 = f(x - t)$; $2H_{0\tau} + 3H_0H_{0\xi} + H_{0\xi\xi\xi} = H_{0\xi\xi}$, $\xi = x - t$, $\tau = \varepsilon t$.

A3.9 From (3.39): $x - \beta y - \frac{(1 + \gamma)M_0^4}{2\beta}\varepsilon y u_0(s) = s =$ constant; $\mathrm{d}y/\mathrm{d}x$ follows.

A3.10 $\Phi_{\hat{y}\hat{y}} = [2\lambda + (1 + \gamma)\Phi_x]\Phi_{xx}$.

A3.11 $\delta = \varepsilon$: $\Phi_{YY} - (\varepsilon M_0)^2\Phi_{xx} = (\varepsilon M_0)^2[\frac{1}{2}(\gamma - 1)(2\Phi_x + \Phi_Y^2)\Phi_{YY} + 2\Phi_Y\Phi_{xY}]$.

A3.12 $\varepsilon^2 D^2 u - Du = 0$ where $D \equiv \frac{\mathrm{d}^2}{\mathrm{d}r^2} + \frac{1}{r}\frac{\mathrm{d}}{\mathrm{d}r} - \frac{1}{r^2}$; $u \sim \frac{A_0}{r} + B_0 r + \varepsilon\left(\frac{A_1}{r} + B_1 r\right)$; near $r = 1$, $1 - r = \varepsilon R, U \sim C_0(\mathrm{e}^{-R} + R - 1) + \varepsilon C_1(\mathrm{e}^{-R} + R - 1)$; near $r = r_0$, $r - r_0 = \varepsilon \rho$, $V \sim r_0 + D_0(\mathrm{e}^{-\rho} + \rho - 1) + \varepsilon[\rho + D_1(\mathrm{e}^{-\rho} + \rho - 1)]$; matching gives: $C_0 = D_0 = 0$; $A_0 + B_0 = 0$; $A_0 - B_0 = C_1$; $A_1 + B_1 = -C_1$; $\frac{A_0}{r_0} + B_0 r_0 = r_0$; $B_0 - \frac{A_0}{r_0^2} = D_1$; $\frac{A_1}{r_0} + B_1 r_0 = -D_1$ from which we can obtain A_0, B_0, A_1, B_1, C_1, D_1.

A3.13 $u \sim \mathrm{e}^{-y} + \mathrm{e}^{-x} + \varepsilon\left\{\mathrm{e}^{-y} - \mathrm{e}^{-1} + \ln\left(\frac{1+\mathrm{e}^{1-x}}{1+\mathrm{e}^{y-x}}\right)\right\}$; with $y = \varepsilon Y$ we obtain $u \sim U_0 = 1 + \mathrm{e}^{-x} + (1 - \mathrm{e}^{-x})\mathrm{e}^{-Y}$; with $x = \sqrt{\varepsilon}X$ we write $V \sim V_0$, where $(1 + \mathrm{e}^{-y})V_{0y} + V_{0XX} + (1 + \mathrm{e}^{-y})\mathrm{e}^{-y} = 0$ with the boundary conditions $V_0(0, y) = 0$, $V_0(X, y) \to \mathrm{e}^{-y} - \mathrm{e}^{-1} + \ln\left(\frac{1+\mathrm{e}}{1+\mathrm{e}^{y}}\right)$ as $X \to \infty$.

A3.14 $t_{0yy} + \alpha(x)t_{0x} = 0$, $\tau_{0yy} = [\alpha(x)\tau_0]_x$; introduce $X = \int_0^x \mathrm{d}x'/\alpha(x')$ then $t_{0yy} = -t_{0X}$ and $\hat{\tau}_{0yy} = \hat{\tau}_{0X}$ where $\hat{\tau}_0 = \alpha(x)\tau_0$; thus we obtain $T \sim \exp(-\pi^2 k)\left\{\exp(\pi^2 X) - \frac{\alpha(0)}{\alpha(x)}\exp\left[-\pi^2 X - h(x)/\varepsilon\right]\right\}\sin\pi y$.

A3.15 $\delta(\varepsilon) \sim 1 - \frac{1}{12}\varepsilon^2$ (for $\sin t$); $\delta(\varepsilon) \sim 1 + \frac{5}{12}\varepsilon^2$ (for $\cos t$).

A3.16 You will obtain $\ddot{y} + 2\alpha\dot{y} + (\alpha^2 + \delta + \varepsilon\cos t)y = 0$.

A3.17 $\delta_1 = 0$; $\delta_2 = -\frac{1}{60}\left(6\lambda^2 - 10\lambda + 5\right)$ (for $\sin t$), $\delta_2 = -\frac{1}{60}\left(6\lambda^2 + 10\lambda + 35\right)$ (for $\cos t$).

A3.18 Introduce $x = \varepsilon^{1/(1+\alpha)}X$, $y = \varepsilon^{-\alpha/(1+\alpha)}Y(X;\varepsilon)$, then $(X + Y)Y' + (\alpha + \beta\varepsilon^{1/(1+\alpha)}X)Y = 0$; with $Y \sim Y_0$ then $Y_0[(1 + \alpha)X + Y_0]^\alpha = (1 + \alpha)^\alpha$ after matching; $y(0;\varepsilon) \sim ((1 + \alpha)/\varepsilon)^{\alpha/(1+\alpha)}$.

A3.19 $Y_0(\xi) = \mathrm{e}^{1/\xi}$; $x_1(\xi) = \mathrm{e} - (2\xi^2 - 2\xi + 1)\mathrm{e}^{1/\xi}$; dominant behaviour of $\xi_0(\varepsilon)$ satisfies $\xi_0 = \varepsilon\,\mathrm{e}^{1/\xi_0}$ and then $y(0;\varepsilon) \sim \xi_0(\varepsilon)/\varepsilon$.

A3.20 $Y_0(\xi) = 1/\xi$; $x_1(\xi) = 1 - \frac{1}{2}(\xi + \xi^{-1})$; $\xi_0(\varepsilon) \sim \sqrt{\varepsilon/2}$ so $y(0;\varepsilon) \sim \sqrt{2/\varepsilon}$.

A3.21 $Y_0(\xi) = (1 + \ln\xi)\mathrm{e}^{-\xi}$ and $x_1(\xi) \sim \ln\xi$ as $\xi \to 0^+$.

A3.22 $Y_0(\xi) = \frac{1}{2}(\xi + \xi^{-1})$; $x_1(\xi) = \frac{1}{4}(\xi - \xi^{-1})$; $\xi_0(\varepsilon) \sim \frac{1}{2}\sqrt{\varepsilon}$ so $y(0;\varepsilon) \sim 1/\sqrt{\varepsilon}$.

A3.23 $Y_0(\xi) = -\xi/(1 + \xi)$ and $x_1(\xi) \to \frac{1}{2}$ as $\xi \to 0$; $\xi_0(\varepsilon) \sim -\frac{1}{2}\varepsilon$ and $y(0;\varepsilon) \sim \frac{1}{2}\varepsilon$. With $y(1;\varepsilon) = 1$, $Y_0(\xi)$ is undefined at $\xi = 1/2$; near this point we have $x = \frac{1}{2} + \varepsilon X$ and $y = \hat{Y}(X;\varepsilon) \sim \hat{Y}_0$ satisfies the equation $\hat{Y} = A\exp\left(-X + 1/4\hat{Y}\right)$.

A3.24 $x(\tau;\varepsilon) \sim \cos\tau + \frac{\varepsilon}{32}(\cos\tau - \cos 3\tau) + \frac{\varepsilon}{1024}(23\cos\tau - 24\cos 3\tau + \cos 5\tau)$ where $\tau \sim \left(1 - \frac{3}{8}\varepsilon - \frac{21}{256}\varepsilon^2\right)t$.

A3.25 $u \sim f(\xi)$ where $\xi \sim x - (1 + \varepsilon\omega_1(\xi))\,t \sim x - [1 + \frac{\varepsilon}{4}f'(x - t)]\,t$.

CHAPTER 4

A4.1 The equation for the amplitude is $A' = -\frac{3}{8}a(1 + a^2)A^3$, and so $x \sim \left[\frac{3}{4}a(1 + a^2)\tau + 1\right]^{-1/2}\cos\left[T + \frac{1}{2a}\ln\left\{\frac{3}{4}a(1 + a^2)\tau + 1\right\}\right]$; if $a < 0$ then amplitude and phase are undefined at $\tau = \tau_0 = -4/[3a(1 + a^2)] > 0$.

A4.2 The equation for the amplitude is $A' = -\frac{3}{8}A^3$ and so $x \sim 3(3\tau + 4)^{-1/2}\cos[T + \frac{1}{2}(1 - \cos\tau)]$.

A4.3 $x \sim \alpha\sin\left[T + \int_0^\tau f(\tau')\mathrm{d}\tau'\right]$; requires $f(\tau)$ and $\int_0^\tau f(\tau')\mathrm{d}\tau'$ to remain bounded as $\tau \to \infty$.

A4.4 $x \sim A_0(\tau)\sin T + B_0(\tau)\cos T + \left(\frac{\alpha}{1-\omega^2}\right)\cos\omega T$ (for $\omega \neq 0, \pm 1$), where $A_0^2 + B_0^2 = \text{constant} = k^2$ (so amplitude is constant) and then the phase is $\phi(\tau) = \left[\frac{3}{8}k + \frac{3}{4}\left(\frac{\alpha}{1-\omega^2}\right)^2\right]\tau + \text{constant}$. For $\omega = 0$ the particular integral is a constant (which is not a problem); for $\omega = \pm 1$ the particular integral is secular. For $|\omega| = 1/3$:

$$2A_{0\tau} + \tfrac{3}{4}B_0(A_0^2 + B_0^2) + \tfrac{3}{2}B_0\left(\tfrac{\alpha}{1-\omega^2}\right)^2 + \tfrac{1}{4}\left(\tfrac{\alpha}{1-\omega^2}\right)^3 = 0;$$

$$-2B_{0\tau} + \tfrac{3}{4}A_0(A_0^2 + B_0^2) + \tfrac{3}{2}A_0\left(\tfrac{\alpha}{1-\omega^2}\right)^2 = 0.$$

For $|\omega| = 3$: $2A_{0\tau} + \frac{3}{4}B_0(A_0^2 + B_0^2) + \frac{3}{2}B_0\left(\frac{\alpha}{1-\omega^2}\right)^2 + \frac{3}{4}(B_0^2 - A_0^2)\left(\frac{\alpha}{1-\omega^2}\right) = 0;$

$$-2B_{0\tau} + \tfrac{3}{4}A_0(A_0^2 + B_0^2) + \tfrac{3}{2}A_0\left(\tfrac{\alpha}{1-\omega^2}\right)^2 - \tfrac{3}{2}A_0B_0\left(\tfrac{\alpha}{1-\omega^2}\right) = 0.$$

A4.5 The equation for the amplitude is $2fA' + (f' + f^2)A = 0$ and so $x \sim \frac{k}{\sqrt{f(\tau)}}\exp\left\{-\frac{1}{2}\int_0^\tau f(\tau')\mathrm{d}\tau'\right\}\sin\left[\varepsilon^{-1}\int_0^\tau f(\tau')\mathrm{d}\tau' + \theta_0\right]$, where k, θ_0 are constants.

A4.6 Write the first term in u as $u_0 = A(\tau)\mathrm{e}^{\mathrm{i}T} + B(\tau)\mathrm{e}^{-\mathrm{i}T}$ and then $2A' + A - 3A^2B = 0; 2B' + B - 3B^2A = 0$ (so $A \propto B$) and so $u \sim \frac{2\mathrm{e}^{-\tau/2}}{\sqrt{1+3\mathrm{e}^{-\tau}}}\cos T$; $v \sim \frac{2\mathrm{e}^{-\tau/2}}{\sqrt{1+3\mathrm{e}^{-\tau}}}\sin T$.

A4.7 $x \sim A_0(\tau)\sin T + B_0(\tau)\cos T$ where $2A_{0\tau} = -\frac{3}{4}B_0(B_0^2 + A_0^2) + \cos\Omega_1\tau$; $2B_{0\tau} = \frac{3}{4}A_0(A_0^2 + B_0^2) + \sin\Omega_1\tau$.

A4.8 $x \sim A_0(\tau)\sin(T + \phi_0) - \frac{\alpha}{8}\cos(3T + \Omega_1\tau)$ where $A_{0\tau} = -\frac{3}{64}\alpha A_0^2\cos(\Omega_1\tau - 3\phi_0)$; $\phi_{0\tau} = \frac{3}{8}A_0^2 + \frac{3}{256}\alpha^2$. Note that the forcing contributes to the term $\sin(T + \phi_0)$, in the form $\frac{1}{3} \times 3T$—a subharmonic.

A4.9 $x \sim A_0(\tau)\sin T + B_0(\tau)\cos T + \varepsilon\left\{A_1\sin T + B_1\cos T + \frac{1}{2}(A_0^2 + B_0^2)\right.$
$\left. -\frac{1}{3}A_0B_0\sin 2T - \frac{1}{6}(B_0^2 - A_0^2)\cos 2T\right\}$;
periodicity requires $A_0 = \text{constant}$, $B_0 = \text{constant}$, and then (except for zero initial data) A_1 and/or B_1 grow linearly in τ for *all* ω_2.
The energy integral is $\dot{x}^2 + x^2 = \frac{2}{3}\varepsilon x^3 + k > 0$ for non-zero initial data: all trajectories are unbounded for $\forall\varepsilon > 0$, so nearly periodic, bounded solutions do not exist.

A4.10 $x \sim a(\tau)\mathrm{cn}(4K(m)T; m(\tau))$ where $a^2 = 32m\theta^2K^2$; $16(1 - 2m)\theta^2K^2 = 1$ with $a^2K^2\theta\int_0^1 \mathrm{sn}^2\mathrm{dn}^2\mathrm{d}T = k\mathrm{e}^\tau$, where k is an arbitrary constant.

A4.11 $x \sim a(\tau) + b(\tau)\mathrm{cn}^2(4K(m)T; m(\tau))$ where $b = 96mK^2\theta^2$; $2a - 1 = 64(1 - 2m)\theta^2K^2$; $a(1 - a) = 32(1 - m)\theta^2K^2$ with $b^2K\theta\int_0^1 \mathrm{cn}^2\mathrm{sn}^2\mathrm{dn}^2\mathrm{d}T = k\mathrm{e}^\tau$, where k is an arbitrary constant.

A4.12 $x \sim A_0(\tau)\sin T + B_0(\tau)\cos T$ where $2A_{0\tau} + (\delta_2 - \frac{5}{12})B_0 = 0$; $2B_{0\tau} - (\delta_2 + \frac{1}{12})A_0 = 0$. Then exponential growth for $\frac{5}{12} > \delta_2 > -\frac{1}{12}$; oscillatory for $\delta_2 > \frac{5}{12}$ or $\delta_2 < -\frac{1}{12}$; linear growth on $\delta_2 = \frac{5}{12}, -\frac{1}{12}$.

A4.13 $x \sim A_0(\tau)\sin T + B_0(\tau)\cos T$ and then the second term is periodic if $\delta_1 = 0$ and the third if

$$2A_{0\tau} = [-\delta_2 - \tfrac{1}{12}(\lambda+1) - \tfrac{1}{60}\lambda^2 - \tfrac{1}{12}\lambda(\lambda+1) - \tfrac{1}{2}]B_0;$$
$$2B_{0\tau} = [\delta_2 - \tfrac{1}{12}(\lambda-1) + \tfrac{1}{60}\lambda^2 + \tfrac{1}{12}\lambda(\lambda-1)]A_0.$$

Thus $\delta_{\pm}(\lambda) = -\frac{\lambda^2}{10} \pm \frac{\lambda}{6} - \frac{1}{3} \pm \frac{1}{4}$.

A4.14 $$x \sim A_0\sin T + B_0\cos T + \varepsilon\left\{A_1\sin T + B_1\cos T + \frac{A_0\sin(1-\Omega^{-1})T}{2(1-2\Omega)} + \frac{A_0\sin(1+\Omega^{-1})T}{2(1+2\Omega)} + \frac{B_0\cos(1-\Omega^{-1})T}{2(1-2\Omega)} + \frac{B_0\cos(1+\Omega^{-1})T}{2(1+2\Omega)}\right\},$$

where A_0, B_0, A_1, B_1 are constants and $\omega_2 = [4\Omega(1-4\Omega^2)]^{-1}$ ($\Omega \neq 0, \pm 1/2$).

A4.15 $\sigma_0 = \omega$; $A_0 = \sqrt{k/\omega}$; $\sigma_1 = \frac{1}{2}\omega^{-1/2}(\omega^{-1/2})''$; $A_1 = -\frac{1}{4}\sqrt{k}\,\omega^{-2}(\omega^{-1/2})''$.

A4.16 $y \sim k\sqrt{\frac{\omega(0)}{\omega(X)}}\exp\left\{-\varepsilon^{-1}\int_0^X \omega(X')\mathrm{d}X'\right\}$ (for a bounded solution as $\varepsilon \to 0^+$); here, $y(0;\varepsilon) = k$.

A4.17 First $\Omega = \sqrt{\lambda\, a(X)}$ and then $y \sim \frac{k}{\sqrt{\Omega}}\sin\left[\varepsilon^{-1}\int_0^X \sqrt{\lambda\, a(X')}\,\mathrm{d}X'\right]$, where k is an arbitrary constant; thus the eigenvalues are given by $\sqrt{\lambda}\int_0^{\varepsilon}\sqrt{a(X)}\,\mathrm{d}X \sim \varepsilon n\pi$. **(a)** $\lambda \sim n^2\pi^2(1 + \frac{1}{2}\varepsilon)$; (b) $\lambda \sim n^2\pi^2(1-\varepsilon)$.

A4.18 $h(x) = (1-x)^{-1}\left[\frac{3}{2}\int_x^1 \sqrt{(1-x')f(x')}\,\mathrm{d}x'\right]^{2/3}$ for $x < 1$;
$h(x) = (x-1)^{-1}\left[\frac{3}{2}\int_1^x \sqrt{(x'-1)f(x')}\,\mathrm{d}x'\right]^{2/3}$ for $x > 1$. N.B. $h \to [f(1)]^{1/3}$ as $x \to 1^{\pm}$. Then $y \sim k\left(\frac{h}{f}\right)^{1/4}\mathrm{Ai}(X)$, where k is an arbitrary constant.

A4.19 For $x > 1$ (bounded solution): $Y \sim \frac{A_0}{[x(x-1)]^{1/4}}\exp\left\{-\varepsilon^{-1}\int_1^x \sqrt{x'(x'-1)}\,\mathrm{d}x'\right\}$;
for $x < 0$: $Y \sim \frac{B_0}{[x(x-1)]^{1/4}}\exp\left\{-\varepsilon^{-1}\int_x^0 \sqrt{x'(x'-1)}\,\mathrm{d}x'\right\}$; for $0 < x < 1$:
$Y \sim [x(x-1)]^{-1/4}\left[C_0\sin\left(\varepsilon^{-1}\int_0^x \sqrt{x'(x'-1)}\,\mathrm{d}x'\right)\right.$
$\left. + D_0\cos\left(\varepsilon^{-1}\int_0^x \sqrt{x'(x'-1)}\,\mathrm{d}x'\right)\right]$.
Near $x = 0$: $Y \sim E_0|x|^{-1/4}\mathrm{Ai}(-\varepsilon^{-2/3}|x|^{3/2})$;
near $x = 1$: $Y \sim F_0|x-1|^{-1/4}\mathrm{Ai}(-\varepsilon^{-2/3}|x-1|^{3/2})$.

A4.20 First $\mathrm{d}X/\mathrm{d}x = \varepsilon^{-1}\sqrt{f(x)}$ then with $Y \sim Y_0$ we obtain $Y_{0XX} + X^3Y_0 = 0$ and so $y \sim X^{1/2}\mathrm{J}_{\pm 1/5}(\frac{2}{5}X^{5/2})$.

A4.21 For $-1 < z < 1$: $\psi \sim A_0[1-z^2]^{-1/4}\sin\left(\varepsilon^{-2}\int_{-1}^z \sqrt{1-z'^2}\,\mathrm{d}z' + \alpha_0\right)$;
for $z < -1$: $\psi \sim B_0[z^2-1]^{-1/4}\exp\left\{-\varepsilon^{-2}\int_z^{-1}\sqrt{z'^2-1}\,\mathrm{d}z'\right\}$;
for $z > 1$: $\psi \sim C_0[z^2-1]^{-1/4}\exp\left\{-\varepsilon^2\int_1^z \sqrt{z'^2-1}\,\mathrm{d}z'\right\}$;
for $z = \pm 1 + \delta Z$ where $\delta = (\varepsilon^4/2)^{1/3}$: $\psi \sim D_{0\pm}\mathrm{Ai}(\mp Z)$.

Matching gives $A_0 = 2B_0 = 2C_0$, $\alpha_0 = \pi/4$, $D_{0+} = 2(2\delta)^{-1/4}\sqrt{\pi}\, C_0$, $D_{0-} = (2\delta)^{-1/4}\sqrt{\pi}\, A_0$, leaving $\varepsilon^{-2}\int_{-1}^{1}\sqrt{1-z^2}\,dz + \frac{\pi}{4} = n\pi - \frac{\pi}{4}$, which gives the required result.

A4.22 First $\omega = -k^3$, then $A_{0T} - 3k^2 A_{0X} = 3kk_X A_0 + ikA_0\,|A_0|^2$.

A4.23 First $\omega^2 = 1 + k^2$, $c = \omega/k$, $c_g = k/\omega$, and then

$$-2i\omega A_{01T} + (c_g^2 - 1)A_{01XX} = \tfrac{1}{3}(1 + k^2)(10 + k^2)A_{01}\,|A_{01}|^2.$$

A4.24 As for A4.23, then $-2i\omega A_{01T} + (c_g^2 - 1)A_{01XX} = \frac{1}{3}(2+ik)(5-ik)A_{01}\,|A_{01}|^2$.

A4.25 First $\omega^2 = k^2 - 1$ (>0), $c = \omega/k$, $c_g = k/\omega$, then

$$-2ikc\,A_{01T} + (c_g^2 - 1)A_{01XX} + 8k^4 A_{01}\,|A_{01}|^2 = 0.$$

A4.26 First $c = k^2/6$, $c_g = k^2/2$, then $-2ikA_{01T} + k^2 A_{01XX} - \frac{9}{2}A_{01}\,|A_{01}|^2 = 0$.

A4.27 E.g. $\omega_X = -\theta_{Xt} = -\varepsilon^{-1}\theta_{xt} = -\varepsilon^{-1}k_t = -k_T$, etc.

**In the next five answers, we have $Y \sim A(x) + B(x)\,e^{-X}$:

A4.28 $A' - 3A = 3x$; $B' + 3B = 0$ and so $y \sim \frac{10}{3}e^{3(x-1)} - x - \frac{1}{3} + \frac{1}{3}(7 - 10e^{-3})e^{-3x-X}$ where $X = \varepsilon^{-1}(x + \frac{1}{2}\varepsilon x^2)$.

A4.29 $(1 + 2x)A' - A = -x$; $(1 + 2x)B' + 3B = 0$ and so $y \sim \frac{2}{\sqrt{3}}\sqrt{1 + 2x} - 1 - x + 2(1 - \frac{1}{\sqrt{3}})(1 + 2x)^{-3/2}e^{-X}$ where $X = \varepsilon^{-1}(2 - x - x^2)$.

A4.30 $(1 + 2x)A' + 2A = 0$; $B =$ constant and so $y \sim \frac{3}{1+2x} + e^{-X}$ where $X = \varepsilon^{-1}(x + x^2)$.

A4.31 $(1 + x)^2 A' - A^2 = 0$; $(1 + x)^2 B' + 2(1 + x + A)B = 0$ and so $y \sim \frac{1+x}{2+x} - \frac{1}{8}\frac{(2+x)^2}{(1+x)^4}e^{-X}$ where $X = (3\varepsilon)^{-1}[(1 + x)^3 - 1]$.

A4.32 $A' + A^2 = 0$; $B' = 2AB$ and so $y \sim (1 + x)^{-1} - (1 + x)^2 e^{-x/\varepsilon}$.

A4.33 The use of $X = (2\varepsilon)^{-1}x^2$ gives rise to a solution which is not defined on $x = 0$; use $X = x/\sqrt{\varepsilon}$, then $T \sim T_0 e^x + (T_1 e^{-1} - T_0)e^x\sqrt{\frac{2}{\pi}}\int_0^X \exp(-\frac{1}{2}y^2)\,dy$.

A4.34 With $Y \sim A(x) + B(x)\,e^{-X}$ we obtain $A' + A^n = 0$; $B' = n\,A^{n-1}B$ and so $y \sim [(n - 1)x + 1]^{-1/(n-1)} - n^{-(n+1)/(n-1)}[(n - 1)x + 1]^{n/(n-1)}e^{-X}$ where $X = \varepsilon^{-1}(1 - x)$.

A4.35 With $Y \sim A(x) + B(x)\,e^{-X} + C(x)\,e^{-Z}$ we obtain $A' = (1 + 2x)A^2$; $\frac{2}{1+2x}B' + \left(2A - \frac{3}{(1+2x)^2}\right)B = 0$; $\frac{2}{1+2x}C' + \left(2A - \frac{3}{(1+2x)^2}\right)C = 0$ and so $y \sim \left[\frac{\sqrt{3}}{2}(1 + \sqrt{3}) - x - x^2\right]^{-1} + \left[x + x^2 - \frac{\sqrt{3}}{2}(1 + \sqrt{3})\right](1 + 2x)^{3/4}\left\{\frac{4}{3}(1 + \sqrt{3})^{-2}e^{-X} - 3^{-1/4}(1 + \sqrt{3})e^{-Z}\right\}$ where $X = \varepsilon^{-1/2}\left(\sqrt{1 + 2x} - 1\right)$, $Z = \varepsilon^{-1/2}(\sqrt{3} - \sqrt{1 + 2x})$.

REFERENCES

Abramowitz, M. & Stegun, I. A. (ed.) (1964), *Handbook of Mathematical Functions*. Washington: Nat. Bureau of Standards. (Also New York: Dover, 1965)

Andrews, J. G. & McLone, R. R. (1976), *Mathematical Modelling*. United Kingdom: Butterworth.

Barenblatt, G. I. (1996), *Scaling, self-similarity, and intermediate asymptotics*. Cambridge: Cambridge University Press.

Blasius, H. (1908), Grenzschichten in Flüssigkeiten mit kleiner Reibung, *Z. Math. Phys.*, **56**, 1–37. (Translated as 'Boundary layers in fluids with small friction,' *Tech. Memor. Nat. adv. Comm. Aero.*, Washington no. 1256.)

Boccaletti, D. & Pucacco, G. (1996), *Theory of Orbits 1: Integrable Systems and Non-perturbative Methods*. Berlin: Springer-Verlag.

Bogoliubov, N. N. & Mitropolsky, Y. A. (1961), *Asymptotic Methods in the Theory of Nonlinear Oscillations*. Delhi: Hindustan Publishing.

Boyce, W. E. & DiPrima, R. C. (2001), *Elementary Differential Equations and Boundary Value Problems*. (7th Edition) New York: Wiley.

Bretherton, F. P. (1964), Resonant interaction between waves. The case of discrete Oscillations, *J. Fluid Mech.*, **20**, 457–79.

Brillouin, L. (1926), Rémarques sur la méchanique ondulatoire, *J. Phys. Radium*, **7**, 353–68.

Burgers, J. M. (1948), A mathematical model illustrating the theory of turbulence, *Adv. Appl. Mech.*, **1**, 171–99.

Bush, A. W. (1992), *Perturbation Methods for Engineers and Scientists*. Boca Raton, FL: CRC.

Byrd, P. F. & Friedman, M. D. (1971), *Handbook of Elliptic Integrals for Engineers and Physicists*. 2nd edn. New York: Springer-Verlag.

Carrier, G. F. (1953), 'Boundary problems in applied mechanics' in *Advances in Applied Mechanics III*. New York: Academic Press.

———(1954), Boundary layer problems in applied mathematics, *Comm. Pure Appl. Math.*, **7**, 11–17.

Carslaw, H. W. & Jaeger, J. C. (1959), *Conduction of Heat in Solids*. Oxford: Clarendon.

Chang, K. W. & Howes, F. A. (1984), *Nonlinear Singular Perturbation Phenomena: Theory and Applications*. Berlin: Springer-Verlag.

Christodoulou, D. M. & Narayan, R. (1992), The stability of accretion tori. IV. Fission and fragmentation of slender self-gravitating annuli, *Astrophys. J.*, **388**, 451–66.

Cole, J. D. (1951), On a quasi-linear parabolic equation occurring in aerodynamics, *Quart. Appl. Math.*, **9**, 225–36.

______(1968), *Perturbation Methods in Applied Mathematics*. Waltham, MA: Blaisdell.

Cook, R. J. (1990), 'Quantum jumps' in *Progress in Optics*, XXVIII (E. Wolf, ed.). Amsterdam: North-Holland.

Copson, E. T. (1967), *Asymptotic Expansions*. Cambridge: Cambridge University Press.

Courant, R. & Friedrichs, K. O. (1967), *Supersonic Flow and Shock Waves*. New York: Interscience.

Cox, R. N. & Crabtree, L. F. (1965), *Elements of Hypersonic Aerodynamics*. London: English Universities Press.

Crank, J. (1984), *Free and Moving Boundary Value Problems*. Oxford: Clarendon.

DeMarcus, W. C. (1956, 1957), The problem of Knudsen flow, Parts I, II (1956) & III (1957), *US AEC Rep.* K-1302.

Dingle, R. B. (1973), *Asymptotic Expansions: their Derivation and Interpretation*. London: Academic Press.

Drazin, P. G. & Johnson, R. S. (1992), *Solitons: an Introduction*. Cambridge: Cambridge University Press.

Dresner, L. (1999), *Applications of Lie's Theory of Ordinary and Partial Differential Equations*. Bristol: Institute of Physics Publishing.

Duffing, G. (1918), Erzwungene Schwingugen bei veränderlicher Eigenfrequenz, *F. Vieweg u. Sohn (Braunschweig)*.

Eckhaus, W. (1979), *Asymptotic Analysis of Singular Perturbations*. (Studies in Mathematics and its Applications, Vol. 9.) Amsterdam: North-Holland.

Erdelyi, A. (1956), *Asymptotic Expansions*. New York: Dover.

Ford, W. B. (1960), *Divergent Series, Summability and Asymptotics*. Bronx, NY: Chelsea.

Fowler, A. C. (1997), *Mathematical Models in the Applied Sciences*. Cambridge: Cambridge University Press.

Fraenkel, L. E. (1969), On the method of matched asymptotic expansions. Parts I–III, *Proc. Camb. Phil. Soc.* **65**, 209–84.

Fulford, G. R. & Broadbridge, P. (2002), *Industrial Mathematics: Case Studies in the Diffusion of Heat and Matter*. (Australian Mathematical Society Lecture Series vol. 16). Cambridge: Cambridge University Press.

Georgescu, A. (1995), *Asymptotic Treatment of Differential Equations*. London: Chapman & Hall.

Hanks, T. C. (1971), Model relating heat-flow value near, and vertical velocities of, mass transport beneath ocean rises, *J. Geophys. Res.*, **76**, 537–44.

Hardy, G. H. (1949), *Divergent Series*. Oxford: Clarendon.

Hayes, W. D. & Probstein, R. F. (1960), *Hypersonic Flow Theory I: Inviscid Flows*. New York: Academic Press.

Hinch, E. J. (1991), *Perturbation Methods*. Cambridge: Cambridge University Press.

Holmes, M. H. (1995), *Introduction to Perturbation Methods*. New York: Springer-Verlag.

Hopf, E. (1950), The partial differential equation $u_t + uu_x = \mu u_{xx}$, *Comm. Pure Appl. Math.*, **3**, 201–30.

Ince, E. L. (1956), *Ordinary Differential Equations*. New York: Dover.

Jeffreys, H. (1924), On certain approximate solutions of linear differential equations of the second order, *Proc. Lond. Math. Soc.*, **23**, 428–36.

Johnson, R. S. (1970), A non-linear equation incorporating damping and dispersion, *J. Fluid Mech.*, **42**(1), 49–60.

______(1997), *A Modern Introduction to the Mathematical Theory of Water Waves*. Cambridge: Cambridge University Press.

Kaplun, S. (1967), *Fluid Mechanics and Singular Perturbations*. (P. A. Lagerstrom, L. N. Howard, C. S. Liu, eds.) New York: Academic Press.

Kevorkian, J. & Cole, J. D. (1981), *Perturbation Methods in Applied Mathematics*. (Applied Mathematical Sciences, Vol. 34.) Berlin: Springer-Verlag.

______(1996), *Multiple Scale and Singular Perturbation Methods*. (Applied Mathematical Sciences, Vol. 114.) Berlin: Springer-Verlag.

Kapila, A. K. (1983), *Asymptotic Treatment of Chemically Reacting Systems*. Boston: Pitman.

King, J. R., Meere, M. G., & Rogers, T. G. (1992), Asymptotic analysis of a nonlinear model for substitutional diffusion in semiconductors, *Z. angew Math. Phys.*, **43**, 505–25.

Kramers H. A. (1926), Wellenmechanik und halbzahlige Quantisierung, *Z. Physik*, **39**, 829–40.

Kuo, Y. H. (1953), On the flow of an incompressible viscous fluid past a flat plate at moderate Reynolds number, *J. Math. and Phys.*, **32**, 83–101.

Kuzmak, G. E. (1959), Asymptotic solutions of nonlinear second order differential equations with variable coefficients, *J. Appl. Math. Mech. (PMM)*, **23**, 730–44.

Lagerstrom, P. A. (1988), *Matched Asymptotic Expansions: Ideas and Techniques*. New York: Springer-Verlag.

Lawden, D. F. (1989), *Elliptic Functions and Applications*. Berlin: Springer-Verlag.

Lie, G. C. & Yuan, J-M. (1986), Bistable and chaotic behaviour in a damped driven Morse oscillator: a classical approach, *J. Chem. Phys.*, **84**, 5486–93.

Lighthill, M. J. (1949), A technique for rendering approximate solutions to physical problems uniformly valid, *Phil. Mag.* **40**, 1179–1201.

______(1961), A technique for rendering approximate solutions to physical problems uniformly valid, *Z. Flugwiss.*, **9**, 267–75.

Lo, L. (1983), The meniscus on a needle—a lesson in matching, *J. Fluid Mech.*, **132**, 65–78.

McLachlan, N. W. (1964), *Theory and Applications of Mathieu Functions*. New York: Dover.

McLeod, J. B. (1991), 'Laminar flow in a porous channel' in *Asymptotics beyond All Orders* (H. Segur, S. Tanveer, & H. Levine eds.). New York: Plenum Press.

Mestre, N. de (1991), *The Mathematics of Projectiles in Sport*. Cambridge: Cambridge University Press.

Miles, J. W. (1959), *The Potential Theory of Unsteady Supersonic Flow*. Cambridge: Cambridge University Press.

Murray, J. D. (1974), *Asymptotic Analysis*. Oxford: Clarendon.

______(1993), *Mathematical Biology*. (Biomathematics Vol. 19.) Berlin: Springer-Verlag.

Nayfeh, A. H. (1973), *Perturbation Methods*. New York: Wiley.

______(1981), *Introduction to Perturbation Techniques*. New York: Wiley.

Olver, F. W. J. (1974), *Introduction to Asymptotics and Special Functions*. New York: Academic Press.

O'Malley, R. E. (1991), *Singular Perturbation Methods for Ordinary Differential Equations*. (Applied Mathematical Sciences, Vol. 89.) New York: Springer-Verlag.

Oseen, C. W. (1910), Uber die Stokes'sche Formel, und uber eine verwandte Aufgabe in der Hydrodynamik, *Ark. Math. Astronom. Fys.*, **6**(29).

Pao, Y.-P. & Tchao, J. (1970), Knudsen flow through a long circular tube, *Phys. Fluids*, **13**(2), 527–8.

Papaloizou, J. C. B. & Pringle, J. E. (1987), The dynamical stability of differentially rotating discs—III, *Mon. Not. R. Astron. Soc.*, **225**, 267–83.

Patterson, G. N. (1971), *Introduction to the Kinetic Theory of Gas Flows*. Toronto: University of Toronto Press.

Poincaré, H. (1892), *Les Méthodes Nouvelles de la Méchanique Céleste II* (available New York: Dover, 1957).

Proudman, I. (1960), An example of steady laminar flow at large Reynolds number, *J. Fluid Mech.*, **9**, 593–602.

Rayleigh, J. W. S. (1883), On maintained vibrations, *Phil. Mag.*, **15**, 229–35.

Reiss, E. L. (1980), A new asymptotic method for jump phenomena, *SIAM J. Appl. Math.*, **39**, 440–55.

Roosbroeck, W. van (1950), Theory of the flow of electrons and holes in germanium and other semiconductors, *Bell System Tech. J.*, **29**, 560–607.

Sanders, J. A. (1983), 'The driven Josephson equation: an exercise in asymptotics' in *Asymptotic Analysis II—Surveys and New Trends* (F. Verhulst, ed.). New York: Springer-Verlag.

Schmeisser, C. & Weiss, R. (1986), Asymptotic analysis of singularly perturbed boundary value problems, *SIAM J. Math. Anal.*, **17**, 560–79.

Segur, H., Tanveer, S., & Levine, H. (eds.) (1991), *Asymptotics beyond All Orders*. (NATO ASI Series B: Physics Vol. 284). New York: Plenum Press.

Shockley, W. (1949), The theory of p-n junctions in semiconductors and p-n junction transistors, *Bell System Tech. J.*, **28**, 435–89.

Smith, D. R. (1985), *Singular-perturbation Theory: An Introduction with Applications*. Cambridge: Cambridge University Press.

Stokes, G. G. (1851), On the effect of the internal friction of fluids on the motion of pendulums, *Trans. Camb. Phil. Soc.*, **9**(II), 8–106.

Szekely, J., Sohn, H. Y., & Evans, J. W. (1976), *Gas-Solid Reactions*. New York: Academic Press.

Taylor, G. I. (1910), The conditions necessary for discontinuous motion in gases, *Proc. Roy. Soc.*, **A84**, 371–77.

Terrill, R. M. & Shrestha, G. M. (1965), Laminar flow through a channel with uniformly porous walls of different permeability, *Appl. Sci. Res.*, **A15**, 440–68.

Tyson, J. J. (1985), 'A quantitative account of oscillations, bistability, and travelling waves in the Belousov-Zhabotinskii reaction' in *Oscillations and Travelling Waves in Chemical Systems* (R. J. Field & M. Burgur eds.). New York: Wiley.

van der Pol, B. (1922), On a type of oscillation hysteresis in a simple triode generator, *Phil. Mag.*, **43**, 177–93.

Van Dyke, M. (1964), *Perturbation Methods in Fluid Mechanics*. New York: Academic Press.

______(1975), *Perturbation Methods in Fluid Mechanics (Annotated Edition)*. Stanford, CA: Parabolic Press.

Vasil'eva, A. B. & Stelmakh, V. G. (1977), Singularly disturbed systems of the theory of semiconductor devices, *USSR Comp. Math. Phys.*, **17**, 48–58.
Wang, M. & Kassoy, D. R. (1990), Dynamic response of an inert gas to slow piston acceleration, *J. Acoust. Soc. Am.*, **87**, 1466–71.
Ward, G. N. (1955), *Linearized theory of Steady High-speed Flow*. Cambridge: Cambridge University Press.
Wasow, W. (1965), *Asymptotic Expansions for Ordinary Differential Equations*. New York: Wiley.
Wentzel, G. (1926), Eine Verallgemeinerung der Quantenbedingungen für die Zwecke der Wellenmechanik, *Z. Phys.*, **38**, 518–29.
Whitham, G. B. (1974), *Linear and Nonlinear Waves*. New York: Wiley.

SUBJECT INDEX